The Science and Technology of

Animal Training

James O'Heare

BehaveTech Publishing

Ottawa, Canada

Title: The Science and Technology of Animal Training

Publisher: BehaveTech Publishing, Ottawa, Canada, www.BehaveTech.com

Author: James O'Heare

Cover art and book design: James O'Heare

Copy editing and proof reading: Kamrin MacKnight

Limits of Liability and Disclaimer of Warranty:

ISBN 978-1-927744-06-2

This book is dedicated to Roscoe 1999-2012, my best friend ever!

Thank you to Kamrin MacKnight, J.D., Ph.D., for editing and proofreading the final manuscript.

Thank you also to the following species-specific chapter reviewers, without which, those chapters would not be nearly as useful:

Jacqueline Munera, CDBC, CCBT, CAP2

Amanda Clase, Ph.D., Dip.CBST

Jelena Kallay, Dip.ABT, KPA CTP

Kathleen Kemp, B.Sc., Cert.PDTSC

FOREWORD BY STEPHEN F. LEDOUX

On this space rock, humans are not alone. We share the planet with many other species. For some of these species, particularly those that we describe as companion animals, we have developed an often mutually profound affinity. Cats, dogs, birds—especially parrots—and horses top the list.

Our concerns with managing the causes of behavior, whether the behavior of our companion animals, our children, our fellow humans, or even ourselves, remain ever present. Unfortunately, the value of science in these endeavors sometimes comes as a surprise. Yet science is indeed valuable to our successes, especially in the form of a sound and substantiated systematic approach, such as that found in the natural science of behavior that we call behaviorology. The openness of the researched principles and designed practices of science certainly save us from the pitfalls of chance, accidents, fads, and secret–knowledge systems that arise when we follow uninformed trial and error patterns, and handed–down tricks and intuitions.

James O'Heare's book, *The Science and Technology of Animal Training* recognizes all that and more. It provides a thorough introduction to all the basic aspects of behaviorology that relate to companion–animal training as the best way to understand them and train them. This enables the reader to see *why* one thing works while another fails. Rather than merely following steps in a cookbook of techniques for a simple set of prescribed animal–behavior tricks, this book enables the improving animal trainer to move on to advanced techniques capable of producing complicated behavioral outcomes, all the while emphasizing positive rather than coercive strategies which, again, is how we would treat each other, and our children, to the best effect. As such, this book benefits all levels of readers, from beginners striving to become skilled, to already established professionals.

Finally, by adhering to natural behavior science, this book helps those who love other animals glimpse the value of behaviorology for cleaning up our own human behavior. Humanity faces some big problems, such as pollution, overpopulation, and war. All these problems, and their solution, involve human behavior, and thus benefit by the input of behaviorology. If we fail to solve these problems, then we endanger not only ourselves but our beloved companion animals as well. By becoming more familiar with behaviorology in general, one becomes more capable not only at animal training projects but also at helping create a better world, one that can remain suitable both for us and our animal friends.

Stephen F. Ledoux, Ph.D., DLBC
Author of Running Out of Time—Introducing Behaviorology to Help Solve Global Problems
Canton, NY, USA
2014 March 9

The Science and Technology of Animal Training

PREFACE

The Science and Technology of Animal Training introduces companion animal training and the natural science discipline that informs it—behaviorology. It was written to provide material to cover the course objectives for (a) BEHG 120: Non-Coercive Companion Animal Behavior Training, provided through The International Behaviorology Institute; (b) ABAP 310: Companion Animal Behavior Training, provided through the State University of New York at Canton, and (c) courses provided through the Companion Animal Sciences Institute. It would also be useful for any other courses on the natural science of behavior and the training of companion animals. Furthermore, anyone that is seeking an introduction to the discipline of behaviorology, with an application of the laws and principles of behavior to training dogs, cats, birds and horses would find the book useful. I wrote *The Science and Technology of Animal Training* to be a Masters level course in animal training. I also wrote it to be self-contained in that it requires no previous coursework in behaviorology as prerequisite. It achieves this by presenting the requisite material in the first two chapters of the book. *The Science and Technology of Animal Training Study Questions* booklet is also available.

There are a great many books available on dog training. The vast majority are popular books making dog training accessible at a very basic level to those with a limited repertoire of training skills and knowledge. These books fulfill a valuable role. Part of a problem in the field of animal training however, is that this level of proficiency sometimes becomes the standard for professionals, largely because there is limited access to educational opportunities (including formal courses and programs of study as well as books) that would expand the trainer's repertoire of training behaviors. This level of skill and knowledge may be suitable for the average companion animal guardian but not for anyone who seeks to be qualified as a professional. Anyone engaged in providing professional training services should have at least a technologist level repertoire of skills and knowledge in the basic natural science discipline that informs the field as well as the technology derived from it. A technologist level of proficiency is a trade or vocational level sufficient for the technologist to work unsupervised on a suitable range of objectives within the field but is not at the scientist level, in this case a fully qualified behaviorologist. Qualification as a behaviorologist is not necessary for animal trainers but a technologist level of proficiency should be a basic foundational level for qualification as a professional animal trainer. The content of this book represents a foundational technologist level of coverage in the natural science of behavior and the technology of animal training. And so, one of my major goals with this textbook is to provide an opportunity for establishing what is required for a basic technologist level of competence that would qualify one as a professional animal trainer. However, simply reading (or even studying) this book does not guarantee that the behavior objectives are met. Achieving the behavior objectives ought to be established through a formal course of study that provides the student an opportunity to achieve the objectives with guidance and supervision and demonstrate that proficiency objectively. You can achieve those through the courses listed above, and perhaps others.

Here are a few pieces of advice on how to work through this book. This is a textbook, not a popular literature book. The material is dense and you should read it as such. You should read it actively. That is, the reader should read carefully and reread sections that are not clear and ensure clarity before proceeding. The reader should generate their own examples and practice explaining the material to themselves and others, using the proper terminology correctly. I recommend that the reader implement the exercises in the book as well, in order to help identify and improve upon any weaknesses identified. As is commonly the case with educational materials, you will get out of what you put into the endeavor.

Please note that because this book is prepared as a text for courses in animal training, it utilizes the most current terminology in behaviorology. Behaviorologists are replacing some old, unfortunately phrased terms with new terms that are more appropriate and these terms are therefore currently in transition. For instance, the old "positive" and "negative" categories of reinforcement and punishment are replaced with "added" and "subtracted" reinforcement and punishment. "Discriminative stimulus" is replaced with "evocative stimulus," emphasizing the evocative capacity of the stimulus.

There are topics upon which not all behaviorologists agree, including the proper definitions for certain terms or whether concepts ought to be addressed or not (e.g., "motivation"). Where this might be the case, I identify contentious issues as such and footnotes are included to provide the source or descriptions of the alternative point of view on the topic. If you are using the book as a course text, ensure that you ascertain your instructor's preference on the topic.

The Science and Technology of Animal Training covers all topic points at an appropriate depth for Association of Animal Behavior Professionals (AABP) core areas of competence in "Principles of Behavior" and "Animal Training." Achieving the behavioral objectives established in this book will thoroughly prepare the candidate for the AABP Proficiency Exam in these two core areas of competence. Passing courses provided by AABP Approved Education Providers (including those referred to above) can serve to exempt certification candidates from these two core areas components of the AABP Proficiency Exam.

I will present technical terms in ***bold italicized font***, either where I introduce them, or where I discuss them in detail, so they are easy to find. The table of contents and index also help direct readers to discussions on specific topics. Readers may also find the Association of Animal Behavior Professionals Encyclopedic Glossary of Terms and Abbreviations[1] useful.

I wrote each of the non-human species-specific training chapters (7, 8, 9 & 10) to stand alone with regard to the other species-specific training chapters. Reading all of them may seem a little repetitive on many points but this repetition will promote effective conditioning. If you elect to read only some of these chapters, you will not be missing important material covered in the unread species-specific training material. In addition, on the topic of repetition, each training project within each species-specific chapter will repeat the basic steps in training, including preliminary, acquisition, fluency and maintenance phases of training. This too will seem repetitive, but this will

---

[1] www.associationofanimalbehaviorprofessionals.com/glossary.html

emphasize the systematic nature of applying principles to training projects and better condition the "habit" of taking a systematic approach to training.

Each species-specific chapter will address basic "good manners" behaviors and will not cover socialization and training of the very young animal in all matters (e.g., house training). Nor will the chapters address training for entertainment or service animal purposes. Furthermore, I made no effort to ensure exactly equal coverage in terms of the number of behaviors or number of pages. For various reasons, some species require more coverage than others do. For instance, because dogs are generally the largest and potentially most dangerous animals that we keep in our homes, there is a lot of training to cover. The horse is the largest species we maintain as companion animals, so they too require a significant amount of training. Much of the training we engage in with dogs relates to managing their movements. Horses are large prey animals and therefore much of their training relates to managing their movements and reducing their fear of certain kinds of handling. Birds—at least the ones we generally maintain as companions—are also prey animals and the behaviors we require to train them relate to handling as well. Cats are more readily trainable than many people believe but there tends to be fewer basic training requirements for cats. In each case, the length and depth of coverage is dictated by what is required for foundational "good manner" training for companionship purposes.

In terms of writing, I will use first person for clarity. Where I use the editorial "we," I am referring to the verbal community of behaviorologists and trainers to refer to how we, as a community, address the topic under consideration in general. I try to use active voice where I can, as it is much easier to read. However, I use passive voice where necessarily to (a) avoid implying an agential perspective, which is contrary to a natural science perspective and (b) where I want to refer to the actions of some others as problematic but it is unnecessary to name specific individuals or groups. You will also notice some repetition. I include these repetitions to provide you with numerous opportunities to work with the terms and concepts, often from different perspectives so that they will more likely result in more effective conditioning in you.

Readers familiar with *The Science and Technology of Dog Training* will find a few chapters here that are very similar, as I based this book on the previous book, but I have updated it, as well as expanded it to address multiple species.

As with all my books, this book will remain a work in progress that I will update it with new editions in order to provide an up-to-date and improved current product. I hope you find it useful for your own purposes.

TABLE OF CONTENTS

The Science and Technology of Animal Training

CHAPTER 1.
BEHAVIOROLOGY

Behavioral Objectives

The objective of this chapter is to measurably expand the reader's repertoire of behaviors in relation to describing and relating the history and characteristics of behaviorology as well as the distinctions between it and other disciplines. Upon successful integration of the concepts outlined in this chapter, the reader, where exposed to contingencies to do so,[2] will accurately:

- Define and discriminate among science and natural science

- Identify the basic assumptions of natural science including naturalism and determinism

- Differentiate between behaviorology and other disciplines or branches of disciplines such as psychology, behavior analysis and ethology

The material in this chapter will introduce you to the natural science of behavior, behaviorology and the philosophy of science that informs it. This will help set the stage for the material in the rest of the book.

Behaviorology

Definition and History

Behaviorology, most simply, is the natural science and technology of behavior. More specifically, behaviorology is the natural science and technology of environment–behavior functional relations. The science of behaviorology began in the 1930s, when B. F. Skinner established radical behaviorism, the philosophical foundation of a natural science of behavior discussed here. Skinner operated from within the discipline of psychology, even though his philosophical orientation was incommensurable with psychology. It was called "operant psychology" at that time. Skinner tried, to no avail, to transform psychology into a natural science of behavior, and in the 1970s, the "operant psychology" school of thought became "behavior analysis." This step was a move to distance itself from mainstream psychology, but behavior analysis remained a branch of psychology. After much debate and many years of attempts to make psychology into a natural science, many natural scientists of behavior recognized that they could not change psychology's fundamental transformational paradigm and that remaining associated with the discipline of psychology would compromise their integrity as psychology imposed its influence over the practice of behavior analysis.

---

[2] Readers will note the phrase "exposed to contingencies" throughout the book. This might evoke consideration is some readers of the contingencies we arrange for companion animals. However, this phrase relates to the exposure of the trainer to contingencies that relate to their own behaviors.

Many natural scientists of behavior recognized that complete independence was necessary in order to maintain scientific integrity, but psychology had claimed ownership of behavior analysis and no change in that status was likely to occur. In the 1980s, the entirely independent discipline of behaviorology was founded, with a professional association, educational institution, and peer reviewed journal.[3] Behaviorology is the *only* completely independent natural science of behavior. For an excellent and concise description of the development of behaviorology, see Ledoux (2014) and for a more in-depth description, see Ledoux (2015).

Philosophy of Natural Science

There are different fundamental approaches to studying nature, different "methods of knowing" if you will. Natural science is one such approach. ***Natural science*** is an empirical approach to studying the phenomena of nature based on certain philosophical assumptions. Together, these constraining assumptions go by the name naturalism, which I will discuss below in detail.

Natural Science

Natural science is an empirical approach to studying the phenomena of nature based on certain philosophical assumptions. Together, these constraining assumptions go by the name naturalism. All methods of "knowing" begin with certain fundamental assumptions, and natural science is no exception. By far, the assumptions of natural science have led to the most reliable and productive solutions, when compared to other, less stringently constrained methods of studying nature. ***Naturalism*** is a philosophy of science that holds that *only natural events exist, that there are no non-real or non-natural events, and that all natural events are theoretically measurable in terms of mass, time, distance, temperature and/or charge*. Natural scientists simply do not study non-natural or supernatural events. It is only through careful adherence to these assumptions and constraints that natural science can generate such reliable products such as space shuttles, cures for diseases and personal computers for example. Thus, natural science generates the most robust and reliable products and theories, as compared to other approaches.[4]

A derivative assumption of naturalism is that of **determinism**, which states that any detectable event represents the culmination of a natural history—that is, that all things are part of a continuous sequence of causes and effects and that there can be no intrusion into this sequence by non-natural events. In other words, nothing occurs spontaneously, initiatively or proactively; all events are completely orderly and lawful—that is, caused. This orderly and lawful characteristic of nature allows us to study it. If events were uncaused, reality would be chaotic and unstudyable. Some real events are challenging to measure due to our current technological capacity, but they

---

[3] See www.behaviorology.org
[4] By comparison, for example, psychology is notoriously ineffective at explaining behavior and even less effective in controlling behavior. Likewise, astrology fails to generate effective predictions of behavior.

are still theoretically measurable with a sufficiently capable technological apparatus. If a supposed event is not theoretically measurable, it is not natural, and natural science only studies natural events.

With respect to behavior, naturalism implies that behavior is a passive and a completely caused natural reaction of a body to the environment. Behavior is not an exception to the laws of nature. There cannot be a so-called "free will" as this would imply a non-natural event, force or agent within that could spontaneously and initiatively direct the body to act. Natural science disallows mystical untestable accounts such as the idea of free will. The natural science approach generates a more parsimonious explanation of behavior, leading to more effective and efficient control over behavior. "The naturalistic understanding and acceptance of our fully caused, interdependent nature is directly at odds with the widespread belief ... that human beings have supernatural, contra-causal free will, and so are in but not fully of this world." (Clark, 2008)

It might seem obvious that we only study natural events, and you might be wondering, "okay, sure, we know natural science is not religion, so outside of that, who is proposing non-natural accounts for phenomena?" Psychology explains behavior by the actions of a so-called "mind," a non-natural "thing" or "force" that cannot be observed and directly measured and supposedly directs the actions of the body. Indeed, psychology is "the study of mind and behavior." Moreover, there are those in other disciplines that do not constrain themselves in the same manner as natural scientists (e.g., astrologists) and as such, they cannot claim the status of natural scientists either. Mystical assumptions and accounts preclude any discipline from the status of natural science, regardless of whether they also happen to use some scientific methods to study non-natural events. The label *science*, alone, is reserved for disciplines that utilize scientific methods but otherwise fail to adhere to the basic naturalistic assumptions. These disciplines are often referred to as "soft sciences," which is a *soft* way of saying that the discipline allows for non-natural, or supernatural, events in its accounting for phenomena or that they study non-natural phenomena. Within the realm of behavior, theology and psychology fall into this category. The current comprehensive natural science disciplines include biology, geology, physics, chemistry, and behaviorology. Behaviorology is the only natural science of behavior.

Radical Behaviorism

Behaviorology was founded on a philosophical framework called "radical behaviorism," which was devised by B. F. Skinner in the 1930s.[5] Radical behaviorism simply extends the assumptions of naturalism to the discipline of studying behavior. Although Skinner was employed within an academic psychology department, his subject matter and approach were different from that which is utilized in psychology. Where psychology postulated an inner agent called the "psyche" or "mind" in order to study "mental" processes, Skinner proposed the study of behavior itself, for its own sake, and from a strictly natural science perspective, without reference to hypothetical

---

[5] Note that the word radical, in this use, refers to "fundamental" or "thorough-going" and not to any kind of extremism; it is merely unfortunate terminology selected to distinguish it from other forms of behaviorism.

(i.e., non-real) constructs. Taking his inspiration from the selection causation paradigm used in biology, Skinner improved vastly upon some other "behaviorisms" that existed at the time. Radical behaviorism is hence a philosophy of natural science and places behaviorology among the natural sciences, along with physics, chemistry, biology and geology.

Radical behaviorism is characterized by three fundamental assumptions:

- Behavior is an entirely natural phenomenon, respecting the continuity of events in space and time, which, accumulates as a natural history of fully caused events.

- The emphasis is on analyzing the environment–behavior functional relations, experimental control over dependent variables and the application of that control in culturally beneficial ways.

- Private events such as thinking or emoting are real behaviors occurring in accordance with the same set of laws and principles as more overt behavior.

Clark (2007) put it eloquently: "To be a thorough-going naturalist is to accept yourself as an entirely natural phenomenon. Just as science shows no evidence for a supernatural god "up there," there's no evidence for an immaterial soul or mental agent "in here," supervising the body and brain."

Modes of Causation

Natural science disciplines study the causal relations between real variables.[6] The *dependent variable* refers to the phenomenon to be explained (in our case, behavior) and the *independent variable* refers to the event that is said to cause or explain it (in our case, environmental stimulation). I explore this further in Chapter 2. Different modes of causation make up the relationships between these variables. Physics and chemistry rely heavily on *mechanistic causation*, which deals with what comes immediately before something else and reliably triggers its occurrence—the second thing *depends on* the first. For example, I might say "sit" and that would trigger/cause my dog to sit. Biology and behaviorology rely heavily on *selection causation*, which deals with selection by consequences—that which comes immediately after something influences the future likelihood of that thing occurring again. For example, because I deliver a treat to my dog every time he sits upon my saying "sit," he is more likely to sit in the future, and the consequence that sitting generated *causes* an increase in the likelihood of future sitting behavior.

---

[6] Actually, natural sciences do not always involve the study of causal relations. Cause is not addresses at all in gravitational astronomy, for instance. See Moore (2008, p. 275). Causal relations are, nevertheless, central to the practice of most natural sciences.

Evolutions via Selection by Consequences

Selection by Consequences

Behaviorology recognizes and utilizes both mechanistic and selection causation in explaining and controlling behavior. However, the selection mode of causation is the unique and central feature in explaining and controlling operant behavior within behaviorology. Operant behavior is behavior selected for by consequences. *Selection by consequences* may generally be understood as iterations through cycles of (a) <u>variation</u>, (b) <u>interaction</u> with the environment, and (c) differential <u>replication</u> as a function of that interaction (Moore, 2008, pp. 136–137). Selection by consequences results in three different types of evolution: biological evolution; repertoire evolution; and cultural evolution, which I will describe below.

Biological Evolution

Over 150 years ago, Charles Darwin (1859) elucidated the process of biological evolution. *Biological evolution* involves the selection of traits by consequences within a population across the span of the existence of the population. Within any population, there is variation in traits among the members of that population. Some traits tend to contribute to greater reproductive rates for members sharing those traits while other traits do not. This is often characterized as the environment "selecting for these traits," however, it is important to recognize that "selection" simply refers to differential reproductive rates resulting from the contribution of these traits interacting with the environment. There is no purposeful or goal-directed selecting involved in this process. It is merely a fact that some traits result in higher reproductive rates than others do. The environment selects for traits that contribute to reproductive success and selects against traits that do not contribute to reproductive success. Thus, the genetic material generating reproductively adaptive traits increases in frequency within the population while genetic material that generates reproductively maladaptive traits decreases in frequency within the population over time. This results in the evolution of traits within the population. We might refer to these selective pressures as *contingencies of reproductive success*.[7] To summarize, there is *variation* in the distribution of traits. *Interaction* with the environment involves differential access to resources, survival and ultimately reproductive success within the environment, including the effect that these traits have on the environment and hence the environment's effect on the traits that interact with it. Differential *replication* involves differential reproductive rates for different traits within the population. As the environment changes over time, the selection forces also change and hence the frequency of traits changes.

---

[7] This is sometimes referred to as contingencies of survival, as in "survival of the fittest," but take note that in biological evolution, survival is certainly necessary, but it is insufficient for evolution to take place. Reproduction rates are both necessary and sufficient. Biological evolution requires survival to the extent that one must survive in order to reproduce but, for example, if one survives to a ripe old age but does not reproduce, the survival per se does not contribute to the gene (or allele) frequencies within the population. In addition, "success" simply refers to competitive exclusion of the alternative and does not refer to a purposeful goal or objective.

Repertoire Evolution

About 85 years ago, B. F. Skinner (1938/1991) elucidated the process of *repertoire evolution*, the selection of repertoires of behavior due to consequences within an individual organism across the lifespan of that individual. The term *repertoire* of behavior refers to the full range of behaviors a subject exhibits and can be broken down into categories (e.g., a repertoire of verbal behaviors). An individual's repertoire of behavior evolves moment-by-moment by way of selection by consequences via *contingencies of reinforcement*.[8] Throughout the life of an individual, various behaviors are strengthened by reinforcement, suppressed by punishment, and weakened by extinction on a moment-by-moment basis, as the behaviors are consequated and the relative frequencies of each behavior changes over time. There is *variation* within the distribution of operant behavior. *Interaction* with the environment involves exhibiting behavior that generates changes in the environment that then impact the individual. Differential *replication* occurs through reinforcement, punishment and extinction of operant behaviors within the individual's repertoire. As a result of repeated iterations through this cycle, the individual develops a repertoire of operant behavior. This repertoire is established through selection by consequences. Behaviorologists study behavior at this level of analysis.

Cultural Evolution

At the cultural level of analysis, selection by consequences involves the selection of cultural practices (e.g., traditions, rituals, norms, ethical rules) within a culture/society/community due to the consequences that the practices generate within the span of existence of the culture (i.e., they outlast the life of individuals). This results in *cultural evolution*. There is *variation* within cultural practices within a community. *Interaction* with the environment involves the impact that the practices have on the environment and how that changed environment impacts upon the community. Differential *replication* involves the differential reinforcement, punishment or extinction of practices, due to the efficacy of the practices in solving enduring community-wide problems. Cultural anthropologists study cultures at this level of analysis.

Selection and Causation: Summary and Conclusion

Although behaviorology recognizes and utilizes mechanistic causation (e.g., a stimulus causes/triggers/evokes a behavior), the unique emphasis that Skinner brought to the study of behavior was the selection by consequences mode of causation, and we call this the *selection paradigm*. Just as Darwin had explained the perpetuation of genetically heritable traits within a population through generations (phylogeny), Skinner explained the perpetuation of behaviors exhibited by the individual organism within its lifetime (ontogeny). In biology, the evolution of populations is explained by natural selection, wherein the environment selects for and against certain genetically

---

[8] Contingencies of reinforcement, in this broad usage refer to contingencies featuring reinforcement punishment or extinction.

coded traits simply due to the traits that tend to result in greater reproductive success. Similarly, in behaviorology, the environment selects for and against behaviors simply by whether the behavior generates effective, ineffective or aversive consequences. As mentioned above, cultural selection refers to the selection of cultural practices among a community of individuals. All three of these selection levels are selection by consequences for the rate of genetic change, behavioral adaptation, and cultural practices, respectively. Fraley and Ledoux (2002, p. 41) described selection by consequences this way:

> The consequences of the past behaviors are said to have *selected* the behaviors that now occur, and the selection paradigm takes its name from that interpretation. But in each instance of behavior, the body is assumed to behave *in the only way that it can behave* under the existing circumstances—an assumption that respects the deterministic natural science of philosophy that informs behaviorology. No explanatory appeal is made to a redundant psychological self that would decide or choose the behavior to be exhibited by the body.

Behaviorology involves the experimental analysis of environment–behavior functional relations, as well as the establishment and application of an efficient and effective technology for controlling behavior. Because of the completely naturalistic character of behaviorology and its emphasis on environment–behavior functional relations, behaviorology is highly effective and efficient in controlling behavior (e.g., training).

More information about behaviorology is available at www.behaviorology.org. In order to provide a means to effectively differentiate between other disciplines, I will discuss branches of disciplines and general approaches that assess behavior below in detail.

Psychology

Psychology is an eclectic aggregate of disparate disciplines, defined as the study of the "mind" and behavior (American Psychological Association, 2012). In some "schools of thought," the mind is the primary emphasis of study, and in others, the activities of the mind are said to explain or elucidate overt behavior. The mind is a hypothetical construct representing consciousness, originating in the brain and manifesting in thought, perception, emotion, will, memory, and imagination (American Heritage Dictionary of the English Language, n.d.). These manifestations of the "mind" are referred to as mental, or cognitive processes, and are the primary topic of study for psychology's most prominent school of thought: cognitive psychology. Whereas behaviorology addresses behavior, not as manifestations of or the "will" of a "mind" but rather as a fully caused natural phenomenon, psychology implies or suggests that the "mind" refers to a free-willing inner agent that may spontaneously, or semi-autonomously, "choose" what behavior to "perform," with or without consideration of the environment. This emphasis on mental processes represents the ***transformational paradigm***, prominent in psychology. Those operating under a transformational paradigm observe input (stimulation), and output (behavior), and hypothesize about how the inputs are transformed inside the organism (by the "mind"), to generate output, the emphasis of interest being placed, not on the behavior for its own sake, but rather the transformational process referred to as mental or cognitive.

E.A. Vargus (1991, cited in Fraley & Ledoux, 2002, pp. 40–41) described the transformational paradigm of psychology eloquently:

> An event occurs, described in any number of ways. The meaning of the event inheres in the action of the organism. The organism perceives, interprets, assesses, integrates, and processes its perceptions and cognitions, and then stores the results of its own actions. It then (or later) engages in performance with respect to that event—or, rather, the transformed nature of that event. In psychology's paradigm, some aspect of the world incites the organism to take action; but before that action occurs, the organism engages in a series of operations, typically called mental or cognitive, that determines the significance of the events and thus determines the nature of the action. In the transformation paradigm, the organism itself, through structures and processes inherent in it, is the agency of its action.

When psychologists study behavior, they commonly seek to explain the behavior, at least in part, by reference to mental processes. These mental processes are said to explain the behavioral output. This is, of course, circular reasoning—the mental processes are inferred from the behavior to be explained. Even where mental processes are said to be completely natural, the reference to a "mind" as a hypothetical thing is suggestive of, and perpetuates the notion of, a free-willing agent. In any event, these mental processes are said to explain the behavioral output. This reliance on a non-natural thing/force/entity cannot benefit natural science. Indeed, this reliance negates the discipline of psychology from being a natural science at all.

In recent decades, psychologists have tended to deemphasize the suggestion of a free-willed inner agent, but they have not abandoned the concept. A more recent trend in psychology is to equate the "mind" with the brain and to study the physiology of mental processes (and behaviors) but this reduces psychology to a branch of physiology. Moreover, the physiological processes are not the cause of behavior, but rather behaviors themselves. In the context of explaining and controlling behavior, the concomitant physiological behaviors are not nearly as informative, accessible or useful as the environment–behavior functional relations studied in behaviorology. Physiology studies *how* behaviors occur, while behaviorology studies *why* behaviors occur.

Behavior Analysis

Many readers are familiar with behavior analysis. Behaviorology and behavior analysis share quite a bit of ground. In fact, behavior analysis and behaviorology shared a common history until behaviorologists declared separation and independence from psychology in the 1980s. There is little difference between what I present in this book and many books on behavior analysis.[9] However, there are some important distinctions between behaviorology and behavior analysis. ***Behavior analysis*** is a natural science of behavior. An "asterisk" is required however to qualify this statement, and the reason the qualification is necessary is the reason that behaviorology is necessary. Behavior analysis is not an independent comprehensive discipline but rather a branch of psychology (APA, n.d.; APA, 2013; Ledoux, 2012a). Psychology, the parent organization that claims behavior analysis as a branch or division of its own discipline, is not a natural science, which presents a credibility problem for natural scientists of behavior operating under the behavior analysis label. Behaviorology, on the other hand, is absolutely not influenced or controlled by psychology and is absolutely not any kind, type or branch of psychology, a discipline that is incommensurable with a natural science of behavior. This independence allows behaviorologists to work free from psychology, and the transformational paradigm, and instead, work solidly in a natural science environment. Being independent of any other discipline also means that behaviorology is a comprehensive discipline and behaviorologists must be educated completely and consistently in behaviorology and natural science principles of behavior itself. Behaviorologists work cooperatively with behavior analysts, where appropriate, but behaviorology remains distinct from behavior analysis in order to remain completely independent and separate from psychology.

Some behavior analysts have studied completely within the natural science of behavior and operate completely as natural scientists of behavior without being controlled in any way by psychology. This would be the case for many graduates of the University of North Texas programs in behavior analysis that are provided by a behavior analysis department separate from the psychology department for instance. There are a great many behavior analysts in this position and they might consider, if qualified, operating under the label behaviorologist instead of behavior analyst in that case since clearly what they are engaged in is more closely aligned with behaviorology. Another recent event of interest is the expansion of certification of behavior analysts by the Behavior Analyst Certification Board at the B.A., M.A. and PH.D. levels (Ledoux & O'Heare, 2015, p. 283). University departments separate from psychology and requirements for more comprehensive behavior analytic curricula are promising developments within behavior analysis. However, until behavior analysis declares unequivocal independence from psychology, behaviorology must stand to protect the existence of an uncompromised natural science of behavior.

---

[9] Behaviorological texts are typically more careful to avoid agential language than behavior analytic texts. Agential language involves the use of terms and grammatical structures that assume, suggest or imply an autonomous inner agent (i.e., the mind). This language pervades most or all other verbal communities, but is not scientifically accurate, perpetuates a nonscientific perspective on behavior and the uncritical acceptance of it. Behaviorologists are generally more careful to avoid such language and it is often possible to distinguish between a behavior analytic work and a behaviorological work in this manner. I heard it put eloquently, though quite colloquially, that behaviorology is "hard core."

Ethology

Ethology is a branch of biology that studies nonhuman behavior patterns that evolve in "natural" habitats, either in the species as a group, smaller population groups or individuals, with an emphasis on behavior patterns that do not depend on, or are know to depend on, operant conditioning. Ethology studies behavior from the phylogenetic level of analysis. That is, ethologists are interested in innate behavior and how it evolved, rather than studying the functional relationship between the behavior of an individual animal and the environment (ontology). Examples of the kinds of behavior patterns that ethologists describe and catalogue include "fixed action patterns," "modal action patterns" and "reaction chains." The innate behavior of individual organisms and the conditioning of such behaviors reflect functional relations at the level of the individual and hence is the rightful purview of behaviorology.

Medical Paradigm

Prior to the appropriate expansion of behaviorology (as the natural science of explaining and changing environment–behavior relations) into the animal training field, veterinarians helped clients resolve problematic companion animal behaviors. They applied a similar approach to dealing with problematic behaviors as they did for disease processes—by labeling and categorizing them, and drawing on ethology and psychology to devise treatment protocols, mainly through trial and error. The labeling and categorizing of behavior patterns, and the rather haphazard approach, generated a hodgepodge of handed-down anecdotal treatment protocols. The behavior was left unexplained.[10] This approach is referred to as the *medical paradigm*. Many non-veterinarian "applied behaviorists," with a more extensive background in ethology, have adopted a largely medical paradigm. With the emergence of a natural science of behavior, the training of companion animals and assessment and resolution of problematic behaviors can now be handled systematically, and in a principled manner, based on the laws and principles of behavior.

Orientation of This Book

This book presents a behaviorology perspective. There is no speculation about what animals may or may not understand, desire, or want, or what might be occurring inside a so-called "mind" between stimulation and their behavior. There is no discussion of other fictitious constructs such as "dominance." Nor will I label problem behaviors with

---

[10] Sometimes, the label or category was put forth as the explanation for the behavior, causing a nominal fallacy (i.e., is the mistaken assumption that naming something explains it).

names to provide supposed explanation for them. Instead, the focus of this book is on the actual functional relations that the behaviors of concern are a part of.

Continuing Education

Ledoux (2015) provides an excellent and comprehensive introduction to the history and emergence of behaviorology.

Moore (2008) provides an in-depth treatment of radical behaviorism, although it ignores the role that behaviorology plays in the natural science of behavior.

CHAPTER 2. PRINCIPLES OF BEHAVIOR

Behavioral Objectives

The objective of this chapter is to measurably expand the reader's repertoire of behaviors in relation to describing and relating the principles of behavior. Upon successfully integrating the concepts outlined in this chapter, the reader, where exposed to contingencies to do so, will accurately:

- Define behavior, stimulus, and conditioning

- Identify the cause of behavior at a level of analysis appropriate to training

- Distinguish between operant and respondent conditioning

- Identify the terms in the 3-term operant contingency and explain their relationship to one another

- Describe the physics of environment–behavior relations

- Define and relate added and subtracted reinforcement, added and subtracted punishment, extinction, and analyze behavior scenarios to determine which principles operate on them

- Identify the variables that affect the strength of operant conditioning

- Define the schedules of reinforcement, under what circumstances each is useful and what effect they have on conditioning

- Define operant generalization and discrimination as well as the roles they play in training

- Explain how to transfer stimulus control

- Define differential reinforcement and its variants

- Define respondent conditioning and its role in the operant contingency and training

- Define habituation, potentiation and sensitization processes as well as respondent extinction

- Graph behavior objectives, establish a baseline on the current strength of the behavior, and continue to track the behavior throughout a training project.

In this chapter, I introduce the basic laws and principles of behavior, providing a solid foundation for the rest of the material in this book. The goal of animal training is to bring specific behaviors under stimulus control. To do this, we utilize certain tactics derived from basic principles of behavior. Therefore, we require an introduction to these basic principles. From these basic principles, we derive strategies and procedures to efficiently and effectively generate specific behaviors and bring them under stimulus control.

Environment–Behavior Relations

In this first major section, I will introduce some key concepts, distinctions and principles of behavior that I will elaborate upon as the chapter progresses. In this section, I emphasize the dependent variables (behavior) and independent variables (environmental stimulation) that behaviorology studies.

Behavior

Behaviorologists study the functional relations between behavior and the environment. In this section we will explore behavior, and in the next section, stimulation. **Behavior** is any measurable, neurally mediated reaction of a body part to stimulation. As a natural science, the dependent variable (i.e., behavior) must be measurable. If the dependent variable is not at least theoretically measurable, then it is non-natural and outside of the scope of what natural science can study. The reaction might be neuromuscular or solely neural. In other words, it can involve the movement of body parts by innervated muscles or glands, or it can be solely neural, including consciousness-related behaviors (e.g., those behaviors commonly referred to as thinking, recognizing, comprehending and/or visualizing). The environmental stimulation that causes the reaction can come from anywhere outside of the body part that is reacting to it, be that outside of the body or within it, which is one reason that we specifically identify the behavior of a body part in the definition of behavior. Being neurally mediated means that the energy transferred from the environment causes neural behaviors to occur that then mediate the reactions of other body parts.[11] The word reaction, as opposed to the word response highlights the reactive and passive (rather than proactive and initiative) nature of behavior, in that it is fully caused and not autonomously initiated by a supposed inner agent that may "will" the body to act. Behavior is simply the body's reaction to stimulation.[12]

Because behavior is a neurally mediated reaction, we can use the "dead-body test" to help make instances and non-instances of behavior more apparent. A dead body may "displace space through time," a common element of some older definitions of behavior, but it cannot do so in ways that are mediated neurally. Falling due to gravity is not an instance of behavior, whereas jumping, walking and thinking are. For example, in jumping off a bridge, the jumping is a behavior, but the falling is not. Thinking and visualizing/imagining are real behaviors, even though they are solely neural behaviors and do not involve muscular movement. Emotional responses are also real behaviors of the respondent type (as opposed to operant), involving the release of chemicals into the bloodstream by glands, as I will discuss below.

---

[11] The behaviorologist's interest is with the functional relationship between the sources of the stimulation and the body part's reaction to it. Physiologists study the mediating neural behaviors, or, the *how* of the behavior; behaviorologists study the *why* of behavior.

[12] People may also exhibit the covert verbal behaviors of "thinking about" the stimulation but this stimulation was also evoked. With the organism, as it is structured at that time, the behavior that was evoked or elicited (including the private thinking behaviors, emotional responding and the outwardly evident behaviors) was the only thing that could have happened (determinism). On occasion, it may seem otherwise, because we exhibit thinking behaviors and awareness of its supplementary role, but these thinking behaviors are also fully caused. Due to the limitations of our nervous system, we fail to exhibit awareness of all of the various links in the causal chain all the way back through time, but this is a limitation on our behavior, not an indication of free will. Consequently, it may *seem* as though individuals are "choosing" when in fact, certain body parts are simply (and complexly) reacting to stimulation in exactly the only way those bodies could.

Stimulation

We often frame the environment in discrete units referred to as stimuli. A **stimulus** is any of various measurable kinds, qualities or intensities of energy changes at receptor cells from a source outside of the body (ectovironmental) or from within it (endovironmental)(Ledoux, 2014). Behavior is influenced by stimuli that occur in close temporal proximity to that behavior. A behavior is triggered by a stimulus that occurs immediately before it and is consequated by stimuli that occur during or immediately following the behavior. We often refer to these "changes in energy" patterns as a "transfer of energy" between the environment and the body. We say that the change in energy patterns at receptor cells is the result of a transfer of energy from the environment to the body. It is merely a different perspective on the same process.

Categories of Behavior

Functional Classification: Operant versus Respondent Behavior

Behaviorologists recognize and differentiate between two fundamentally distinct kinds of behavior—operant behavior and respondent behavior. **Operant behavior** is behavior that is selected, maintained and brought under stimulus control by the consequences that it generates. Examples of operant behaviors are sitting, walking and speaking. The other major type of behavior that behaviorologists are interested in is respondent behavior. **Respondent behavior** is behavior that, (a) once the antecedent stimulus reaches a certain threshold, completely compels a bodily reaction, which cannot be prevented with supplemental stimulation and (b) is insensitive to the consequences that it may generate. It includes such behavior as blinking when a puff of air impacts the eyeball, salivating when food is placed in the mouth, and the release of various chemicals into the bloodstream (from glands) referred to as emotional behaviors. Some respondent behaviors are innate, the response portion of the "reflex" relation. However, neutral stimuli can come to also elicit a response similar to the unconditioned response. We refer to this as a conditioned response. The environment *elicits* respondent behaviors and *evokes* operant behaviors. This emphasizes the distinction between these two fundamentally different kinds of behavior.[13] To summarize, the key distinction between operants and respondents, respondents are *not* interruptible once elicited, operants *are* interruptible once evoked, and whereas respondents are essentially unaffected by consequences, operants are driven by consequences.

---

[13] Take careful note that respondent behavior is always and only *elicited* and operant behaviors are always and only *evoked*. To say that operants are elicited or that respondents are evoked would be incorrect.

Unconditioned versus Conditioned Behavior

Behaviorologists are largely interested in operant behaviors. However, a full appreciation of behavior requires an appreciation of the distinction between conditioned and unconditioned behavior. There are both unconditioned and conditioned forms of respondent behavior. In this section, I will explore the distinction between these unconditioned respondent behaviors and all other behaviors (including operants), which are conditioned. Each species evolves to exhibit certain unconditioned behaviors. The species' anatomical structure mediates these unconditioned responses to specific stimuli. Biologists refer to these eliciting stimuli → respondent behavior relations as *reflexes*. These *Unconditioned behaviors* are behaviors that have not been conditioned but occur because the body is so structured by biological evolution to react automatically and inevitably to certain kinds of stimulation. Generally, these behaviors are (a) relatively uniform across the species, (b) topographically stereotypical, and (c) elicited by very specific forms of stimulation. Many unconditioned behaviors, such as eye blinking when a puff of air touches the eyeball, or salivating when food enters the mouth, are discrete and simple in scale. Some are slightly more complex, including the startle response that involves retracting a limb from a heat source or the opposition reflex of opposing pressure applied to certain body parts such as the head (such as for instance when a barber tries to move your head and your body pulls against it).

In some cases, certain innate reactions comingle with operant behaviors in what ethologists refer to as "*fixed action patterns*" or "*modal action patterns*." Ethologists refer to these as "species-typical" or "species-specific" behaviors or behavior patterns. Respondent and operant behaviors are always occurring, including simultaneously. Contingencies involving innate species-typical respondent behaviors can interact in operant contingency processes. Unconditioned responses to certain stimuli that automatically function as reinforcers are instances of species-typical behavior. Examples of these species-typical behavior patterns include feeding, mating, nesting, social greeting rituals and attack patterns. These more complex behavior patterns involve both unconditioned responses and operant behaviors. For instance, in a fixed or modal action pattern involving ritualized greeting behaviors, contact with a conspecific elicits the entire set of behaviors in that pattern. These types of behavior patterns are elicited by discrete stimuli, differentiating them from reaction chains discussed below. Historically, our ethologist colleagues have catalogued these behavior patterns, in what they call *ethograms*.

Reaction chains are chains of species-typical behaviors, with each reaction segment of the chain being elicited by environmental stimulation (Moore, 2008, pp. 85–86). In contrast to fixed or modal action patterns, in a reaction chain involving predatory hunting behaviors of certain canine species (including some dogs), the eyeing, stalking, pouncing, grabbing and shaking/holding behaviors are each triggered by different stimulus conditions made available by the previous segment in the chain. The sight of a wounded prey animal for instance, may trigger eyeing and crouched slow-approach behaviors, and then the closeness of the prey animal may trigger the pouncing and grabbing segments. Finally, the stimulus provided by the prey in the mouth triggers the shaking and holding behaviors. It is important to note that although these species-typical behaviors involve some unconditioned components, they generally involve operant behavior components, and as such, they are, to a degree, interruptible and the consequences they generate will influence the future likelihood of those behaviors occurring again.

It is appropriate to appreciate the role of unconditioned reactions, but I caution the reader to avoid ethological labels when analyzing behavior episodes. Terms such as "dominance," "submission," "displacement," "cutoff behavior," or worse, terms that are not even ethological such as so-called "calming signals" all provide a false sense that the label renders the behavior explained and this militates against proper analysis. Labels *do not* explain the behavior and they *certainly* do not help control it.

Traits are Not Behavior, Nor are They Causes of Behavior

To further understand what behavior is, it is important to make a distinction between traits and behavior. Laypeople may refer to traits such as stubbornness, rudeness, spitefulness, selfishness, aggression or hostility, to describe an animal's behavior, but although these descriptors might be evoked when exposed to contingencies to render a general verbal description of a tendency, they are not behaviors per se. People even sometimes use these labels as explanations for behavior. For example, someone may say that a dog bites because he or she is aggressive. These traits are not even behaviors; they certainly fail as explanations of behavior, as they involve only one variable in the stimulus–response sequence. Animals do not bite because they are "aggressive" or possess some trait called "aggression" or "hostility." This fallacy (referred to as the "nominal fallacy,") might persuade some people to believe that the behavior is then explained, but it fails to adequately provide a dependent variable for the observed behavior, let alone functionally related dependent and independent variables. In the above example, note that the dog is said to bite because he or she is aggressive but he or she would also be said to be aggressive because he or she bites. Thus, the reasoning is circular and counterproductive, as there is really just one variable that has been given two names. The nominal fallacy is very common in the animal behavior field. As another example, some people attempt to explain certain behaviors by saying that the dog is "dominant." The animal exhibits the behavior because he or she is dominant, and he or she is dominant because he or she exhibits the behavior. Notice only one variable again—the dependent variable. The word "dominant" is simply another word for the behavior. That is another circular explanation, along the same vein as described above for aggression. Valid explanations require two distinct variables, the dependent variable (the behavior) and the independent variable (the stimulus that caused the behavior); they must be distinct events. It is important to ensure that actual behavior and the functional relations of which it is a component are the currency of the animal training profession and not traits, speculations or assumptions about behaviors or fictitious hypothetical constructs like "mind" or "dominance."

Functional Relations and Contingencies

Nature is composed of an unbreakable continuity of completely natural events that accumulate in a natural history. In natural science, the event (e.g., the response) that is being explained or accounted for (the thing being caused) is referred to as the dependent variable, and the event (e.g., the stimulus) that causes it is referred to as the independent variable.[14] Natural science examines the functional relationship between real dependent and independent variables as they occur in nature. In behaviorology, the dependent variable is the behavior under consideration, and the independent variable is the environmental stimulation necessary and sufficient to produce the behavior. Therefore, behaviorologists study the functional relations between behavior (the dependent variable) and the environment (the independent variable).[15] A *functional relation* refers to a reliable relationship between a dependent variable and an independent variable. Thus, when we refer to a functional relation, it is the relationship between two events; one depends on the other to occur. Since we cannot observe every single interaction between a particular dependent and independent variable, this term is more appropriate and justifiable than the term "causal relation." I will discuss the basic functional relations studied by behaviorologists below.

You have likely been exposed to the phrase *ABCS of behavior. A* stands for antecedent, *B* stands for behavior and *C* stands for consequence (a type of postcedent stimulation). The *A* and *C* represent the environment and the *B* represents the behavior that is functionally related to *A* and *C*. Antecedent stimuli evoke the behavior or influence the effectiveness of the evocative stimulus (S^{Ev}), and the consequences increase or decrease the likelihood of the behavior being evoked on subsequent occasions. The ABCs of behavior are a useful acronym for evoking the concept of the three-term contingency, which is of fundamental importance to the natural science of behavior even if it slightly over-simplifies the formula. However, for now, the ABCs of behavior ease us into the complexity of the formula.

A *contingency* is a description of a functional relation in episodes of behavior and will be a major focus of our analysis of behavior. The phrase functional relation emphasizes the relation between variables/terms, whereas the word contingency refers to the terms/variables and the functional relation between them. The term *contingency analysis* is used to refer to both (a) the *procedure* involved in analyzing episodes of behavior, including describing functional relations, and (b) the descriptive notation of the contingency—the *product* of the analysis.

Let us elaborate on the slightly over-simplified ABCs of behavior by introducing a bit more complexity. First, let us look at a slightly broader formulation of the three-term contingency than provided in the ABCs. While the basic three-term contingency is A $\rightarrow$ B $\rightarrow$ C, the broader formulation is A $\rightarrow$ B $\rightarrow$ P. In this sequence of events, A is the antecedent stimulus, B is the behavior (and will be represented by a response class), and

---

[14] You can remember which is which with the verbal supplement that INdependent variables are the variables the trainer INfluences and DEPENDent variables are the variables that DEPEND on the other variable.

[15] There are several uses for the word causal. For instance, there are mechanistic and selection causation and these are both quite different kinds of relationships between two events. The term functional relation (as opposed to causal relation) is used to ensure greater clarity and precision.

P is the postcedent stimulus (as a prompt, *ante* means before and *post* means after). Antecedent stimuli are those that occur immediately before a given behavior and postcedent stimuli are those that occur immediately following that behavior. The arrow indicates the functional relation. Thus, the antecedent stimulus *functionally controls* the behavior, which in turn *functionally controls* the postcedent stimulus. There are other ways to frame the relationship. For example, the postcedent stimulus is *contingent upon* the occurrence of the behavior and the behavior is *contingent upon* the occurrence of the antecedent stimulus. Alternatively, the postcedent stimulus *depends* on the behavior, which *depends* on the antecedent stimulus. Arrows in symbolic notations of contingencies stand in for the italicized words above. Once we confirm that an antecedent stimulus has evoked the behavior, we refer to it as an evocative stimulus.[16] Once a postcedent stimulus is functionally related to the behavior and changes the likelihood of future instances of the behavior, we refer to it as a consequence.[17] Thus, operant behaviors are components of a three-term, antecedent → behavior → consequence, contingency, often referred to as a stimulus–response–stimulus contingency. In contrast, respondent behaviors are components of a two-term antecedent → behavior contingency, often referred to as a stimulus–response contingency.

With regard to respondent behaviors, the antecedent stimulus *elicits* the behavior. This is a mechanistic cause. The body is so structured, by biological evolution, as to make certain reactions inevitable, automatic and uninterruptible when the organism is exposed to certain kinds of stimuli. The number and kind of stimuli that can come to elicit the same reactions also expands via conditioning. We call behaviors that occur automatically without any previous conditioning **unconditioned responses**, and behaviors that occur only after conditioning **conditioned responses**. These are all respondent behaviors and inevitable, once elicited. To be clear, there is no free-willing agent that "chooses" to "perform" behavior or "anticipates" the unconditioned stimulus or that "learns" about such "associations." An elicited behavior is completely inevitable and the organism is a passive functioning physiology (i.e., body) that is structured in a way that it will simply exhibit the behavior when exposed to the stimulus. The organism does not "learn to associate" stimuli or "learn to perform" behaviors. The organism merely mediates the behavior.

With regard to operant behaviors, the mechanistic cause is also the antecedent stimulus that *evokes* the behavior. Antecedent stimulation triggers an operant behavior because of a history of consequences generated by that behavior in the presence of the antecedent stimulus (specifically, reinforcement in the case of increasing and maintaining the rate of behavior). This is selection causation. The consequence restructures the nervous system in a way that changes the organism's receptivity to the antecedent stimulus in the future, making the behavior more or less likely to occur in reaction to it. With each trial through the contingency, the consequence changes the structure of the organism. We refer to this as operant conditioning. This cybernetic loop constitutes the physics of the selection causation process discussed in detail below. In a given trial of a consequated behavior, it is important to realize that the consequence

---

[16] The older term "discriminative stimulus" (S^D) is acceptable as well. However, the newer term evocative stimulus is now more common within behaviorology and we consider it the proper term.

[17] I will introduce other categories of postcedent stimulation (e.g., selectors) below.

observed in the trial did not cause or affect the behavior in *that* trial, as the future cannot cause the past. Mechanistically, a response occurs because an antecedent stimulus occurred and evoked the response. Through the mechanism of selection causation, a response occurs because past instances of the response class resulted in reinforcing restructuring of the nervous system (i.e., conditioning) that results in an increased likelihood of the response class across subsequent exposures of the animal to the stimulus. More specifically, the behavior → reinforcer contingency occurs just after the antecedent stimulus → behavior contingency, and thereafter, the behavior is more likely to occur when the antecedent stimulus occurs.

Some people distinguish between respondent and operant behavior based on the argument that respondent behavior is "involuntary," whereas operant behavior is "voluntary." This is inaccurate, because it implies that some behaviors are not fully caused by the environment, but rather that some behaviors (i.e., respondents) are reflexive and that others (i.e., operants) are free-willed. This is another fallacious influence from psychology to guard against. There are no inner agents that can spontaneously or semi-autonomously initiate behavior and no behavior is "voluntary" in the proper sense of that word. Some people believe that because operants are interruptible by exposure to new streams of energy between evocation and reacting, and because awareness behaviors and private verbal behaviors (e.g., thinking) participate in the evocation of behavior, that they "choose," "will" or "initiate" that behavior. This is, however, an illusion. More accurately, this belief behavior is based on a false inference. The verbal behavior participating in the contingency is also fully caused and being interruptible by newly introduced streams of energy does not indicate free will. Behavior is not "performed"; it simply, inevitably, naturally happens. Just as we do not say that a radio voluntarily performs the sounds it emits, nor do we say organisms voluntarily perform behaviors that they emit. The radio merely mediates the sounds and living things also simply mediate behavior—we are not exceptions to the laws of nature just because the belief that we are an exception generates reinforcement for many people. These "beliefs" are behaviors too, and can be maintained by reinforcement.

Conditioning

Conditioning is a behavior change process wherein energy pattern changes at receptor cells cause a cascade of neural firings that results in a micro restructuring of relevant parts of the nervous system, which results in a change in receptivity to stimulation in the future and hence a change in behavior. In short, conditioning is a behavior change process wherein stimulation results in small-scale changes to the nervous system that results in a change in behavior. As a result of conditioning, the antecedent–behavior functional relation is strengthened, suppressed or weakened, whether by a simple pairing as in respondent conditioning or by consequating the behavior as in operant conditioning.

Stimulation, as described above, refers to any of various measurable kinds, qualities or intensities of energy changes at receptor cells. When we say that the antecedent–behavior relation is strengthened, suppressed or weakened, this is in reference to the effects of various different conditioning processes. With respect to operant behavior: reinforcement *strengthens* the antecedent–behavior relation, resulting in maintenance of, or increase in, the rate of that behavior; punishment *suppresses* the antecedent–behavior relation, resulting in a decrease in the rate of that behavior; and extinction *weakens* the antecedent–behavior relation resulting in a decrease in the rate of that behavior.

With respect to respondent behavior, pairing neutral (or conditioned) stimuli with unconditioned (or established conditioned) stimuli will *strengthen* the likelihood or magnitude of the conditioned response, while presentation of the conditioned stimulus without pairing it with the unconditioned stimulus will *weaken* the likelihood or magnitude of the conditioned response (more on these processes below). Procedurally, when a neutral stimulus and unconditioned stimulus impact the organism at approximately the same time, that stimuli have been **paired**; the stream of energy from each of the two stimuli impinge upon the organism contiguously.[18]

In the case of either respondent or operant conditioning, conditioning involves a transfer of energy from the environment to the organism, resulting changes in energy patterns at the receptor cells and then in small-scale restructuring of the organism's nervous system. This restructuring then changes how the body reacts to the environment. These streams of energy include the photons impacting the optic nerve, which are seen; pressure waves entering the ear, which are heard; pressure on the body, which is felt; and molecules entering the nasal cavity, which are smelled and the like, all of which are then processed by the brain through malleable neural networks. I will discuss this in detail in the next section.

The term "learning" is commonly used in psychology, including the branch of psychology called behavior analysis. However, behaviorologists avoid this term in favor of the technical term **conditioning**. The term "learning" is suggestive of a cognitive operation that individual agents "perform" in an active manner. The grammar of its use commonly supports such an implication. A psychologist might say "a student learned certain things in class that day," whereas the behaviorologist might say "conditioning occurred" or that the "contingencies conditioned a student to exhibit certain behaviors that day." In passive voice, there is no implication of free will, and in active voice, the environment (i.e., contingencies) rather than the student is the subject, thereby avoiding agential implication. Behaviorologists favor the word conditioning as a technical term in order to avoid such agential implications and careful use of grammar to avoid implying that organisms are engaging in an initiative "learning" process.

Training is simply arranging the circumstances that will result in conditioning toward the end of bringing specific behaviors under stimulus control. The objective of a training project is to change behavior as efficiently and effectively as possible toward established objectives. Trainers establish specific behavior objectives and systematically identify and apply standard training strategies, in order to condition the behavior and bring it under the control of specified stimuli. They apply the appropriate principles in formulating procedures or protocols that will achieve the objectives, and then carry them out. They achieve all of this by controlling the causes of behavior—the

---

[18] Note that the most effective conditioning occurs when the unconditioned stimulus comes immediately after the neutral stimulus with only a very small interval between them, or when the conditioned stimulus begins occurring and ends just after the unconditioned stimulus begins occurring, with a little overlap. Other arrangements are less effective.

environment. The change made to the environment causes the resulting change in behavior.

The Physics of Behavior

Informed by this appreciation of the variables studied by behaviorologists, I will describe the physics of how these variables interact in more detail below.

The environment is awhirl with streams of energy (stimuli) of various kinds in various quantities. Photons (detected as light) are ubiquitous in the environment, as are pressure waves (such as those detected as sounds). A vast number of molecules enter the olfactory system through the nose and we feel pressure from anything touching the body. While not all of the streams of energy that impact upon the nervous system cause large-scale reactions, at any given moment, a body is likely reacting to multiple sources of stimulation, with only a fraction of the stimuli also simultaneously evoking awareness behavior. In other words, we are not even aware of all of the stimuli that impinge upon us, nor the behaviors that they elicit and evoke. Furthermore, behavior is usually multiply controlled rather than being influenced by a simple, single, discrete stimulus. Moreover, just as we are reacting to one set of stimuli, other stimuli are already eliciting and evoking additional reactions.

The statement that a stimulus impacts upon or affects an organism, refers to the stream of energy emanating from or deflecting off of something in the environment that impinges upon the organism's sensory nervous system, resulting in energy pattern changes at receptor cells triggering a cascade of neural firings within the nervous system. In this way, energy is transferred from the environment to the nervous system.[19] This transfer of energy triggers a cascade of neural firings that induces greater energy expenditures involved in and required for neuromuscular behavior. We call this incoming energy stream contacting the body stimulation and the resulting bodily reaction to it behavior. Thus, the stimulation evokes the behavior. The energy transferred to the body during stimulation is adequate to trigger the behavior, but it does not fuel it; the potential energy inside the body, derived from nutrition, fuels the behavior. The process described thus far constitutes the antecedent → behavior contingency component of the three-term contingency. Three examples follow. Example 1: A Pepsi® can on a table deflects ambient light and the photons impact upon the photoreceptor cells in the retina of a person's eyeball. The light physically changes compounds within these cells and electro-chemical signals are transmitted through successive neurons in the retina, to the optic nerve fibers, and eventually the cerebral cortex. Due to the current structure of that nervous system (as structured by biological evolution and previous conditioning), energy expenditure processes are mediated and a reaction is triggered. In this case, the hand of the body reaches out, grabs and squeezes the can until it is crushed. Example 2: Light emanating from the sun impinges upon the sensory nervous system causing a cascade of neural firings and an expenditure of

---

[19] The energy is actually *presumably* emanating from or reflecting off something in the environment because our only contact with nature occurs within our nervous system. We presume that the environment outside of our reactions to it exists, but this must necessarily remain just a useful assumption. For a detailed discussion on this fascinating topic, see Fraley (2008; 2015) and Ledoux (2014).

energy from the body. The reaction of the body is for the eyelids to squint. Example 3: Pressure waves emanating from a fire truck siren impact upon the receptors in the auditory system, which results in a cascade of neural firings associated with the sense of hearing, and an expenditure of energy from the body in the form of the tact "fire truck."[20]

The behavior (i.e., the reaction) generates a stream of energy that impinges upon the environment, thereby causing a change in that environment in some way, notably, a change in the streams of energy that constitutes the environment. The body then detects this change as the, now different, stream of energy contacting the organism's nervous system. The transfer of energy from the environment to the body again induces a cascade of neural firings through the nervous system and causes small-scale changes to the nervous system (a micro-restructuring) that then change the way that the organism reacts to the environment thereafter. We refer to this postcedent stimulation as the consequence. Along with the behavior that generated it, this consequence constitutes the behavior → consequence contingency in the three-term contingency. The process of behavior causing a change in the environment or change in how the body contacts the environment that is then detected by the body is referred to as a *cybernetic loop*.[21] Together with the contingency described in the previous paragraph, this entire process constitutes the basic three-term contingency. The body's reaction to stimulation is behavior and the thereafter-changed nervous system that results in a change in how the body reacts to the environment in the future is *conditioning*. Let us revisit each of our three previous examples, describing the transfers of energy and reactions to it.

Light deflecting off an empty Pepsi® can impacts a sensory nervous system causing a cascade of neural firings and an expenditure of energy from the body. The reaction of the organism is to reach out and crush the empty can, causing light to deflect off of the, now crushed, can impacting the nervous system, again causing a cascade of neural firings and small-scale structural changes in the nervous system, possibly resulting in a greater likelihood of can crushing behavior across subsequent exposure to empty cans.

Light emanating from the sun impinges upon a sensory nervous system causing a cascade of neural firings and an expenditure of energy from the body. The reaction of the organism is for the eyelids to squint, causing changes in the stream of energy (light) impacting the nervous system, again causing a cascade of neural firings and small-scale structural changes to the nervous system resulting in a greater likelihood of eyelid squinting behavior across subsequent exposure to strong sunlight. Notice that this example differs from the previous one in that squinting simply changes how energy impinges upon the nervous system whereas can crushing causes a change to the can, which the body then detects.

---

[20] A tact is "a verbal operant in which a response of given form is evoked (or at least strengthened) by a particular object or event or property of an object or event." (Skinner, 1957/1992, pp. 81–82)

[21] Occasionally, framing the transfer of energy in terms of "streams of energy" can evoke confusion, such as when the stream of energy is difficult to envision, as in the squinting example. Another way to frame this process is by referring to "changes in energy patterns of receptor sites" instead of to streams of energy. By framing the process as changes in energy patterns at receptor sites, we can more readily conceptualize the transfer of energy involved in examples such as eye squinting. In that example, you can see that squinting changes the pattern of energy at the receptor cells in the eye. Energy is transferred throughout the process and from this transfer of energy, patterns of energy at receptor cites are changed.

Pressure waves emanating from the fire truck siren impact upon a sensory nervous system causing a cascade of neural firings and an expenditure of energy from the body. The reaction of the organism is a verbal tact "fire truck," causing a stream of energy (pressure waves) impinging upon another nervous system (i.e., a parent) that reacts with the verbalization "yes, that's right." This results in a stream of energy (pressure waves) traveling back to the nervous system, again causing a cascade of neural firings and small-scale structural changes to the nervous system resulting in a greater likelihood of the verbal tact "fire truck" occurring upon subsequent exposure of the individual to pressure waves produced by fire truck sirens.

These are rather simplified representations of the very complex relationships between stimuli and behaviors exhibited by biological organisms. The impact of other factors adds to the complexity of behavioral responses. For instance, it is likely that not all empty Pepsi® cans would evoke can-crushing behaviors in all people. There are other forms of stimulation occurring, which we refer to as "context" (and technically, function-altering stimulation). The body may not reach out and crush empty cans outside of one's own house; perhaps only cans destined for the over-crowded recycling bin may evoke the crushing behavior. In addition, once in a while, the can might leak a bit of sticky Pepsi® onto the hand or a tear in the aluminum might cut the hand, introducing aversive contingencies to the mix, competing with the reinforcing contingencies. Being exposed to an empty Pepsi® can destined for the over crowded recycling bin might involve several concurrent contingencies, some competing with others. This example and indeed the others described above may evoke the recognition of various other complexities in addition to those in the simplified examples presented above. It is truly, and complexly, amazing, yet all completely natural.

Operant Conditioning

In this section, I will explore operant conditioning in detail, with a focus on elaborating on the various kinds of operant contingencies. *Operant conditioning* is a behavior change process wherein behaviors become more or less likely to occur on subsequent occasions of the occurrence of the evocative stimulus, due to the historic consequences that it has generated. In other words, when a behavior generates a consequence, the organism is conditioned such that the evocative stimulus becomes more or less likely to evoke that behavior. The functional relation between the evocative stimulus and the behavior is strengthened, suppressed, or weakened depending on the effects of the consequence. This discussion calls for a review of contingencies, this time in greater depth.

Operant Contingencies

At this point, you should be familiar with a number of concepts and terms that I will apply in the following sections. This should prove useful in expanding your understanding of the principles of behavior as I introduce more terms and concepts.

Terms Used to Describe Operant Contingencies

The three-term contingency, as introduced above, is a basic formula for beginning to analyze episodes of operant behavior. The three terms in the three-term contingency are:

- the antecedent stimulus;

- the behavior; and

- the postcedent stimulus.

With this foundation established, a more in-depth elaboration is possible and in order.

Antecedents

Antecedent stimuli comprise all stimuli that occur immediately before a behavior. We refer to stimuli that does not evoke or otherwise influence behavior as a neutral stimulus (S^N) with respect to that particular behavior. That is, a neutral stimulus might occur before a behavior, even frequently, but it is either coincidental or otherwise simply not functionally related to the behavior. Once we have confirmed a functional relation (that is, that the stimulus actually does evoke the behavior), we refer to the antecedent stimulus as the *evocative stimulus* (S^{Ev}). The older term, *discriminative stimulus* (S^D), is acceptable as well. There are often numerous stimuli sharing control over behavior, even though we usually simplify the analysis by identifying the stimulus that exerts the most control. A stimulus becomes evocative of a behavior after a history of reinforcement following that evocative stimulus–behavior contingency. If an evocative stimulus–behavior contingency is no longer reinforced, it will become extinguished and once it does, the evocative stimulus becomes a neutral stimulus once again—it is a specific kind of neutral stimulus referred to as an S-delta (S^Δ). If, on the other hand, the behavior is suppressed (even completely) by punishment, we still refer to the antecedent stimulus continues as the evocative stimulus because punishment merely suppresses the behavior and the stimulus remains evocative. I will describe these processes in greater depth below.

I will explore other classes of antecedent stimuli in the section to come, including various kinds of function-altering stimuli (i.e., stimuli that are not evocative but they influence the likelihood of the evocative stimulus evoking the behavior).

Behavior

As discussed, *behavior* is any measurable, neurally mediated reaction of a body part to a stimulus, be the reaction neuromuscular or solely neural. Whereas the term behavior is general, a *response* is a particular occurrence of a behavior. No two responses will ever be precisely the same in every way. Thus, a response is a one-time event. By definition, one cannot repeat a one-time event and therefore we cannot track it quantitatively. If walking in general is an example of a behavior, walking on one particular occasion is a response. A *response class*, with respect to operant behavior, refers to a set, group or variety of response forms that may differ topographically but share the common function of contacting the same reinforcer.[22] We define response classes functionally rather than topographically. We refer to different kinds or forms of responding within a response class as a ***response class member*** or ***response class form***. Some examples will help bring these terms under appropriate stimulus control. If forward progress were the reinforcer maintaining walking behavior, we would say that the behavior that achieves progress in forward motion is a response class. There may be numerous types of walking, with some being quick and others slow, or some responses characterized by long strides and others by short strides. For convenience, we ignore these variations and refer to each response to as a member of the response class "walking." In other examples, an even wider range of response class forms might achieve forward motion, including such forms as running, skipping, crawling, galloping, cantering and trotting. Each of these is a response class form because they are specific forms of the response class of achieving progress in moving forward. This wider response class would not be identified as "walking," because walking is defined more narrowly to include motion involving at least one foot on the ground at all times (e.g., in dogs and horses, a four-beat gait with two or three feet on the ground at all times). In this case, we might invoke the word "locomotion," to represent this wider response class. The point here is that a response is a single instance of behavior. Thus, we utilize the notion of a response class in order to conveniently clump a number of *functionally* related responses together and treat them as repeatable behaviors. We can further clump different forms of that response class together as response class forms.

These can be some challenging distinctions to master if they are unfamiliar, particularly if one previously considered some of them as synonyms. To summarize, we define a response class functionally and it is composed of potentially multiple response class forms. The response class forms are the various types or forms of the responses that we may observe within the class. So, for one response class involving drinking water from a glass, response class forms include taking a small sip of water from the glass, drinking a large quantity of water over multiple sips, and "chugalugging" the entire contents of the glass at one time. We can consider all of these forms to be the target behavior by defining the response class as "drinking water from a glass." In this manner, we can study multiple instances of behavior over time.

Postcedents

Postcedent stimuli comprise everything that occurs immediately after the behavior. Many things occur after a behavior, and only some of them are functionally related to the behavior. Once a postcedent stimulus is confirmed to change behavior, whether it was generated by the behavior or occurred coincidentally, it is called a ***selector***. This stimulus selects for behavior. A selector that changes behavior coincidentally, rather than actually being generated by the behavior, is called a ***coincidental selector***, and the behaviors maintained by coincidental selectors are referred to as ***superstitious behaviors***. Superstitious behaviors are usually short-lived because the reinforcement is merely coincidental and therefore, usually on a sparse schedule. We refer to a selector that influences behavior, and is not merely coincidental, but is actually generated by the behavior as a ***consequence***. A consequence can be generated by the behavior in the

---

[22] A response class, with respect to respondent behavior refers to a set of responses that share a common eliciting stimulus.

sense that it changes the environment directly (e.g., you squeeze a cola can and the can becomes crushed), or it can be mediated by someone else, as long as it occurs *contingently* upon occurrence of the behavior (e.g., you provide a treat to a dog each time he or she sits). Figure 1 provides a visual depiction of these relations and categories. Take some time to examine all of the pathways presented. Consequences can be divided into different types, depending on (a) whether the consequence increases or decreases the likelihood of the behavior, (b) whether the stimulus was added or subtracted immediately following the behavior, and (c) whether the stimulus achieves the change as a result of previous conditioning or not. Another postcedent behavior change process involves a lack of consequence where once there was a consequence. We can more easily appreciate all of these processes by framing them in terms of the contingencies they are components of, the topic that I will address below.

Contingency Analyses

By ascertain the basic three-term contingency that operates in any particular episode of behavior, the relationship between the behavior and its environment is investigated and described, thereby explaining the cause of the behavior event (identifying the dependent and independent variables and their functional relations). The evocative stimulus → behavior contingency describes the mechanistic cause, and the behavior → consequence describes the selection cause. The principles of behavior discussed below elaborate how these relations operate. Whether you are explaining a behavior event that has occurred or planning a training project, the contingency analysis is an important foundation for these activities.

For now, this is a rather simplistic treatment of the topic. There can be more than three relevant terms in a contingency and there often are. Furthermore, with respect to a particular response class, there may be several contingencies that operate simultaneously on the behavior, some reinforcing and others punishing. We determine the rate of responding (or the likelihood of a behavior being evoked) by the net result of the contingencies, as well as the history of those contingencies. Furthermore, both operant and respondent contingencies occur simultaneously and often influence each other. I will discuss these complexities in more detail below. For now, let us keep exploring the three-term contingency and how to depict it.

Before we examine contingencies further, it is useful to discuss the symbolic notation used to depict functional relations in written contingencies. We use arrows (i.e., →) to indicate a functional relation. If there is no functional relation, no arrow is used. If a previously operational functional relation existed but no longer does, then we use an arrow with a slash (/) through the arrow to indicate this (i.e., –/→). In this situation, the functional link has been broken. In cases where a term has not yet been identified, a question mark (i.e., ?) is used to hold its place.

I will provide a general representation of the three-term contingency before any specifics have been determined or identified below. Note that we have not yet inserted arrows, because we have not yet confirmed the specific functional relations.

Antecedent – Response class – Postcedent

Once the actual antecedent stimulus that evokes the behavior has been determined as well as the postcedent stimulus that affects the future likelihood of the behavior, we may depict the specific contingency in a manner similar to this example:

$$\text{"Sit"} \rightarrow \text{Jake sits} \rightarrow \text{Treat}^{(+R)}$$

This example involves an added reinforcement consequence. Note that we identified the specific consequating stimulus (i.e., treat), as well as the conditioning outcome (i.e., $^{(+R)}$) that identifies the specific kind of functional relation involved. I will describe the various types of reinforcement and punishment in detail below.

When analyzing a behavior episode, start with the target behavior. In any given situation, there might be multiple behaviors exhibited, some operant and some respondent. If you are analyzing an operant behavior, start with identifying the *particular* response class you seek to explain. This clarity makes the analysis much simpler and cleaner; failing to do so is the source of much confusion when novice trainers seek to analyze an episode of behavior.

$$? \quad - \quad \text{Jake sits} \quad - \quad ?$$

You may proceed to identify the evocative stimulus and the consequences, once you have identified the target behavior. I will illustrate the process here by identifying the evocative stimulus next, but depending on the circumstances, you may identify the consequence first, particularly if the evocative stimulus is currently unclear.

Usually, what occurs just before the behavior is what evokes it. However, if a few different salient events occur right before the behavior, you may have to choose one as the most likely candidate for evocative stimulus, taking into consideration the subject's reaction to each of the various relevant stimuli in the past. There may be one obvious stimulus package, but you may be unsure about which features of that stimulus are evocative and which coincidental. For instance, if a male postal carrier evokes barking behavior at the door, you may not know whether all of the features of that stimulus package are evocative together, or whether just maleness or just the postal carrier uniform are evocative or even if simple approach to the door is the actual specific evocative stimulus in question. Hypothesized functional relations can be tested and confirmed or refuted with simple experiments of testing each stimulus individually, controlling for the others, until you identify the one that evokes the behavior.[23] At first, you may not be able to determine exactly what is evoking the behavior and you may leave a "?" in that term position, and move on to identify the consequence. We would use a dash to represent a hypothesized functional relation and an arrow for a confirmed one. Let us say that we have confirmed that Jake sits reliably when someone says, "sit."

$$\text{"Sit"} \quad \rightarrow \quad \text{Jake sits} \quad - \quad ?$$

---

[23] We call this experimental assessment process a "functional analysis."

Next, we assess the postcedent environment for a hypothesized consequence. Usually, it will be the event occurring immediately after the behavior. However, as with antecedents, you might need to hypothesize the most likely consequence for the contingency and test it experimentally to confirm it. Let us say that someone often gives Jake a treat right after he sits and if we stop giving treats, the rate of the behavior declines. In that case, we replace the dash with an arrow. Furthermore, the stimulus is added (as opposed to subtracted) and the change involves an increase in responding (as opposed to a decrease). Therefore, we include the type of consequence as added reinforcement (+R).

$$\text{"Sit"} \rightarrow \text{Jake sits} \rightarrow \text{Treat}^{(+R)}$$

Of course, there may indeed be other added reinforcers and/or subtracted reinforcers participating in the control of this behavior. If more than one reinforcer is likely important in maintaining this behavior, you should write two contingencies, one above the other and ideally ranked by importance. For example, when Jake sits, sometimes it may be consequated with a treat and sometimes with a quick game of tug-of-war. It is likely useful to identify both sources of reinforcement in this case. You could identify which reinforcer is most effective experimentally, but this is rarely necessary. You may also find a situation in which the behavior is maintained by both added and subtracted reinforcers. This might occur if the guardian demands the sit in an aggressive manner and then delivers a treat upon occurrence of the behavior. The sitting behavior may function to escape the aversive social contact and to enhance contact with the treat. Let us assume for the current example a relatively simple arrangement. We depict the confirmed contingency as above.

We can use the contingency analysis both to analyze situations for the causes of observed behavior and to help plan a contingency in order to train a behavior. For example, you will likely begin most training projects by assessing the current proficiency of the subject with respect to certain behaviors as well as the client's training behaviors; it will be useful to be able to identify the ongoing contingencies. Then, in your training plan, you may seek to arrange for different contingencies. For instance, in the aversive example above, once you have identified the concurrent aversive contingency, you can plan to eliminate the subtracted reinforcement contingency, leaving only the added reinforcement contingency intact to proceed. This analysis process keeps the focus on the functional relations and the actual causes of the behavior. This improves clarity, both in explaining behavior and in planning how to change it, which also improves efficiency and effectiveness.

Relations Within the Three-Term Contingencies

The three-term contingency is the basic starting point to describe the functional relations between the environment and the target behavior in any given behavioral episode. It is the basic *starting point* because we may need to incorporate other terms that participate in controlling the target behavior into the contingency analysis. You cannot start with less than the three terms in the three-term contingency for a complete accounting, no matter how simple the episode of behavior.

The three-term contingency is made up of two two-term contingencies. I described this above, but I will address it here in greater depth. The first two-term contingency is the evocative stimulus → behavior contingency. This describes the mechanistic cause of the behavior, the stimulus that evokes the behavior and the response class it evokes. The second two-term contingency is the behavior → consequence contingency. This describes the selection cause of the target response class and the consequence that it generates. These two contingencies are integrally related. The behavior → consequence contingency functions to strengthen, suppress or weaken the evocative stimulus → behavior contingency. As a verbal short cut, you can say that it functions to strengthen, suppress, or weaken the evocative capacity of the evocative stimulus, as long as you appreciate that it does not change the behavior itself, but rather the body's reaction to the evocative stimulus. Occasionally, one might refer to the strengthening, suppressing or weakening of behavior. This is a verbal shortcut, perhaps a misleading and inappropriate one, depending on how loose we are to get with terms in making use of verbal shortcuts. To be clear, the functional relation between the evocative stimulus and the behavior is changed and this occurs structurally within the body. To reiterate, conditioning creates changes to the structure of the body of the organism via stimulation from the environment, which changes the organism's receptiveness to the evocative stimulus on subsequent presentations of the stimulus. Across repeated trials through the contingency, you can observe the maintenance or changes in the likelihood of the evocative stimulus evoking the behavior, and you can observe the consequences selecting for or against future instances of the behavior. I have depicted these processes diagrammatically in Figure 1 (which also includes the different kinds of consequences that result in these different kinds of conditioning). Notice how the behavior → consequence contingency functions to change (strengthen, suppress or weaken) the evocative stimulus → behavior contingency on subsequent trials.

Externalization of Contingencies

It is counterproductive to speculate about what an animal may or may not "understand" or "desire" etc., or even what they might be thinking, although thinking is real behavior. It is more productive to trace private events back through the causal chain to a link that is publicly accessible, something in the environment that can we can manipulate in order to control the target behavior. In other words, we observe a behavior (dependent variable) and, instead of speculating about the thought processes or especially any "mental transformations" as the independent variable causing it, we trace the causal chain back to a point that we can observe, measure and influence. For example, a mat might evoke approach and lying down behaviors, and there is no need to look for thinking behaviors that might come between the seeing of the mat (stimulation) and the approach and lying down behaviors. No doubt a number of neural and physiological behaviors are also elicited and evoked inside the dog but they are simply other behaviors, either simultaneously evoked, or are links in the causal chain occurring before the going to the mat behavior. Instead of speculating about these, we identify the mat as the stimulus functionally related to the approach and lying down behavior. We refer to this process as ***externalizing the contingency*** as it involves identifying stimuli and responses that are external and publicly accessible.

Frequently, students of behavior refer to the animal coming to "realize," "know" or "understand" something. However, it is important to remember that we infer this from the functional relation between the behavior and the environment. Indeed, we may take the operant behavior as indirect evidence of this supposed "understanding," but it is not generally a productive/pragmatic course of analysis. It is less speculative and presumptuous, and more scientific, to avoid the inference and instead discuss the functional relation from which the inference was made. Therefore, if an animal begins responding reliably to an evocative stimulus, we can discuss the functional relations involved in this instance of stimulus control. This is preferable to speculating about what the subject "knows," "realizes" or "understands." There is a place for discussing knowledge and understanding of course, but for most purposes in training companion animals, it is best to refer to the directly observable and measurable components of the behavioral episode. Knowledge and understanding are not the target behaviors that we as trainers are concerned with in training.

Expanding the Contingency Analysis

When another variable becomes important in controlling behavior, it can be advantageous and appropriate to include a fourth term in the contingency analysis. Whether further terms are included in the contingency analysis depends on how important these terms are in controlling the target behavior, and our interest in examining these other variables, which might provide greater explanatory power, in a given behavioral episode. One can include fourth, fifth or more terms in the antecedent, behavior or postcedent positions of the contingency. I will discuss these complexities in other sections.

Postcedent Principles, Processes and Procedures

Law of Effect

The *law of effect* states that consequences select for or against operant behavior—some consequences will tend to strengthen behavior, while others will tend to suppress it. We refer to behavior-strengthening consequences as reinforcers and the behavior-suppressing consequences as "punishers." Originally, the law of effect was formulated in terms of stimuli that generated a "satisfying state of affairs" and stimuli that generated an "annoying state of affairs." We no longer formulate laws and principles of behavior by presumptions or speculation about unverifiable private experiences. Rather, we identify whether the consequence does, in fact, increase or decrease the likelihood of that behavior upon exposure to the evocative stimulus. For example, it would be inappropriate to say that reinforcement involves following a behavior with pleasant stimuli. It is better if we frame this sentence without the word "pleasant." The law of effect is the foundational law of behavior for operant conditioning.

A reiteration of some important points may be in order. It is often said that consequences strengthen, suppress or weaken the behavior. This is fine when exposed to contingencies to simplify the process with verbal shortcuts. But, as discussed above, it is not as strictly accurate as saying that consequences provide the conditioning history that creates changes within the organism that increase or decrease the probability of the evocative stimulus evoking the behavior on subsequent occasions of its presentation. It does not strengthen the behavior directly, but rather causes changes in the organism's body that thereby alter the effectiveness of the evocative stimulus on subsequent occasions—conditioning occurs physically within the organism.

I will discuss five kinds of consequences—three that decrease the rate of behavior and two that increase the rate of behavior. We refer to the two that increase the future likelihood of behavior as reinforcement and two of the three that decrease it as punishment. We can say that reinforcement increases the likelihood of behavior or that it strengthens the behavior and they express the same thing. We can also say that punishment decreases the likelihood of a behavior or that is suppresses the behavior and they express the same thing. These four principles involve changes to the postcedent environment that change the future likelihood of that behavior occurring. They are different from the fifth kind, extinction, which involves an absence of postcedent environmental change for previously reinforced behaviors. In other words, withholding reinforcement results in a subsequent decline in responding and weakens the behavior. We can say that extinction decreases the likelihood of a behavior or that it eliminates or extinguishes the behavior. In reinforcement and punishment procedures, a stimulus can be added (i.e., presented or increased in magnitude) or subtracted (i.e., withdrawn or decreased in magnitude). This means that you can have added reinforcement (+R) and subtracted reinforcement (–R), and added punishment (+P) and subtracted punishment (–P). The diagram below depicts these relations.

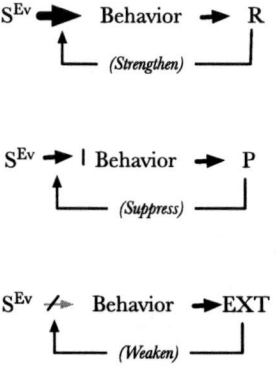

Figure 1. This diagram shows how each of the three kinds of consequences (reinforcement, punishment and extinction) influences the evocative stimulus → behavior contingency. Reinforcement strengthens the relation (indicated by the enlarging of the functional relation arrow), punishment suppresses it (indicated by the barrier in front of the functional relation arrow, and extinction weakens it (indicated by the shrinking and fading functional relation arrow).

Historically, scientists used the terms "positive" and "negative" to refer to the addition or subtraction of consequating stimuli respectively. These terms have been confusing to many because positive and negative have an established connotation as "good" and "bad," respectively. The notions of negative punishment and negative reinforcement are particularly confusing. Positive merely means added, and negative merely means subtracted. In behaviorology, the trend is to replace positive and negative with added and subtracted respectively, because they promote clarity and reduce confusion (see Ledoux, 2002b).

Each of these types of consequence (i.e., +R, –R, +P and –P) can be further broken down into unconditioned or conditioned forms. This means that you can have unconditioned added reinforcement, conditioned added reinforcement, unconditioned subtracted reinforcement, conditioned subtracted reinforcement, unconditioned added punishment, conditioned added punishment, unconditioned subtracted punishment and conditioned subtracted punishment. I will discus these in greater depth below. At this point, just note that in these cases, unconditioned consequences function to change behavior without any previous conditioning to establish them as such. Conditioned consequences on the other hand become effective only after we utilize a pairing procedure to condition them. Although we use capital letters by default when the status of the consequence as conditioned or unconditioned is unspecified, we use capital letters to specify unconditioned consequences and lower case letters to represent conditioned consequences (i.e., +r, –r, +p and –p) when contingencies require that level of specificity.

Figure 2 is a flow chart that illustrates how to analyze the postcedent stimulation of any particular operant conditioning episode and identify the principles operating on the behavior in question.

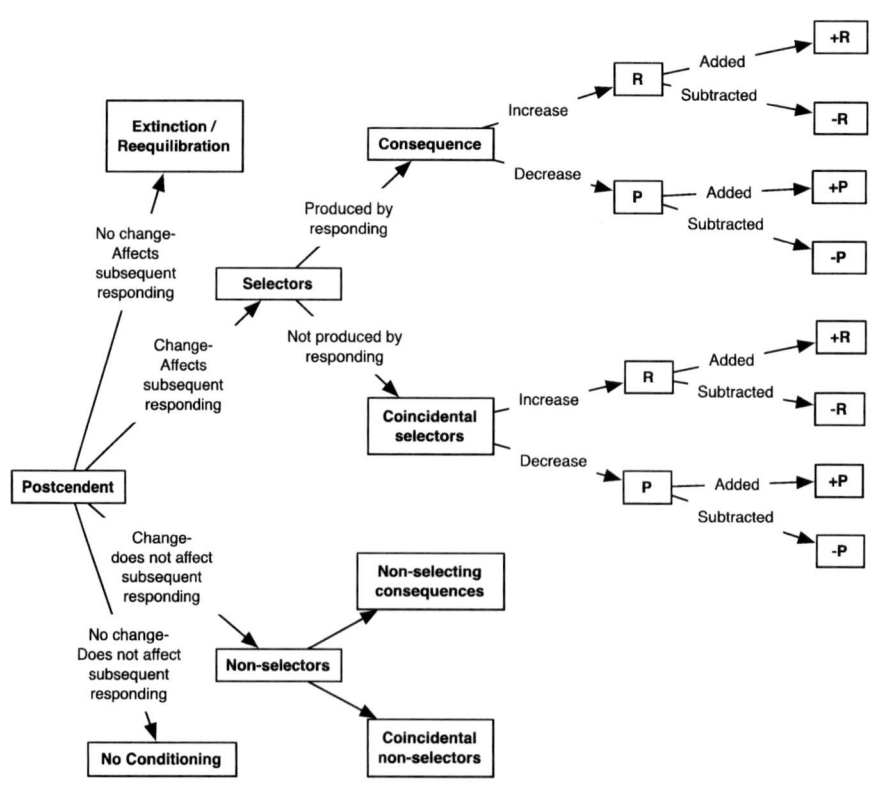

Figure 2. Flow chart depicting various operant processes and principles. Adapted and modified from Ledoux (2002b). I will not explore non-selectors further in this diagram because they are of little interest to our current purposes.

One of the easiest ways to explore the relations among the types of consequences is in the contingency diagram depicted in Figure 3. Many readers are likely to be more familiar with a table showing just the four quadrants of operant conditioning. The main problem with the traditional quadrants table is that it addresses only changes in behavior resulting from postcedent changes and ignores extinction, the fifth basic principle of operant conditioning that involves withholding the reinforcer for a behavior and its subsequent decrease in strength. In other words, it is a table of consequences and ignores changes in behavior that result from the withholding of previously available consequences. Figure 3 accommodates extinction along with the other basic principles.

The four principles of added and subtracted reinforcement, and added and subtracted punishment occupy the four corners of the diagram because they involve changes to the postcedent environment (addition or subtraction of stimulation—the horizontal axis in the figure) as well as an increase or decrease in the rate of responding (the vertical axis). In the case of reinforcement, the increase strengthens the evocative capacity of the evocative stimulus. In the case of punishment, the decrease suppresses the evocative capacity of the evocative stimulus. Extinction, on the other hand, involves no functional postcedent change in the environment (from the antecedent environment) for a behavior with a history of reinforcement. It therefore exists only directly on the vertical axis line; because the rate of behavior decreases, we find extinction at the bottom of the figure. Do not attribute the effects of extinction to punishment (which would require a left–right axis position in the figure). In the case of extinction, the decrease weakens the evocative capacity of the evocative stimulus.

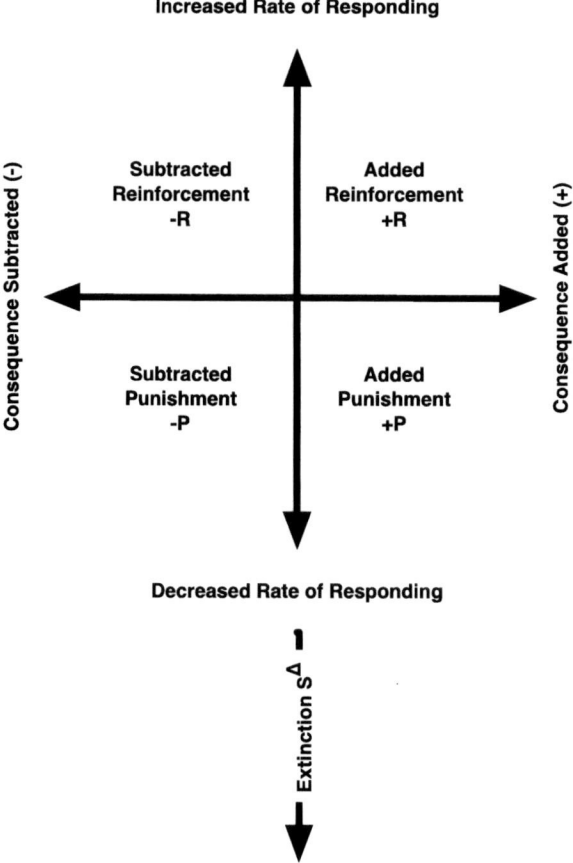

Figure 3. Diagram relating basic principles of behavior. Adapted from Fraley (2008) with additions adapted from Ledoux (2014).

I will discuss each of the principles below.

Contingencies of Reinforcement[24]

Reinforcement is a behavior change process in which a change in stimulation, during or immediately following a response, causes changes to the structure of the

---

[24] Most behaviorologists refer to all contingency types as "contingencies of reinforcement," including contingencies of punishment and extinction. We generally refer to them this way because all of the contingencies address behavior and reinforcement in some way. Reinforcement contingencies obviously involve the reinforcement of behavior. Punishment contingencies involve the suppression of previously reinforced behaviors and extinction contingencies involve the elimination of previously reinforced behaviors. Stipulating that all of the contingencies are traditionally referred to as "contingencies of reinforcement," the headings refer to the particular kinds of "contingencies of reinforcement" in order to ensure clarity and reduce confusion.

organism's nervous system, which result in an increase in the likelihood of that behavior occurring upon subsequent exposure to the stimulation. That is, the behavior → consequence[R] contingency results in a *strengthening* of the evocative stimulus → behavior contingency. On subsequent occasions of its occurrence, the evocative stimulus is *more* likely to evoke the operant behavior. The consequence strengthens the relation between the behavior and whatever stimuli were present at the time. There are two basic types of reinforcement: added reinforcement and subtracted reinforcement.

Added Reinforcement

Added reinforcement (+R) is a behavior change process in which the addition of stimulation during or immediately following a response, causes changes to the structure of the organism's nervous system, which result in an increase in the likelihood of that behavior on subsequent occasions. That is, the behavior → consequence[(+R)] contingency results in a *strengthening* of the evocative stimulus → behavior contingency. On subsequent occasions of its occurrence, the evocative stimulus is *more* likely to evoke the behavior because of that previously added consequence. A valuable way to conceptualize +R is by listing the necessary and sufficient conditions, as follows:

- **Behavior** occurs.

- Stimulus is **added** during or immediately following behavior.

- Subsequent **increase** in likelihood of behavior on subsequent occasions of exposure to the stimulus.

We can divide added reinforcement into conditioned and unconditioned added reinforcement (i.e., there can be a conditioned added reinforcer or an unconditioned added reinforcer, depending on whether or not the reinforcer was effective without previous conditioning to establish it as a reinforcer before its use). We use both unconditioned reinforcers (also referred to as primary reinforcers) and conditioned reinforcers (also referred to as secondary reinforcers) in training.

Unconditioned reinforcers are capable of reinforcing behavior without any previous conditioning to establish them as reinforcers. The body is structurally arranged through biological evolution such that the stimulus simply is a reinforcer. These types of reinforcers commonly involve biologically significant stimuli such as food, air, water or sex. However, unconditioned reinforcers such as these identified stimuli are not *always* reinforcing, and in fact, they can become aversive. If an organism is satiated with respect to the stimulus, the body is so structured as to prevent that stimulus from acting effectively as a reinforcer, and indeed, it may constitute an aversive stimulus until the satiation subsides. For example the last thing you want after a large meal is food and access to it certainly will not reinforce any behavior, other than perhaps escape behaviors, but that is an example for subtracted reinforcement, which I will discuss next. On the other hand, if we deprive the organism with respect to the stimulus, the stimulus should act effectively as a reinforcer. Because we can rarely deliver unconditioned reinforcers with precise timing sufficient to reinforce fleeting behaviors, we make use of conditioned reinforcers in training.

Conditioned reinforcers, which are discussed below, function to reinforce behavior but only because it has been previously paired with an unconditioned reinforcer or an already established conditioned reinforcer. Once a conditioned reinforcer (usually a click sound in animal training) is established, we can make use of the conditioned reinforcer to reinforce specific behaviors, particularly fleeting ones. We administer the click immediately following the behavior followed itself within a second or two by the unconditioned reinforcer. We often characterize the conditioned reinforcer as "bridging the gap" between the behavior and the unconditioned reinforcer, and some trainers refer to the conditioned reinforcer as a "bridge." This allows for reinforcement of a precise behavior and sufficient time for the trainer to deliver the unconditioned reinforcer to the subject.

Laypeople commonly use the word "reward," but this is *not* a principle of behavior and should *absolutely not* be used in place of reinforcer or added reinforcer. It is best to avoid this term altogether. A reward is something provided to someone for good or bad behavior in recognition of that behavior. Note that this does not apply to "good" behavior only. Furthermore, the "reward" does not necessarily reinforce behavior—it is simply presented in recognition of the behavior (Peirce & Cheney, 2013, p. 5). Furthermore, the term "reward" is usually framed in terms of rewarding the individual, which is semantically meaningless if you interpret the word reward as a synonym for reinforce. If you are exposed to the contingencies to express something about reinforcement, it is better to use the proper terms: reinforcement or added reinforcement. The word reinforcement is a well-established household term in everyday language, so there is no need for concern that laypeople will fail to respond appropriately to its use.

Subtracted Reinforcement

Subtracted reinforcement (–R) is a behavior change process in which the subtraction of an ongoing stimulus, during or immediately following a response, causes changes to the structure of the organism's nervous system, which result in an increase in the likelihood of that behavior on subsequent occasions. That is, the behavior → consequence$^{(-R)}$ contingency results in a *strengthening* of the evocative stimulus → behavior contingency. On subsequent occasions of its occurrence, the evocative stimulus is *more* likely to evoke the operant behavior, because of that previously subtracted stimulus. Subtracted reinforcement involves the strengthening of escape behavior. The necessary and sufficient conditions are as follows:

- **Behavior** occurs.

- Ongoing stimulus is **subtracted** during or immediately following behavior.

- Subsequent **increase** in likelihood of behavior on subsequent occasions of exposure to the stimulus.

We can further divide subtracted reinforcement into conditioned and unconditioned subtracted reinforcement (i.e., there can be a conditioned subtracted reinforcer or an unconditioned subtracted reinforcer, depending on whether or not the reinforcer was effective without previous conditioning to establish it as a reinforcer before its use). For example, if you squeeze a dog's ear and only let go when the dog sits, the release of pressure on the ear will function as an unconditioned subtracted reinforcer and after some number of trials, your reaching for the dog's ear will act as a conditioned subtracted reinforcer. Such coercive training procedures are fraught with problematic side effects, which is why added reinforcement based procedures are preferred, but they do illustrate the principle of subtracted reinforcement. Subtracted reinforcement is unique among the basic principles of behavior in that the consequence is a subtracted (reduced or eliminated) version of the evocative stimulus that evoked the behavior. In a subtracted reinforcement contingency, the evocative stimulus is an aversive stimulus, and the behavior functions to reduce or eliminate (i.e., subtract) contact with that stimulus. Although subtracted reinforcement strengthens escape behavior, we usually refer to escape from an unconditioned subtracted reinforcer as *escape* and escape from a conditioned subtracted reinforcer is usually referred to as *avoidance*. The unconditioned subtracted reinforcer is avoided, in a manner of speaking, but the contingency really involves escape from the conditioned subtracted reinforcer—it is escape in both instances.

Some Potentially Confusing Distinctions

All reinforcers strengthen behavior but there can be value in making certain distinctions regarding the manner in which certain reinforcers occur. There are also a few common distinctions and terms that refer to the same thing that frequently cause confusion. In this section, I will explore a few common distinctions, including some potentially confusing ones.

Under certain circumstances, it may become useful to make a distinction between whether reinforcement directly or indirectly generates behavior. In behaviorology, an *intrinsic reinforcer* is one that is produced *directly* by the behavior. For instance, if you close your hand around a can and the crushing of the can reinforces your can-crushing behavior, the can becoming crushed is an intrinsic reinforcer. An *extrinsic reinforcer* is one that is produced *indirectly* by the behavior, arbitrarily mediated by another organism. For instance, if you give a treat to a dog after he or she sits and that increases the rate of sitting behavior, giving the dog the treat was an extrinsic reinforcer.

Confusion arises when the term intrinsic is also used to refer to *endoreinforcement* (also known as *endogenous reinforcers, automatic reinforcers*)—that is, reinforcers that are generated *solely within the body* of the subject, such as when a behavior causes the release of certain chemicals (e.g., adrenaline, corticosteroids, etc.) into the bloodstream. The opposite of this usage would involve extrinsic reinforcers, also known as *ectoreinforcement* (also known as *exogenous reinforcers*), which refers to reinforcers that are generated *outside* of the subject's body.

To avoid confusion, I generally use the terms *trainer-mediated reinforcers* and *nontrainer-mediated reinforcers* in this book to differentiate between reinforcement generated by the behavior or reinforcement that is mediated by a trainer. Trainer-mediated reinforcement is also commonly referred to as *contrived reinforcement* and nontrainer-mediated reinforcement is commonly referred to as *"natural" reinforcement*. This terminology is problematic since contrived reinforcers are, of course, perfectly natural. Since various authors use the terms intrinsic and extrinsic to refer to either of these two different things, these terms should evoke further exploratory behaviors to determine the intended usage in that source.

Variables Influencing Effectiveness of Reinforcement

For reinforcement to be effective, certain conditions must be met. Although there can be many such variables, a few of them are particularly important. I will describe these below.

Contiguity

Contiguity refers to the interval of time between the behavior and the reinforcer. The smaller the interval, the more effective reinforcement will be. Behavior is continuous; behaviors are always occurring and consequences are influencing their future likelihood. In the interval between the target behavior and its reinforcer, other behaviors can occur, even barely noticeable or private ones. The longer the interval between the target behavior and the reinforcer, accumulation of additional behaviors becomes more likely and it is less likely that the reinforcer will actually reinforce the target behavior. Furthermore, if the reinforcer does not occur immediately after the behavior, the behavior and the reinforcer may not actually become neurologically correlated (i.e., conditioned). Schlinger and Blakely (1994) tested reinforcer delays in three subject groups: one with a delay of zero seconds, another with a delay of four seconds, and another with a delay of 10 seconds. They found *dramatic* reductions in conditioning as the delay increased, with little to no conditioning having taken place in the 10–second delay group. Contiguity is the most important of the variables discussed here.

Contingency

Contingency, recall, refers to a functional relation between behavior and the environment. The greater the degree of contingency (i.e., the higher the correlation between the behavior and reinforcer), the more effective conditioning will be. In other words, if a behavior is reinforced each time it occurs and not otherwise, conditioning will proceed quickly and efficiently. If the correlation is not as strong, conditioning will occur more slowly, with the rate depending on the degree of correlation. Contingency is particularly important in the initial acquisition phase of training (i.e., in the initial stage of conditioning). As the behavior becomes stable, the benefits of reinforcing intermittently rather than continuously come to outweigh the benefits that accrue from reinforcing each time the behavior occurs. I will discuss this in detail below.

Reinforcer Characteristics

In general, smaller, more frequently delivered reinforcers are more effective than fewer, larger reinforcers. However, in a particular trial, a larger reinforcer is generally more effective than a smaller one. "*Jackpotting*," a term common in animal training, refers to the procedure of delivering a higher magnitude reinforcer in relation to the magnitude of the reinforcer provided in other trials. In the jackpotting procedure, the trainer provides reinforcers of higher quality (i.e., generally more effective) or higher quantity for particularly "good" responses. There is some evidence to suggest that jackpots increase responding in the session in which they are used (Weatherly, McSweeney, & Swindell, 2004), although there is also a post reinforcement decline in responding when a high-magnitude reinforcer is delivered. This indicates that jackpots might actually be disruptive, perhaps due to it more quickly inducing satiation and/or as an emphasized post reinforcement pause schedule effect (schedule effects are discussed below). There is also evidence that the increase in responding itself is minimal (Bond, 2007). It may be more productive to simply maintain consistent reinforcement criteria. I will discuss jackpotting in detail below.

Task Characteristics

Different species are genetically disposed such that certain behaviors are easier to train than others are. Evolution results in bodies so structured that certain behaviors readily contact highly effective reinforcement, often generated inside the subject's body. For example, a chicken tends to begin pecking at an experimental apparatus, regardless of trainer-mediated reinforcement contingencies placed on it, although manipulating the contingencies can influence the rate of the behavior. This concept helps explain why some behaviors are more challenging to train than others are. For example, waiting for (rather than approaching) a reinforcer, or looking away from (rather than at) a reinforcer, or holding back (rather than pulling ahead) on a leash to contact something are all examples of challenges to training due to what some might call "counterintuitive" contingencies or what might more accurately be described as contingencies with competing biologically established counter-contingencies.

Concurrent Contingencies

Concurrent contingencies are always operating in a particular situation. That is, there are always various stimuli differentially controlling the target behavior, as well as other behaviors that might displace it at any given time. The behavior that results in any particular instance is the net result of all of the contingencies vying for control over the behavior. I will discuss concurrent contingencies in greater depth below.

Motivative Operations

Motivative operations are procedures that establish a function-altering condition that influences how effective consequences will be and hence how strongly evocative the evocative stimulus will be. Motivative operations include establishing operations and abolishing operations. *Establishing operations* involve *deprivation* with respect to the reinforcer, which tends to cause an *increase* in responding. *Abolishing operations* involve *satiation* with respect to the reinforcer, which tends to cause a *decrease* in responding. A deprived body tends to exhibit behavior that contacts that reinforcer and a satiated body tends not to exhibit those behaviors (Laraway, Snycerski, Michael, & Poling, 2003). I will discuss motivative operations further below.

Contingencies of Punishment

Punishment (P) is a behavior change process in which a change in stimulation, during or immediately following a response, causes changes to the structure of the organism's nervous system, which result in a decrease in the likelihood of that behavior on subsequent exposure to the stimulus. That is, the behavior $\rightarrow$ consequence$^{(P)}$ contingency results in *suppression* of the evocative stimulus $\rightarrow$ behavior contingency.[25] On subsequent occasions of its occurrence, the evocative stimulus is *less* likely to evoke the behavior. The consequence suppresses the relation between the evocative stimulus and the behavior. Note that a punitive contingency does not make any changes to the ongoing reinforcement contingency maintaining the behavior. Because of this, we make the distinction that punishment *suppresses* behavior, whereas extinction *eliminates* or *weakens* it.

Added Punishment

Added punishment (+P) is a behavior change process in which the addition of stimulation, during or immediately following a response, causes changes to the structure of the organism's nervous system, which result in a decrease in the likelihood of that behavior on subsequent occasions. That is, the behavior $\rightarrow$ consequence$^{(+P)}$ contingency results in *suppression* of the evocative stimulus $\rightarrow$ behavior contingency. On subsequent occasions of its occurrence, the evocative stimulus is *less* likely to evoke the operant behavior because of that previously added consequence. The necessary and sufficient conditions are as follows:

- **Behavior** occurs.

- Stimulus is **added** during or immediately following behavior.

- Subsequent **decrease** in likelihood of behavior on subsequent occasions of exposure to the stimulus.

We can further divide added punishment into conditioned and unconditioned added punishment (i.e., there can be a conditioned added punisher or an unconditioned added punisher, depending on whether or not the punisher was effective without previous conditioning to establish it as a punisher before its use). For example, being hit with a belt will often act as an unconditioned punisher and after some number of trials, the mere sight of the belt may become a conditioned punisher. Behavior under relatively strong punitive controls will often be temporarily suppressed. However, the evocative

---

[25] Some clarity is in order here. You may read elsewhere that reinforcement strengthens behavior whereas punishment weakens it. It is accurate to say that reinforcement *increases* the likelihood of behavior and punishment *decreases* it. However, this is not the same as strengthening versus weakening the evocative stimulus $\rightarrow$ behavior contingency or the evocative capacity of the evocative stimulus; a decrease in the likelihood of the behavior does *not* necessarily mean that the evocative stimulus $\rightarrow$ behavior relation is weakened. Reinforcement maintains behavior and punishment of behavior leaves the reinforced evocative stimulus $\rightarrow$ behavior contingency unaddressed. Thus, behavior is *suppressed* by punishment rather than eliminated or weakened. Consequently, we say that reinforcement *strengthens* behavior, extinction *weakens* behavior, and punishment merely *suppresses* behavior. I discuss these nuances are in detail in the following section.

stimulus does not become a neutral stimulus, unlike the situation with extinction. If the stimulus were neutral, the behavior would not reemerge at the pre-punishment rate as soon as we discontinue the punitive contingency, which it tends to do as the reinforcement contingency remains in effect.

Subtracted Punishment

Subtracted Punishment (–P) is a behavior change process in which the subtraction of a stimulus, during or immediately following a response, causes changes to the structure of the organism's nervous system, which result in a decrease in the likelihood of that behavior on subsequent occasions of exposure to the stimulus. That is, the behavior → consequence$^{(-P)}$ contingency results in *suppression* of the evocative stimulus → behavior contingency. On subsequent occasions of its occurrence, the evocative stimulus is *less* likely to evoke the operant behavior because of that subtracted consequence. The necessary and sufficient conditions are as follows:

- **Behavior** occurs.

- Stimulus is **subtracted** during or immediately following behavior.

- Subsequent **decrease** in likelihood of behavior on subsequent occasions of exposure to the stimulus.

We can further divide subtracted punishment into conditioned or unconditioned subtracted punishment (i.e., there can be a conditioned subtracted punisher or an unconditioned subtracted punisher, depending on whether or not the punisher was effective without previous conditioning to establish it as a punisher before its use). For example, if a subject exhibits a non-criterion behavior and you say "Oops," snatch the treat back and turn away without interacting with the subject for several seconds, the removed access to the treat and your social attention, assuming these are reinforcers, will likely act as an unconditioned subtracted punisher. After some number of trials, the vocalized "Oops" will become a conditioned subtracted punisher. The emphasis should be on utilizing added reinforcement by setting the subject up for success. However, if one uses subtracted punishment, they can use a conditioned subtracted punisher for the same reasons we utilize a conditioned added reinforcer.

Variables Influencing the Effectiveness of Punishment

For punishment to be effective, it should be:

- Abruptly intense

- Immediate (contiguous)

- Consistent (contingent)

Generally, an *abrupt* and *intensely* aversive punisher, delivered *immediately* upon occurrence of the behavior, *consistently*, each time the behavior occurs will be more effective in suppressing behavior than a punishment procedure failing to achieve any of these criteria. Notwithstanding these variables and their impact on the effectiveness of punishment, the side effects of aversive stimulation are significantly problematic.

Specific to punishment, it is important to also manage the reinforcement contingency. This is a separate procedure. Punishment will be less effective if the reinforcement for the behavior remains available. Of course, if one is going to address the reinforcement contingency with extinction and do so effectively, one might as well simply reinforce an alternative behavior rather than punish the problematic behavior. By addressing the reinforcement contingency, we instate extinction and therefore punishment should not be necessary. Furthermore, if the punishment is successful, then we actually protect the behavior against extinction, because the behavior does not occur and therefore there is no opportunity to extinguish it. Meaning, if there are no trials of the behavior failing to generate that reinforcer, then extinction does not occur. Thus, the behavior is suppressed, rather than eliminated. This is just one of the reasons why punishment is generally not the most productive approach to reducing behavior excesses (i.e., problematic behaviors).

Contingencies of Extinction

Operant behavior is a function of the reinforcer that maintains it. **Extinction** (EXT) is a behavior change process in which no functional postcedent environmental changes occur, which causes changes in the structure of the organism's nervous system, which result in the strength of the previously reinforced behavior decreasing across subsequent occasions of exposure to a stimulus. Extinction is a procedure in that it involves withholding the reinforcement and is a process in that it involves a reduction in the likelihood of the target behavior. Unlike reinforcement or punishment, which involve changes in the postcedent environment (i.e., adding or subtracting stimulation), extinction of a behavior that already has a history of reinforcement involves no postcedent change in the environment. In other words, nothing of significance is added to or subtracted from the environment when the behavior occurs. In extinction, the focus is on the reinforcer not occurring; the behavior fails to produce a change in the environment that it generated in the past. Unlike punishment, extinction actually changes the contingency of reinforcement that was maintaining the behavior. It therefore weakens the evocative capacity of the evocative stimulus and eliminates behavior, rather than merely suppressing it through superimposing a punitive contingency over a reinforcing contingency. We can refer to the decrease in the rate of the behavior as the **extinction curve**, in reference to the tracking of the rate of the behavior as a line on a graph. The necessary and sufficient conditions are as follows:

- **Behavior** occurs.

- Reinforcer does *not* occur.

- Subsequent **decrease** in likelihood of behavior on subsequent occasions of exposure to a stimulus.

Following instatement of an extinction procedure, the rate of the behavior may initially increase briefly. We call this spike in the graphed line an **extinction burst**. There may be a series of extinction bursts during the extinction process, although they will become gradually less frequent and prominent. An extinguished behavior tends to be more readily reconditioned if a reinforcer is reintroduced at some point (presumably because some structures in the subject's nervous system remain susceptible to the contingency). While the stimulus continues to evoke the behavior under an extinction procedure, it continues to be an evocative stimulus. Once it no longer evokes the behavior at all, it becomes a neutral stimulus. Once a behavior has become extinct, the subject may occasionally exhibit, in appropriate contexts, instances of the behavior as a probe. This has been referred to as "spontaneous recovery," however there is nothing spontaneous about it (or anything else for that matter). Following instatement of an extinction procedure, the form of the behavior becomes more variable. It is from these variable forms of the response class that shaping can take place, refining the form of the behavior.

To end this section, let us follow a brief advanced tangent as a teaser for further study. Traditionally, extinction is defined as a process by which a previously additively reinforced[26] behavior is followed by an abrupt prevention of the reinforcement and a subsequent decline in responding, as discussed above. One could logically apply extinction to subtracted reinforcement, and to added and subtracted punishment as well. In fact, the *"learned helplessness"*[27] phenomenon is consistent with being a side effect of extinction of behavior maintained by subtracted reinforcement. Furthermore, extinction does not need to be an abrupt process; we can gradually alter the postcedent environment to become increasingly similar to the antecedent environment. The word extinction becomes troublesome when we apply it to punishment because the word "extinction" generally connotes a decline in something but instead, extinction of punished behavior results in an *increase* in the rate of that behavior. We use the term *reequilibration* for this broader application of extinction, as applied either abruptly or gradually, and to reinforcement or punishment, because the behavior returns, in all cases, to its previous equilibrium rate, during the conditioning process. Interested readers can refer to Fraley (2008, pp. 438–472).

Schedules of Added Reinforcement

A schedule of added reinforcement sets the rule that determines which responses, among a series of responses, will be reinforced. We use different schedules of reinforcement because each schedule generates a characteristic effect on the rate of the behavior. I will review the basic schedules of reinforcement below.

Continuous Reinforcement

Continuous reinforcement (CRF) sets the rule that we add reinforcers after each occurrence of the behavior. A continuous reinforcement schedule produces a steady

---

[26] Please note that the term for this grammatical use of the principle of stimulus subtraction (or addition) has not been established yet with consensus. I will use the word "subtractively" but one might use "subtractional" and "subtractionally," though they seem needlessly wordy. Generally, behaviorologists have thus far bypassed this issue of grammar by changing the sentence structure to allow for the use of subtracted and added.

[27] "Learned helplessness" is a behavioral phenomenon in which a subject that cannot effectively escape aversive stimulation, ceases attempts at escape (Overmier & Seligman, 1967; Seligman, Maier & Geer, 1968; Seligman, 1975).

reliable rate of responding with relatively little variation in form and is particularly useful in the initial, acquisition phase of training. After the acquisition phase, it is best to transition to an intermittent schedule. Behaviors maintained on a continuous reinforcement schedule are less resilient than those maintained on intermittent schedules. Thus, behaviors maintained on a continuous reinforcement schedule become extinguished more rapidly when reinforcement is not forthcoming. In contrast, on certain intermittent schedules, behavior can become very persistent even on a very sparse schedule of reinforcement. As there typically is very little variability in the form of the behavior on continuous reinforcement schedule, there is little opportunity for shaping to fine tune the form of the target behavior. Furthermore, satiation can occur quickly, if we reinforce every response.

Vending machines operate on a continuous reinforcement schedule. Every time one inserts money, the machine dispenses the appropriate product. We can exemplify the schedule effect discussed above by the likelihood of a person's money-inserting behavior if we switched the continuous reinforcement schedule to an extinction schedule, or even any intermittent schedule. In the scenarios where we switched the schedule to an extinction or intermittent schedule, we would quickly extinguish the money-inserting behavior. This is not the case with slot machines, which operate on an intermittent schedule, as discussed below.

Figure 4 shows the typical effect of a continuous reinforcement schedule on conditioning.

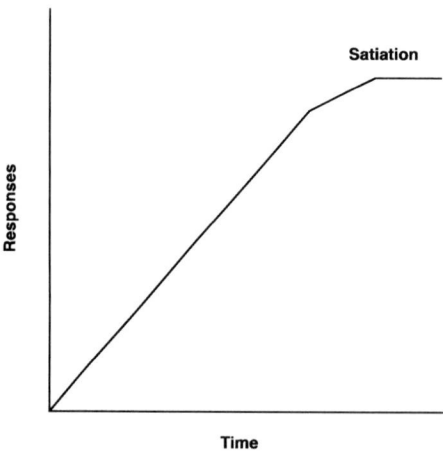

Figure 4. Typical conditioning curve associated with continuous reinforcement.

Intermittent Reinforcement

An *intermittent reinforcement* schedule sets the rule that added reinforcers follow some, but not all, occurrences of the behavior. Behavior on an intermittent schedule is less susceptible to extinction and the form is more variable than behaviors produced under continuous reinforcement. I will discuss six basic intermittent schedules, three fixed and three variable, below.

Fixed Ratio

A *fixed ratio* (FR) schedule sets the rule that reinforcement will be added following the final response after a fixed number of responses have occurred. A number is included in the description of the schedule to identify the number of responses that generates the reinforcement. Therefore, reinforcing after, say, four responses would be FR-4. On an FR schedule, the subject usually responds at a high rate, but responding wanes after each delivery of reinforcement—we refer to this as a *post reinforcement pause*. Because of this post reinforcement pause, we rarely use FR schedules in training. Trainers usually move from a continuous reinforcement schedule to an intermittent schedule with a high rate of reinforcement and gradually stretch (or thin) the ratio over time. Avoid resetting to a higher ratio until the behavior has stabilized at the current ratio. Conditioning is a physical process within the body and it takes a certain amount of time for the body to structurally change in response to stimulation. If you thin the schedule of reinforcement before the body has had a chance to complete the change at the current ratio, you may actually strain the schedule and the behavior may fail to stabilize at all, and in some cases may extinguish.

Figure 5 shows the typical effect of an FR schedule on conditioning.

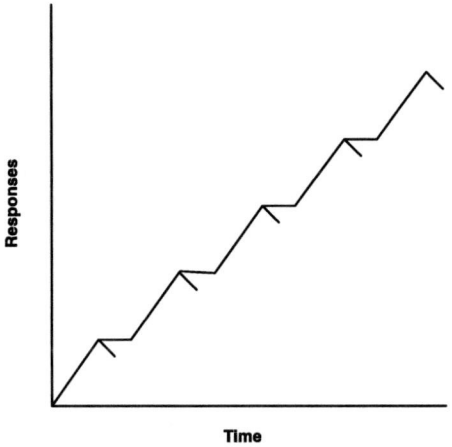

Figure 5. Typical conditioning curve associated with a fixed ratio schedule of reinforcement.

Variable Ratio

A *variable ratio* (VR) schedule sets the rule that reinforcement will be added following the final response after a specific, but variable, mean average number of responses has occurred. Variable means that we deliver reinforcement in a seemingly random manner around the mean specified in the training plan. Avoiding getting too far away from the mean. Responding under a VR schedule usually occurs at a high rate and with less of a post reinforcement pause than under a FR schedule. In addition, behaviors on VR schedules are highly resistant to extinction.

The schedule effects for the VR schedule are very important for training. Engineers design slot machines to operate on a VR schedule because this schedule maintains the highest rate of responding in relation to the number of reinforcers provided. Unlike the vending machine, the slot machine delivers its payout on a seemingly random schedule, which is in fact a VR schedule. As you would expect, subjects on VR schedules may exhibit the target behavior vast numbers of times before reinforcement is delivered (persistence), assuming the ratio is stretched gradually enough.[28] You can **stretch a ratio** of reinforced to unreinforced behaviors, but if you stretch it too far, too quickly, you may strain the ratio. **Ratio strain** can occur when one thins a ratio schedule too quickly or when one stretches the ratio too far at once.

Because of these schedule effects, a VR schedule is the most useful basic schedule of reinforcement to transition to after continuous reinforcement. As with FR schedules, ensure that the subject's body has adjusted to the current change, as evidenced by stabilization of the behavior, before thinning the schedule.

Figure 6 shows the typical effect of a VR schedule on conditioning.

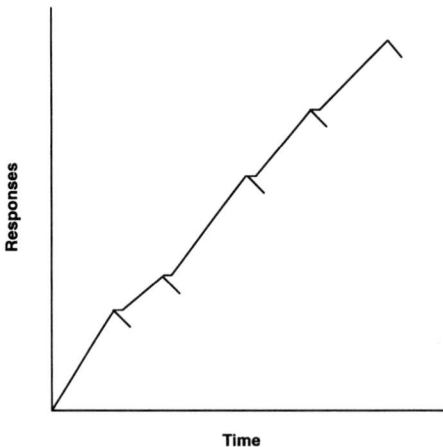

Figure 6. Typical conditioning curve associated with a variable ratio schedule of reinforcement.

Fixed Interval

A *fixed interval* (FI) schedule sets the rule that we provide reinforcement immediately after the first response exhibited *after* a specific interval of time has passed. It is important to note the two necessary conditions for interval schedules: the specified interval must pass, and then and only then, we reinforce the *next* response to occur. A number and unit of time are included in the schedule description to identify the time interval involved. For example, we would depict reinforcing for the first response after 2 minutes as FI-2min.

---

[28] In animal training circles, there is frequent talk about thinning a schedule of added reinforcement. "Thinning the schedule" refers to stretching the ratio of reinforced to unreinforced behaviors. While the phrase "thinning the schedule" is more common among animal trainers, the phrase "stretching the ratio" is more common in the general disciplinary vernacular.

Increases in the rate of responding are moderate under FI schedules, and the subject ceases responding after we deliver each reinforcer, in what is generally called "scalloping." Because of this scalloping schedule effect and only moderate responding, we rarely use FI schedules in training. This schedule is addressed here primarily for completeness, and because other useful schedules (discussed below) are based on the FI schedule.

Let us follow a tangent to discuss an advanced topic related to the use of fixed interval schedules. ***Adjunctive behavior*** is operant behavior that appears intermittently while the subject exhibits other operant behaviors. This behavior often seems unrelated to the prevailing contingencies operating at the time. Adjunctive behavior is a schedule effect. It is most common in fixed schedules, particularly fixed interval schedules and occurs during the initial part of the interval when reinforcers are not available. You can think of this as behavior that intrudes during the post reinforcement wane in responding that is associated with fixed schedules of reinforcement. During the interval when reinforcers are unavailable, other antecedent stimuli that exert weaker control than the stimuli controlling the primary behavior briefly become more evocative and can become prepotent in controlling behavior. As a cyclic process occurring along with a primary behavior and its reinforcement during training repetitions, adjunctive behaviors can be reinforced; they can even become excessive. When you are training and observe other behaviors between cycles of reinforcement availability, you should consider the possibility that they are adjunctive behaviors. You should control for these other antecedent stimuli and instate a VR schedule in order to eliminate the adjunctive behavior.

Adjunctive behavior is distinct from situations in which some antecedent stimulus simply becomes more strongly evocative than the antecedent stimulus that had previously prevailed. In these cases, the change in behavior is due not to a schedule effect but rather to a change in the relative evocative capacity of different stimuli.

Intruding behaviors such as we see in episodes of behavior involving adjunctive behaviors or when other contingencies prevail over current contingencies may seem strange, "abnormal," or mysterious, and have sometimes been referred to as "displacement" behaviors. However, these behaviors are not mysterious or abnormal at all—a careful analysis of the contingencies will reveal the functional relations operating in a perfectly normal and natural law-abiding manner.

Figure 7 shows the typical effect of an FI schedule on conditioning.

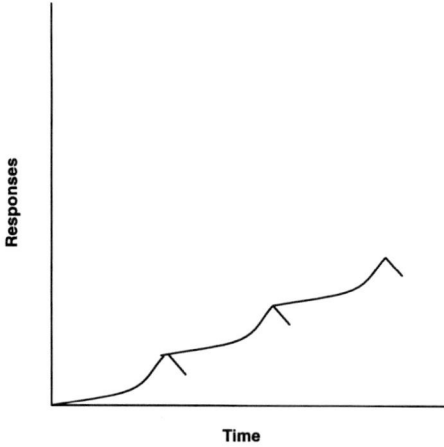

Figure 7. Typical conditioning curve associated with a fixed interval schedule of reinforcement.

Variable Interval

A ***variable interval*** (VI) schedule sets the rule that we deliver reinforcement on the first occurrence of the target behavior after a specific, but variable, interval of time has passed. As with the FI schedule, the specified interval must pass, and only after that do we reinforce the next response. Again, a number and unit of time are included in the schedule description for specific instances. For example, a VI-30sec. would indicate a variable interval schedule of 30 seconds.

Increases in rate of responding are moderate but steady under VI schedules. VI schedules lack the scalloping observed with the FI schedules, but the rate remains merely moderate, compared with behavior under a ratio schedule.

Figure 8 shows the typical effect of a VI schedule on conditioning.

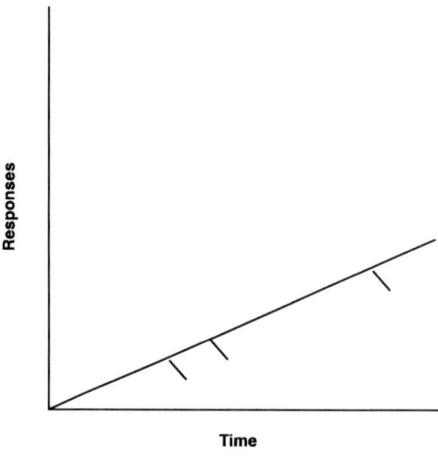

Figure 8. Typical conditioning curve associated with a variable interval schedule of reinforcement.

Fixed Duration

A *fixed duration* (FD) schedule sets the rule that the behavior must be exhibited continuously for a specified period of time, at which point reinforcement is delivered. A number and unit of measure are included in the schedule description to identify the duration. For example, we would depict a behavior maintained on a fixed duration schedule of 30 seconds as FD-30sec.

Variable Duration

A *variable duration* (VD) schedule sets the rule that the behavior must be exhibited continuously for a specific, but variable amount of time around a mean average. Again, a number and unit of measure are included in the schedule description for specific instances of the behavior, such as VD-30sec. In this example, we reinforce the behavior only if it occurs continuously for a seemingly random duration around a mean average of 30 seconds.

Extinction

An extinction (EXT) schedule sets the rule that we reinforce no responses. This schedule may initially result in an increase in responding, referred to as an *extinction burst*, but responding then gradually decline until the behavior no longer occurs at all. If the behavior was previously on a variable ratio schedule, the extinction curve may be quite lengthy; if it was on a fixed schedule, particularly continuous reinforcement, extinction typically occurs much more quickly.

Limited Hold

A *limited hold* (LH) is not a schedule *per se* but rather a schedule extension that we may incorporate into schedules of reinforcement. It sets the additional rule that the reinforcer is only available for a specified period of time after the evocative stimulus occurs. In other words, once the behavior to be reinforced is cued, the behavior must then be exhibited within a specified period of time in order for it to be reinforced. If the behavior is not exhibited within the specified time period, an occurrence after that interval is not reinforced. This is a latency reduction schedule.[29] For instance, if an FR-

---

[29] There is some disagreement with respect to what limited hold is or ought to be. Skinner (1957, p. 729) originally defined limited hold as "A short period during which a reinforcement arranged by an interval schedule is held available." This is still how many sources continue to define it (e.g., Ledoux, 2014; Cooper et al. 2007; Pierce and Cheney, 2013). However, some sources (Burch & Bailey, 1999, pp. 41–42; Catantia, 1998/2013) define it as I use it here. In Skinner's original use, once the behavior has been exhibited, *then* the reinforcer is available only for a specified interval of time. If the subject does not collect the reinforcer within that interval, that reinforcer is no longer available until it becomes available again according to whatever schedule is being used. In accordance with Burch and Bailey (1999) and Catania (1998/2013), I will use the term limited hold to refer to the latency, the period of time between the evocative stimulus and the response. For animal training, the concept of limited hold as a latency reduction schedule is particularly useful. However, take note that many prominent behaviorology and behavior analysis texts define the limited hold in accordance with Skinner (1957).

5, LH-8sec. schedule is set, this means that we will reinforce every fifth response if it occurs within 8 seconds of the evocative stimulus. Similarly, if an FI-1min., LH-10sec. schedule is set, this means that the first criterion response to occur after 1 minute is reinforced, but only if the behavior occurs within 10 seconds after the evocative stimulus was presented.[30]

Complex Schedules

There are times when we implement multiple schedules simultaneously. The most common example is the use of a basic ratio or interval schedule along with a fixed duration schedule. For example, let us say that "sit" is on a fixed ratio three (FR-3) schedule and you combine that with a fixed duration 10–seconds schedule (FD-10sec.), this would mean that every third sit response that occurs continuously for 10 seconds would be reinforced. That would be indicated as FR-3, FD-10sec. Let us say "sit" is on a VR-6 schedule and you combine that with a FD-8sec. schedule; this would mean that every sit that occurs and is maintained for eight seconds around a varying but average of every six responses would be reinforced. In practice, a limited hold extension is frequently used.

Combining ratio or interval schedules with duration schedules provides a complex schedule that relates to a single response class. When you combine a ratio and a duration schedule, you are specifying criteria for reinforcement of the one response class under consideration. There are a number of useful complex schedules characterized by simultaneously establishing schedule rules for *different* behaviors (or in some cases differentially reinforce occurrence of a target behavior under certain circumstances to the exclusion of the *same* behavior under different circumstances). We refer to these schedules "differential reinforcement" and I will describe them below.

Differential Reinforcement

Differential reinforcement encompasses a number of compound procedures involving targeting one behavior (i.e., response class or response class form) for reinforcement and another for extinction (Cooper et al. p. 470; Miltenberger, 2008, p. 587; Chance 2009, p. 224; Burch & Bailey, 1999, p. 153).[31] We use differential

---

[30] The duration of other responses in the sequence are usually held to the same standard. In other words, if every fifth response is scheduled for reinforcement and there is a limited hold of 8 seconds for that fifth response, trainers usually require that 8 seconds of reinforcer availability for each response in order for it to be counted as a criterion-meeting response. Therefore, if five responses occurred in the sequence but one of them occurred after 8 seconds, we identify that response as a non-criterion response and there were only four criterion responses counted. The limited hold applies to the reinforced response in the sequence, but it is common practice to require the limited hold criteria for all responses in the sequence.

[31] Not all sources define differential reinforcement in the same way. I define it here broadly as targeting one behavior for reinforcement and another behavior for extinction. This is the broadest definition available for differential reinforcement supported in the literature. Some credible sources define differential reinforcement more narrowly as targeting a member of a response class with reinforcement and targeting another member of that *same* response class with extinction (e.g., Ledoux, 2014). Other sources support this narrower definition as well. Indeed Cooper and colleagues (2008) define differential reinforcement in the body of the text (p. 470) broadly in accordance with how I use it, and in their glossary (p. 693) in accordance with this narrower definition. In the broader definition, the reinforcement-targeted behavior and extinction-targeted behavior do *not* need to necessarily be members of the same response class. Recall that

reinforcement extensively in training companion animals. Typically, we use it either to replace specific problem behavior of concern, or to train a new specific behavior in place of other, non-criterion, behaviors that the subject might otherwise exhibit.[32] There are several variations of differential reinforcement, each of which may be appropriate under different circumstances. We define these variants by the relationship between the target behavior and the extinguished non-criterion behaviors. The following procedures are some of the more commonly utilized differential reinforcement procedures.

The following three differential reinforcement procedures, differential reinforcement of incompatible behavior, differential reinforcement of alternative behavior and differential reinforcement of other behavior involve differentially reinforcing one behavior to the exclusion of a *different* behavior. The next three differential reinforcement procedures, (which might better be termed differential reinforcement-like procedure) differential reinforcement of high rate of responding, differential reinforcement of low rate of responding and differential reinforcement of diminishing rates differ in that they differentially reinforce occurrence of a target behavior under certain circumstances to the exclusion of the *same* behavior under different circumstances. Whether these three procedures are in fact instances of differential reinforcement *per se* is debatable but placing them elsewhere, considering the fact that "differential reinforcement" occurs in their names, would invite confusion. They are, at the least, a different kind of differential reinforcement procedure. I will present the final differential reinforcement procedure, differential reinforcement of excellent responding, as an alternative to the arguably sloppy "jackpotting" procedure common in animal training.

Differential Reinforcement of Incompatible Behavior

Differential reinforcement of incompatible behavior (DRI) involves arranging added reinforcement for a target behavior that is mutually exclusive with respect to the behavior that the subject would otherwise exhibit in that context. For example, we use a DRI procedure when we train a dog to sit to greet people rather than jumping up on them, by reinforcing sitting and failing to reinforce jumping up. Both behaviors are members of the same response class in this example because they share an effect on the environment. If the subject cannot exhibit the extinction-targeted behavior at the same time as the reinforcement-targeted behavior, the procedure is DRI.

Differential Reinforcement of Alternative Behavior

Differential reinforcement of alternative behavior (DRA) involves providing added reinforcement for a behavior that, while compatible with the behavior that the subject

---

members of the same response class may differ topographically but they share the same effect on the environment. Let us say that there is a problem behavior maintained by a reinforcer that we deem inappropriate to use in training. So, the problem behavior is targeted for extinction in the presence of the relevant evocative stimulus, a different behavior (that necessarily must now be of a different response class) is chosen to replace it, and an unrelated, but still effective, reinforcer is used in the presence of the same evocative stimulus. Under the narrower definition, this would not technically be an instance of differential reinforcement, because the two behaviors are not members of the same response class, although it would certainly be a differential reinforcement-like procedure. However, it would be differential reinforcement by the broader definition. As indicated above, you may find both definitions in the literature.
[32] We also use differential reinforcement in discrimination training.

would otherwise exhibit in that context, is still a specific, different behavior. We use a DRA procedure when the behavior we are training is not mutually exclusive with respect to other behaviors that are likely to compete with it.

Differential Reinforcement of Other Behavior

Differential reinforcement of other behavior (DRO) involves reinforcing any behavior other than the target problem behavior.[33] DRO is often erroneously referred to as "differential reinforcement of zero responding." This is an error because it implies reinforcement of no responding. One can only reinforce behaviors—one cannot reinforce the absence of behavior.

Differential Reinforcement of Low Rate

Differential reinforcement of low rate of responding (DRL) sets the rule that a fixed interval of time is set after which the next response is reinforced only if no responses occurred during that interval. If a response has occurred during that interval, you reset the clock on the interval. One may use DRL to reduce the rate of a behavior that is unacceptable at the current rate of responding but would be acceptable at a lower rate. The difference between FI and DRL is that, under the FI schedule, responding during the interval is permissible but simply goes unreinforced, whereas, under the DRL schedule, responding during that interval resets the interval clock. For example, imagine you have a stopwatch and have established a DRL-1min. schedule for a subject who vocalizes more frequently than you would prefer. You push the start button on the stopwatch and wait. If the subject vocalizes during that 1 minute, you push the reset button when they do so, and the minute starts counting from zero again. On the other hand, if the subject did not vocalize during that interval, you reinforce the next response that occurs.

You can include an LH component to DRL schedules. For example, this would involve starting the clock and, assuming you pass the interval without a response, reinforcement for a response after that interval is only available for a specified period of time. If no response occurs during that window of opportunity, the schedule has ended and the session is over or you reset the clock to run another trial.

We use the DRL schedule for excessive but acceptable behaviors. Alternatives exist, such as extinction procedures to eliminate the problematic behavior altogether or a preclusion procedure, where contact with the evocative stimulus is prevented. These options are not always appropriate, such as when it is not possible to limit contact with the evocative stimulus.

---

[33] The primary use for DRO schedules are in behavior change programs instated to resolve problem behaviors that are affected by strong aversive emotional behaviors, the after-effect of which would be referred to as fear, panic, or anxiety. In these cases, the DRO procedure allows for a very high rate of added reinforcement, which is thought to achieve respondent counter-conditioning as a byproduct of the operant conditioning taking place. In other words, as you reinforce operant behaviors in the presence of a feared stimulus, for instance, the added reinforcers are paired with that stimulus and the problem stimulus comes to elicit emotional reactions similar to those elicited by the treats or whatever other added reinforcer was used.

Differential Reinforcement of High Rate

Differential reinforcement of high rate of responding (DRH) sets the rule that a fixed interval of time is set after which the next response is reinforced only if a specified number of responses have already occurred during that interval. We use DRH when a behavior is occurring but not frequently enough. Unlike with DRL, in which no response is permitted during a fixed interval, DRH requires a specific minimum number of responses during the interval. In the DRH schedule, you are required to set the interval and the number of responses that must be surpassed in that interval.

Here is an example of what DRH looks like for a subject who vocalizes but not enough in the specific context under consideration. You set a DRH 10/1min (10 responses during a one–minute interval). You press the start button on the stopwatch and then start counting responses. The trial is over after the one–minute interval. If the subject did not exhibit at least 10 vocalization responses before the one–minute is up, you either end the session, or push the reset button and restart the one–minute interval for a new trial. If the subject did reach at least 10 vocalization responses in the one–minute interval, we reinforce the next response after that interval.

Differential Reinforcement of Diminishing Rates

Differential reinforcement of diminishing rates (DRD) is a schedule of reinforcement in which reinforcement is delivered at the end of a predetermined interval if the number of responses is less than a criterion that is gradually reduced across time intervals based on the behavior of the individual (Cooper, et al., 2007). This differs from DRL in that it gradually sets a lower rate of responding as you progress through the training. Once the rate of the behavior has decreased to the set criterion, the next criterion rate is set, and so on, such that the rate of behavior gradually and incrementally decreases.

Differential Reinforcement of Exceptional Responding—And Alternative to "Jackpotting"

Differential reinforcement of exceptional responding is not a formally recognized schedule of reinforcement. The reason I have included it here is that it is an alternative to the "jackpotting" procedure commonly used by animal trainers. ***Differential reinforcement of exceptional responding*** (DRE) sets the rule that during intermittent schedules of reinforcement, particularly "exceptional" responses are reinforced even if they momentarily deviate from the set intermittent schedule.[34] The reinforcement is qualitatively and quantitatively the same as with other instances of its delivery. In contrast, in a VR schedule, we might not always reinforce these particular responses. Under the DRE schedule, you can set the additional overriding schedule rule that, regardless of the schedule, we *do* reinforce exceptional responses. DRE is likely to be less disruptive and more productive than "jackpotting," but it still contributes to conditioning in a similar way.

---

[34] One reason it is likely not a formally recognized schedule is that the term "exceptional" is just as vague for its use with DRE as it is for jackpotting. Vagueness is *not* a desirable quality in a schedule of reinforcement and this schedule is *only* presented as a better alternative to jackpotting.

In contrast to a DRE schedule, the "jackpotting" procedure sets the rule that higher magnitude reinforcers will be provided for certain particularly "exceptional" responses However, it is important to note that higher magnitude reinforcement can promote quicker conditioning, but this is not linear. With each increase, the rate of beneficial effect decreases. Also, note the phenomenon referred to as ***negative incentive contrast***, in which decreased responding occurs when going back to the lower magnitude reinforcer from the higher magnitude reinforcer (Nation & Woods, 1980). Jackpotting can be disruptive for these reasons and may not be as beneficial in training as some believe it to be. Rather than being explicitly clear and consistent with regard to specific criteria you are working on in a training project at any given time, by jackpotting, you merely refer to some vague "exceptional" occurrence and target that for higher magnitude reinforcement. By adjusting the criterion to reinforce this "exceptional" occurrence, you are deviating from the consistency of the criterion. If you jackpot for significant criterion leaps, you may not be able to maintain the new criterion in successive trials and this may become disruptive to conditioning. The effectiveness of jackpotting should not be taken as a given, as more research is warranted in clarifying and analyzing the effectiveness of jackpotting procedures.

Managing Schedules of Reinforcement

Some of the schedules of reinforcement discussed above are more useful than others are in different circumstances and at certain points during a training project. This discussion could also give the impression that we use schedules in a rather static manner, but we generally use them in applied settings in a more dynamic way, particularly in the ratio stretching process.[35]

Schedules of reinforcement are vitally important in training because the schedule effects that they generate produce the rate of responding and resilience characteristics required to achieve the training objectives. In addition, providing continuous trainer-mediated reinforcement is usually not feasible in the long run. Selecting a schedule of reinforcement in a training project depends on the effect that we need. Under most common training situations, begin with continuous reinforcement and then quickly transition to VR, LH (with FD if needed). Then, gradually stretch the ratio, resetting the ratio as soon as the behavior becomes stable at the current level.[36]

We rarely use interval and FR schedules in training, mainly because of the post-reinforcement pause and adjunctive behaviors that they tend to generate.

When we initiate training for fluency and improving the duration, distance and distraction characteristics of the behavior, these parameters are addressed one at a time. When starting work on each of these factors, we relax the schedule of reinforcement, usually going back to continuous reinforcement, and then we re-thin it. The schedule of reinforcement typically changes throughout the project, depending on the circumstances. We instate a duration schedule where appropriate and a limited hold

---

[35] References to "stretching the ratio" and "thinning the schedule" refer to the same thing, gradually reducing the overall number of reinforced responses among a series of responses.

[36] This progressive thinning of the ratio schedule has been referred to as a ***Progressive Ratio Schedule*** (Peirce & Cheney, 2013)

extension to reach an acceptable *latency* (i.e., the time interval between the evocative stimulus and initiation of the behavior).

Since non-criterion behaviors occasionally occur during training, we usually use a compound procedure throughout. Differential reinforcement procedures establish rules for meeting reinforcement criteria and an extinction schedule for failure to meet the reinforcement criteria. Instead of EXT, –P can be used to target non-criterion behaviors, although this would no longer meet the definition of a differential reinforcement procedure.

Trainers rarely prepare a detailed plan for the progression of their schedules through a training project, for at least two reasons. First, the choice of schedule is based on actual responding progress during the training session and indeed throughout the project. Second, training typically progresses quickly, and modifications must often be made quickly. Furthermore, the type of progression through schedules is usually the same, as discussed above. For example, in training "sit," you might plan for continuous reinforcement for the first several trials and then, once the behavior is smooth and reliable, switch to a VR-2, LH-10sec., FD-1sec. schedule. You might then quickly increase the ratio, depending on rate of responding, to ensure at least 90% accuracy, and gradually reduce the LH and increase the FD, using EXT for non-criterion responses. Because increasing the ratio depends on ensuring at least 90% response accuracy, that component of the schedule is adjusted "on the fly," once responding meets that success rate (at least 9 out of 10). Then, once you introduce a change in the criteria, such as increased, distance or distraction levels, you will bring the schedule back to continuous reinforcement and re-thin it. Once that is smooth, you will typically introduce another feature and instate continuous reinforcement again, re-thinning again until you have met all appropriate criteria, at which point you gradually thin the schedule for the last time, resting at a point that will maintain the behavior.

The Premack Principle and Activity Reinforcers

The *Premack principle* states that the opportunity to engage in a higher probability behavior can reinforce occurrence of a lower probability behavior. That is, if an individual is more likely to exhibit one behavior than another is, the opportunity to exhibit the behavior that they are most likely to exhibit can actually reinforce occurrence of the behavior that they are less likely to exhibit. It is sometimes colloquially referred to as "grandma's rule"—if you eat your broccoli, you can then eat your cake.[37] If eating cake is a higher probability behavior than eating broccoli, you can use the opportunity to engage in cake eating as a reinforcer for broccoli eating. We can refer to this as an *activity reinforcer* as well.

---

[37] The Premack principle is not to be confused with "bribery." Bribing, in nontechnical terms, refers to promising something to someone in order to get him or her to do something wrong (www.macmillandictionary.com). Activity reinforcers do not involve bribes. The reinforcer is not present or promised, or provided before the behavior it is to reinforce and it is not provided as a reinforcer sample or cue. We provide the reinforcer *after* the behavior; the only difference in this case is that the reinforcer is the opportunity to engage in some other activity. Moreover, of course, trainers are not trying to generate "bad" behaviors.

In animal training, the animal contacts a more effective reinforcer while exhibiting a higher probability behavior, and the opportunity to engage in the behavior that generates this more effective reinforcer can be used to reinforce some other, less probable behavior. A consequence of this principle in animal training is that you do not always have to use treats to reinforce behaviors. This might be done at first in the acquisition phase of training but there are many reinforcers available to animals on a daily basis, and the opportunity to engage in behaviors that function to contact these effective reinforcers in the real world can be used to reinforce an unrelated behavior. For example, you can reinforce walking at your side on a loose leash with the opportunity to sniff a fire hydrant, or reinforce sitting while you open the door with getting to run outside. The "trick" is to make the opportunity to engage in the higher probability behavior contingent on exhibiting the lower probability behavior. The Premack principle helps us appreciate that we can use a wide range of things other than treats in order to reinforce behaviors. This is particularly important in the maintenance phase of training.[38]

Antecedent Principles, Processes and Procedures

The previous major section mainly related to principles and processes that influence subsequent behavior by directly affecting the postcedent environment. In this section, I will describe principles, processes and procedures that influence behavior by affecting the antecedent environment. Primarily, this includes stimulus control principles, including prompting the behavior, so that we may bring the behavior under stimulus control, the expansion or contraction of the range of stimuli that evoke the behavior, and transferring stimulus control from one antecedent stimulus to another. After this section, more advanced topics are addressed, including some further antecedent conditions such as function-altering stimulation.

Stimulus Control: Generalization and Discrimination Training

Stimulus control refers to a range of phenomena relating to antecedent control of behavior. **Stimulus control** exists when the likelihood of a response class form varies with respect to some property of the antecedent environment (Moore, 2008, p. 98). I will discuss the two basic processes of generalization and discrimination in this section,[39] as well as the procedure for transferring stimulus control.

---

[38] The notion of activity reinforcers is arbitrary in many, if not most or all cases. After all, while a treat is a reinforcer, the opportunity to eat a treat is an activity. There is usually a way of specifying an activity that is associated with contacting a reinforcing stimulus. However, the Premack principle is useful in evoking attention to everyday reinforcers over more contrived trainer-mediated reinforcers.

[39] There is a current trend in behaviorology toward replacing the term "discrimination" with the term evocation. However, the term evocation refers to the triggering of an operant behavior whereas discrimination refers to a pattern of behavior or process by which we narrow the range of evocative stimuli. Therefore, while I have replaced "discriminative stimulus" with "evocative stimulus," I use the term "discrimination" in reference to the narrowing of the range of stimuli that evoke a behavior and to the training that is used to carry that out. This should help promote appropriate discriminative responding to

Generalization and discrimination are inversely related, and involve an increase or decrease in the range of stimuli that evoke a behavior respectively. As Moore (2008, p. 100) points out, generalization and discrimination are not things individuals "do" but rather a process that occurs or a function of a pattern of behaving. Hence, while it is appropriate to refer to a pattern of responding to stimuli as an instance of generalization or discrimination, it would be inappropriate to refer to a subject generaliz*ing* or discriminat*ing*.

Generalization

Generalization, specifically, *stimulus generalization*, with regard to operant behavior,[40] is the process by which the range or set of evocative stimuli triggering an operant behavior is increased.[41] That is, additional antecedent stimuli also come to evoke the behavior in question. We increase the evocative capacity of an antecedent stimulus with reinforcement. The more similar other stimuli are to a stimulus involved in the original training, the more likely they too will evoke the behavior—this results in an increase in the number of stimuli that may evoke the behavior. For example, if a training procedure is carried out and a 1000 Hz tone becomes the evocative stimulus for a parrot to raise a wing and after conditioning, one finds that a 950 Hz tone or a 1050 Hz tone also sometimes evokes the behavior, it is be said that generalization has occurred. Typically, the less similar the stimulus becomes, the less likely it is to evoke the behavior in questions. For instance, the 1000 Hz tone may evoke the behavior 99% of the time and the 950 Hz tone or the 1050 Hz tone may evoke the behavior 80% of the time. An 800 Hz tone or a 1200 Hz tone may evoke the behavior 60% of the time. The distribution usually forms a common bell curve form when graphed.

We can promote generalization by trainer-mediated procedures or by nontrainer-mediated events. Because reinforcement tends to strengthen the evocative capacity of stimuli with shared or similar properties, these similar stimuli will come, through repeated trials, to take on stimulus control over the behavior in question. The less similar a stimulus is—that is, the fewer shared salient properties there are between the stimuli—the less likely the stimulus is to take on stimulus control. For instance, if you train a subject to sit when cued with a vocal "sit," and the word "sip" also comes to evoke sitting, that is an indication that generalization has occurred. In this example, you are likely to find that the word "down" does not evoke sitting because "down" and "sit" do not share enough properties to be strengthened by reinforcement along with the "sit" stimulus.

Response generalization, another form of generalization, involves an increase in the range of response class forms evoked by a stimulus. For example, if walking to contact a reinforcer is effective, other response class forms such as skipping, crawling and running may also become likely. We refer to this increase in the likelihood or rate of other response class forms as response generalization because the range of response class forms that are evoked increases. You can promote this by reinforcement of other response class forms, or you can discourage it by differentially reinforcing only the very

---

both evocation and discrimination concepts. However, take note that other behaviorological literature may use the term evocation in place of "discrimination" in both of these uses.

[40] Generalization is also applicable to respondent conditioning.

[41] When we use the word generalization alone, it generally refers to stimulus generalization.

specific response class form you want and extinguishing similar response class forms, a process referred to as discrimination, which I will discuss next.

Discrimination

Discrimination is the process by which the range or set of evocative stimuli triggering an operant behavior decreases. ***Discrimination training*** is a procedure by which we decrease the range of evocative stimulus properties by differentially reinforcing responding only to a specific set of stimulus properties while extinguishing responding to similar stimulus properties. We refer to the specific stimulus targeted in discrimination training, of course, as the evocative stimulus. We target similar stimuli that might otherwise evoke the target response class form, but do not fall within the narrow property range of the evocative stimulus for extinction, and refer to them as S-delta, symbolized S^Δ. To compare this with the example used for generalization, if a training procedure is carried out and a 1000 Hz tone becomes the evocative stimulus for a parrot to raise a wing and after conditioning, one finds that a 950 Hz tone or a 1050 Hz tone also sometimes evokes the behavior, it would be said that generalization has occurred. Assume the trainer's objective is for the parrot to raise a wing when exposed to the 1000 Hz tone but not to a 950 Hz tone or a 1050 Hz tone. Discrimination training would involve arranging trials in which we expose the parrot to 950 Hz tones, 1000 Hz tones and 1050 Hz tones. We will only reinforce responding (i.e., wing raising) to the 1000 Hz tone and not responding to the other stimuli. Through repeated trials, responding to 950 Hz and 1050 Hz tones will decrease and become extinct, thereby reducing the range of stimuli that will evoke the behavior. We refer to this narrowing of the range of evocative stimuli as discrimination and the training carried out to achieve it as discrimination training. Obviously, there are limits to this process. When the difference between two stimuli becomes indistinguishable by the nervous system in question, discrimination cannot take place.

Here is another example involving a common training situation. Let us say you train a parrot to wave a foot when you say "wave" and you reinforce each instance of this. You might then carry out discrimination training by saying "waze" and failing to reinforce if the bird waves. You could present other similar stimuli and do the same thing, with trials of "wave" mixed in, which you *do* reinforce responding to. You will find that discrimination occurs. Instead of a wide range of similar stimuli evoking the behavior, only that narrowly defined stimulus evokes the waving behavior and other similar stimuli no longer do.

To take the above example a step further and illustrate both generalization and discrimination training, you might promote stimulus generalization by reinforcing when *any* person says "wave" at the same time as you carry out discrimination training to restrict the evocative stimulus to the exact word "wave" and not similar-sounding words. We use generalization and discrimination training to ensure that the "right" stimuli evoke a behavior and the "wrong" stimuli do not.

The ***differential outcome effect*** is a robust behavioral phenomenon whereby discrimination training occurs more quickly and the discrimination is more accurate when different behaviors contact different reinforcers. For example, if you are training a dog to sit, stand, and down, and you are randomly evoking each behavior, your training will be more efficient and effective if you follow each of the three behaviors with different reinforcers rather than just the same one. For instance, sit might generate kibble, down might generate a piece of veggie burger, and stand might generate a piece of apple. This makes the three contingencies more distinct than if the consequences were the same in all cases, and this greater distinction may be the reason for the effect. It can be particularly useful when working on inter-stimulus discrimination training.

Transferring Stimulus Control

Once a behavior is more likely to occur immediately after we present a specific stimulus, the behavior is under **stimulus control**—the antecedent stimulus comes to control the behavior. The transfer of stimulus control from an established evocative stimulus to, what is at that time, a neutral stimulus is important for training, because trainers are often required to change what evokes a behavior. For instance, it is common to transfer stimulus control from the motion used to lure or prompt a behavior to a hand signal or a vocal cue.

We may conveniently categorize stimuli based on the sensory system they impinge upon. To transfer stimulus control between stimuli that share the *same* sensory modality, simply take advantage of stimulus generalization and gradually make the old established stimulus, seem gradually more like the new one through repeated trials. The subject will respond to the new stimulus because it shares similar properties (i.e., generalization) with the previous stimulus and through repeated trials, all of the features of that new stimulus are strengthened until the new stimulus takes on stimulus control. We call this a **prompt fading** procedure. For example, when transferring stimulus control from a luring motion to a hand signal that is not similar enough to allow immediate generalization, gradually make the old luring motion look more like the new hand signal until the new hand signal reliably evokes the behavior (i.e., takes on stimulus control). On each successive trial, we make the motion look just a little bit less like the lure motion and just a little bit more like the hand signal motion.

To transfer stimulus control between stimuli that involve *different* sensory modalities, it is not possible to take advantage of an incrementally shifting generalization process as easily and we rely on a pairing[42] procedure instead. In this case, we use the following contingency sequence:

New stimulus → Old stimulus → Behavior → Reinforcement

We call this a **prompt delay** procedure. Through repeated trials, the new stimulus takes on stimulus control over the behavior. For example, when transferring stimulus control from a hand signal to a vocal cue (such as "sit"), these are different sensory system categories—one stimulus is visual and the other auditory. We carry out the following sequence: "sit" → hand signal → subject sits → reinforcement. You can test

---

[42] **Pairing** refers to more than one stimuli impinging upon the subject's nervous system at approximately the same time. This pairing process results in certain physical changes within the subject's nervous system that cause certain shared reactions to one or the other stimuli later. Some authors (e.g., Moore, 2008) refer to this pairing process as **correlating** the stimuli. It is best to avoid the older term "associating" as it tends to imply agential participation.

whether the procedure has been successful by delaying presentation of the old stimulus and observing whether the behavior occurs or not.

In both the prompt fading and the prompt delay procedures, the range of stimuli that will evoke the target behavior increases and hence generalization occurs. You can encourage this generalization process by allowing reinforcement of behavior evoked by similar but not identical stimuli. Alternatively, you can discourage it through discrimination training, which I discussed in a previous section.

Prompting

A *prompt* is any antecedent stimulus, other than the designated primary stimulus, that contributes to evoking the target behavior. The *primary stimulus* is the currently non-evocative or weakly evocative stimulus designated to become the evocative stimulus for the behavior after conditioning. Recall that when reinforcement occurs, it increases the evocative capacity all of the antecedent stimuli that are present, including any non-evocative stimuli that the trainer might seek to install as the primary evocative stimulus, as well as any prompts they might use and even some perhaps unintended stimuli that are present coincidentally. We use the prompt in order to contribute to generating the behavior in the presence of the primary stimulus, so that the primary stimulus will take on stimulus control (along with other present stimuli). Reinforcement affects the evocative capacity of the prompt and the primary stimulus. Once the primary stimulus takes on stimulus control over the behavior, the prompts are gradually *faded*, leaving the primary stimulus with a strong capacity to evoke the behavior on its own. Fading the prompt as soon as practicable is important because with each trial, it too, along with the primary stimulus, takes on stimulus control over the behavior. If the prompt becomes too well conditioned, it will require more extensive discrimination training (including extinction trials) to eliminate. The unfaded prompt can become what we call a function-altering stimulus (to be discussed below), resulting in the behavior becoming *prompt dependent*.

Establishing and strengthening the capacity of a stimulus to evoke behavior is a major part of what training achieves. The challenge in these situations is to initially generate the behavior, so that it can be reinforced and thereby strengthen the capacity of antecedent stimuli to evoke the behavior. In this case, we can use shaping of successive approximations of the behavior. This is an advanced procedure, introduced in a later chapter. There are times when shaping is preferable but if all else is equal, and prompting is possible, prompting is the quicker means to generate the behavior. If we can use other stimuli or an accumulation of stimuli in combination to evoke the target behavior, it is not necessary to shape the behavior. Once we evoke the behavior, we may reinforce it. This reinforcement strengthens the capacity of all antecedent stimuli present at that time to evoke the behavior.

Sometimes, you will be able to present the primary stimulus and prompt stimuli together, both contributing to generating the behavior, and once you have sufficiently reinforced both, you may gradually faded the prompt. In many cases though, your prompts will do the vast majority of the evocative work, so to speak, and you will then transfer stimulus control to a newly introduced stimulus once it has taken on enough stimulus control to evoke the behavior reliably. Therefore, we can use prompts along with the primary stimulus as a supplementary cue or you can use prompts on their own and introduce the primary stimulus once we can generate the behavior reliably. Often, in this latter case, the prompt is actually evoking attending behavior or approach/contact behavior and as we manipulate the prompt stimulus, the subject exhibits the target behavior as he or she moves to attend, orient, approach and contact the prompt. Some examples of each approach may help in understanding these principles.

For example, you might train a naïve puppy to come to you by saying "here," which tends to exert weak evocative control merely because you are looking at the puppy and vocalizing something, which often evokes approach behavior. But, it may be too weakly evocative to ensure quick and reliable approach every single time, so you prompt the behavior by supplementing the "here" with waving motions and high pitched noises that, together with the primary stimulus, effectively generate the behavior. These additional prompts are then gradually faded out. This is a prime example of prompts used as supplementary evocative stimuli.

On the other hand, when training a dog to sit, you may utilize a treat in your hand to generate **targeting** behavior (i.e., attending, orienting, approaching and contact behaviors) and the dog follows the treat with his or her nose, which, as you move it over his or her head, he or she sits and the behavior is generated. In that case, the prompt alone generates the behavior without the primary stimulus. With each trial, the prompt takes on stimulus control of the behavior until it is faded, transferring stimulus control to either an intermediate (e.g., hand signal) or the final cue (e.g., vocal), in either case, the primary stimulus.

Point of common confusion. Many trainers verbalize about prompting or evoking a behavior versus "capturing" it. All operant behavior is evoked. One might talk coherently about trainer-mediated prompts and cues—ones mediated directly by a trainer, versus behaviors that are evoked by some other stimulus in the environment. We might call these nontrainer-mediated evocative stimuli and the behavior they evoke is then reinforced, but they are all evoked by stimuli in the environment. If you "capture" a behavior, that just means that some event not mediated by you evoked the behavior.

Once the primary stimulus becomes capable of evoking the behavior, generalization has occurred in that a greater number of stimuli evoke the behavior. Trainers do not often think of this as an instance of generalization even though it is. At this time, the prompt can be gradually faded. **Fading** often involves presenting the prompt in a slightly less salient or prominent form with each successive trial until we no longer present it at all. In some very common training scenarios, we gradually transform the prompt into what will be the evocative stimulus for the behavior, be it temporary or permanent as described above. For example, when prompting a sit behavior with a luring prompt, the now redundant prompt is gradually made to look less and less like the luring motion and more and more like a hand signal that will then be used to evoke the behavior. We call this kind of fading **prompt fading**.

As mentioned, we can categorize prompts by the sensory system that they impinge upon for convenience. For instance, olfactory prompts involve the transfer of energy from the stimulus to the organism through the olfactory system (smell). Tactile prompts contact the touch sensitive nerves. Visual prompts impact upon the optic nerve in the eye. Aural prompts impact upon the nerves attached to the eardrum.

Sometimes physical manipulation is mistaken for prompting. A tactile prompt might involve a touch but it does not force a behavior. It is better to call the so-called "physical prompt"—such as pushing on a dog's hips to generate a sit a *physical manipulation*. A prompt adds a contributory evocative stimulus, but physical force or manipulation is not evocative—it is just force. The fundamental problem with physical manipulation is that it superimposes a concurrent and conflicting contingency into the situation and is usually distracting and disruptive of training, depending on how aversive it is. When one applies force to an animal's body, the subject may indeed eventually end up in the intended position (or merely allow the movement to be forced) but it often elicits an opposition reflex as well, which means that the opposite muscle group is activated. A physical manipulation that achieves the movement as the animal pushes against pressure may therefore promote a completely different behavior—at best. Applying the "dead body test," it is not even a behavior at all where it is just a matter of a person manipulating the subject's body into a position; the behaviors that are exhibited in this situation are usually counter-controlling behaviors in response to the aversive manipulation. At worst, physical manipulation imposes intensely aversive stimulation, which also tends to disrupt conditioning and promote problematic side effects.

Note that certain kinds of barricades that might be used in training—for example, practicing sit with the dog in a corner so that they cannot back up, or using a chair or leg to lure through to achieve a down—are not prompts. We might also use prompts, but the barricade itself involves use of a *preclusion procedure* and not a prompt procedure. Such preclusion procedures function to prevent the possibility of certain behaviors being exhibited, and can be removed later.

Multiple-Term Contingencies

Thus far, I have discussed the basic two- and three-term contingencies. It is now time to elaborate on four and greater term contingencies. We write contingencies with as many terms as we need to adequately describe functional relations in behavioral episodes under consideration. Three terms are common in the simplest contingency analyses. However, expanding the number of terms can provide added explanatory power, which provides a finer-grained accounting of the behavioral episode when the terms we add represent events that exert significant control over the behavior. We can add terms to the antecedent, behavior or postcedent positions of the contingency. Note that adding terms does not necessarily provide a more complete accounting of the behavior episode; in some cases, it only imposes needless complexity.

Added Consequating Terms

We commonly identify the unconditioned reinforcer alone in our contingency analyses for simplicity purposes even though we frequently use a conditioned reinforcer as well. We do this simply as a matter of convenience—a verbal shortcut. To increase accuracy, we may expand the contingency analysis to include both the unconditioned reinforcer and the conditioned reinforcer. This four-term contingency would be of this form:

$$S^{Ev} \rightarrow Behavior \rightarrow Consequence_1 \rightarrow Consequence_2$$

The specific contingency might look like this:

$$\text{"Sit"} \rightarrow Jake \ sits \rightarrow Click \rightarrow Contact \ with \ treat$$

Added Behavior Terms

Occasionally, a sequence of behaviors becomes important in an analysis, rather than either ignoring the extra behaviors or clumping the behaviors together in a general description of the response class, we can improve accuracy by including each distinct behavior in the contingency analysis. In this case, the contingency might have the following form:

$$S^{Ev} \rightarrow Behavior_1 \rightarrow Behavior_2 \rightarrow Consequence$$

The specific contingency might look like this:

$$\text{"Mat"} \rightarrow Jake \ walks \ to \ mat_1 \rightarrow Jake \ lies \ down \ on \ mat_2 \rightarrow Treat$$

Added Antecedent Terms (Function-Altering Stimuli)

The Function-Altering Stimulus

The other broad category of antecedent stimulation besides the evocative stimulus is the function-altering stimulus (S^{FA}). A *function-altering stimulus* is an antecedent stimulus that causes changes to the nervous system of an organism in such a way that it alters the organism's receptivity to other stimuli, making these other stimuli more or less likely to evoke behavior. The function-altering stimulus does not itself evoke the behavior but rather alters the function of evocative stimuli. Occasionally, an evocative stimulus will evoke a behavior under certain circumstances but not under other circumstances. These "circumstances" are comprised of function-altering stimuli. They are sometimes referred to as "setting events," "occasion setters," "enabling stimuli" or simply "context." Specific behavioral processes such as motivative operations, sensitization, habituation and potentiation are examples of specific kinds of function-altering stimuli. Take note that prompts are not function-altering stimuli. A prompt adds an evocative stimulus, whereas function-altering stimuli only alter the evocative capacity of other stimuli. Consideration of context (function-altering stimulation) will help us achieve a higher degree of explanatory power in our contingency analysis.

Fraley (2008, p. 512) provided an excellent example of function-altering stimulation in which the presence of a fire alarm lever evokes lever-pulling behaviors, but only under certain conditions. In many instances, the lever is a neutral stimulus rather than an evocative stimulus. The presence of flames or smoke (function-altering stimulus) alters the capacity of the lever (the evocative stimulus) to evoke the lever-pulling operant; it does so physically within the nervous system of the behaving organism, just as all stimulation does. Without the presence of the function-altering stimulus, the maintaining consequences do not occur; merely pressing a lever any time you see one would not likely be reinforced, and indeed a punitive consequence would likely suppress it.

Another important example might involve a situation in which a dog sits when we vocalize the word "sit," but only if treats are obviously present (a function-altering stimulus). The vocal cue evokes the behavior reliably, but *only if* the treats are present. You can see why this fourth term becomes important in adequately explaining and predicting behavior. The vocal stimulus is the evocative stimulus, and the presence of the treats is the function-altering stimulus because it functions to alter the capacity of the evocative stimulus to evoke the behavior. When one fails to quickly fade food prompts, the presence of the food as the unconditioned reinforcer becomes a function-altering stimulus and the cue comes to evoke the behavior only if the function-altering stimulus is present; the behavior that the food reinforces becomes what we call ***prompt dependent***.

Notice the phrase "only if" in these examples. The fire alarm lever is evocative of lever-pulling behavior, but *only if* flames are present. The vocalized "sit" is evocative of sitting, but *only if* treats are present. When you encounter a behavior episode that evokes an "only if" descriptor, you know that a function-altering stimulus is present.

In some cases, it can be challenging to determine which stimulus we should assign to the function-altering stimulus position and which should be assigned to the evocative stimulus position in a contingency analysis that explains the functional relation between the two antecedent terms. By general convention, we write the independent variable on the left, and the dependent variable on the right (Fraley, 2008, pp. 515–516). However, the question remains, which is which? Catania (1998, p. 265) provides a good example with an obvious distinction. If someone says the stove is hot, that clearly alters the function of the stove within the organism hearing the warning. The stove may now evoke retreat or avoidance behaviors. The verbal warning is clearly the function-altering stimulus and the stove is clearly the evocative stimulus and not the other way around. However, many cases are not as obvious. Does the lever evoke lever-pulling behavior only if flames are present or do flames evoke lever-pulling only if a lever is present? When the function-altering stimulus and the evocative stimulus coincide in time, as will be the case in dog training, the two stimuli may be separated experimentally; one of the stimuli is evocative for the behavior and the other is not (Schlinger & Blakely, 1987). In our case, if you present flames, this may evoke any number of behaviors (e.g., yelling warnings or running away) and the flames themselves would not evoke lever-pulling behavior (however, they might evoke looking-around-for-levers behaviors). The lever, on the other hand, could very well evoke lever-pulling behavior under various circumstances; the lever is, under appropriate circumstances, evocative (discriminative) of lever-pulling behavior. In this case, the lever is necessarily required for the behavior to occur. Furthermore, notice that in this example, the lever does not alter any other stimulus's capacity to evoke lever-pulling behavior but the presence of flames does. In the case of the treat-dependent sitting dog, the word "sit" evokes sitting behavior *only if* the dog sees that treats are present. The stimulus "sit" is, under appropriate circumstances, evocative for sitting behavior, as opposed to any number of other behaviors and is so in this example. The vocal cue "sit" does not alter the evocative capacity of the treats. The dog does not sit, as opposed to stand, down or anything else, because the presence of treats evokes it. Sit is the evocative stimulus and the treat is the function-altering stimulus because "sit" is evocative for sitting in this relation. The point here is that it will not always be obvious which stimulus is the evocative stimulus, and which is the function-altering stimulus. When in doubt, separate the stimuli, and consider which would be evocative, under appropriate circumstances, as opposed to altering the function of the other.

Below is a general depiction of a four-term contingency featuring a function-altering stimulus:

$$S^{FA} \rightarrow S^{Ev} \rightarrow \text{Behavior} \rightarrow \text{Consequence}$$

We can depict this as in the previous examples, in a way that illustrates how the contingency functions to strengthen the antecedent stimuli. Notice that this form begins with a stimulus that, alone, does not evoke any behavior. This is a neutral stimulus (S^N). In this case, you can see that only when another stimulus is present does the stimulus evoke the behavior, and a reinforcing consequence functions to strengthen the capacity of the evocative stimulus to evoke the behavior *only* when the function-altering stimulus is present. In other words, the reinforcer strengthens both the evocative stimulus $\rightarrow$ behavior and the function-altering stimulus $\rightarrow$ evocative stimulus contingencies.

$$S^N - \text{Behavior} - \text{No Consequence}$$

Motivative Operations

"Motivation," when used colloquially, refers to anything that increases the likelihood of behavior, particularly where the reinforcer is identified (Fraley, 2008). Motivation can act as function-altering stimulation. One might say that "being hungry" motivates eating, "being hot" motivates turning on an air conditioner, or that "being fearful" motivates running away from something for example. "Motivation" is usually invoked when some antecedent condition exists such that an established evocative stimulus becomes more likely to evoke the behavior that resolves that condition. Behaviorology accommodates the notion of motivation more technically, as it relates to operant behavior with the concept of the motivative operation as introduced by Michael (1993) as "establishing operations" and expanded by Laraway and colleagues (2003) as "motivating operations." A ***motivative operation*** (MO) refers to any event that makes changes to an organism's nervous system in a way that alters the effectiveness of a reinforcer and hence the frequency of behavior maintained by that reinforcer (Pierce & Cheney, 2013, p. 486; Moore, 2008, p. 128, citing Michael, 1993). The motivative operation established what we might call a *motivative condition* within the body. Notice that "motivation" (either the operation or the condition it establishes) is not something that an organism "has." The motivative operation is a *procedure* that we may implement that causes certain structural changes within the organism (condition) thereby changing that organism's receptivity to certain other stimuli. The condition it establishes within the body relates to the current structure of the nervous system and not to a thing that the organism "has" or "does." Motivative operations commonly involve generating a reinforcer-deprived or a reinforcer-satiated condition within an organism, which then momentarily alters the effectiveness of that reinforcer and hence the evocative stimulus that would evoke behavior that functions to enhance contact with that reinforcer. For example, if we deprive the body that we refer to as "you" of food (i.e., make you "hungry"), food will be a much more effective reinforcer for behaviors that lead to consuming food (e.g., getting food and eating). On the other hand, if we satiate the body[43] with respect to food (i.e., make you "full"), food will be a much less effective reinforcer for behaviors that lead to consuming food.

We can conceptually divide motivative operations into two broad categories: establishing operations and abolishing operations. **Establishing operations** are operations that result in evocative stimuli being *more* evocative by depriving the organism of the reinforcer in that contingency. **Abolishing operations** are operations that result in evocative stimuli being *less* evocative by satiating the organism with respect to the reinforcer in that contingency (Laraway et al., 2003). Put another way, the body can be satiated with respect to the reinforcer, making the evocative stimulus *less* evocative (i.e., an abolishing operation). Or, the body can be deprived with respect to the reinforcer, making the evocative stimulus *more* evocative (i.e., an establishing operation). Establishing operations and abolishing operations refer to the procedure implemented to establish or abolish the evocative capacity of a stimulus, while satiation and deprivation refer to the condition of the body in relation to the reinforcer. We define **deprivation** by a historical interval during which a body does not contact a reinforcer—the longer the interval the more deprived the body. We define **satiation** as the situation where the body has been in contact with the reinforcer to the point where the presence of the reinforcer is no longer an effective reinforcer. Deprivation and satiation are inversely related. A deprived body is *more* disposed to exhibit behavior that has historically contacted that reinforcer and a satiated body is *less* disposed to exhibit behavior that has historically contacted that reinforcer. We can use establishing operations to make behaviors targeted for reinforcement more likely. Excessive

---

[43] As an advanced tangent, appropriate here only within a footnote, you may have noticed that I did not write "your body." The phrase "your body" is problematic from a natural science perspective. What would be the "you" in this statement? Are "you" a secular body-supervising agent that directs the actions of your possession called the "body"? No, "you" *are* the body. We can conceptualize "you" in a couple of different ways that are consistent with natural science, such as you being the body that evokes pronouns such as "you," or as you being your full repertoire of behaviors, including the various consciousness related behaviors of awareness, recognition, comprehension, visualization and so on. However, "you" are not something that "has" a body. For more detailed discussion of this fascinating topic see (Ledoux, 2014) and Fraley (2008).

deprivation is counterproductive, and indeed abusive, but a mild deprivation state (e.g., "hunger" that commonly occurs before meals) can make training with food reinforcers more efficient at that time.

Fading Function-Altering Stimuli

As with prompts, function-altering stimuli can be faded. Once the training is well under way and the evocative stimulus is effective with a deprived body, you can gradually eliminate the condition of deprivation from your training sessions. Although a satiated body will make training with food reinforcers less effective, the behavior will now be less under control of trainer-mediated reinforcers and more under control of nontrainer-mediated ones. Thus, food becomes less important as training proceeds and the behavior comes to generate endoreinforcers. Professional trainers are quite familiar with the phenomenon of a dog beginning to reject treat reinforcers, apparently eager not to delay a further opportunity to exhibit the behavior—the dog takes the treat and promptly spits it out.

More complex multiple-term contingency analyses exist, but these three- and four-term contingencies will suffice for now. Interested readers may see Fraley (2008) for additional discussion of multiple-term contingencies and function-altering stimulation.

Concurrent Contingencies

Real life is dynamic and complex and behavior is multiply controlled. Rarely is there a single contingency operating on an individual at any given moment. In fact, it is possible that there never is just one contingency in operation, even in a highly controlled laboratory setting. In even the best-managed training environments, there are multiple concurrent and often competing contingencies operating on individuals, with each contingency vying for control over behavior. These contingencies may include different reinforcers available for different behaviors and even punishers that compete with the reinforcers. When more than one contingency is competing for control over behavior, it is sometimes unfortunately referred to "choice." Behavior is of course reactive and passive, not initiative and free-willed. Although, at least with humans, much thought shares in control over overt behaviors, which some may refer to as "choices," but these thoughts are behaviors, which are evoked themselves, and controlled by environmental stimuli. So-called "choice arrangements" are better framed as *concurrent contingencies*, to reflect a more accurate analytical perspective on the topic. The reason it is useful to appreciate concurrent contingencies is that, when training, it is important to control as many of the contingencies as possible to ensure that the trainer can prompt and reinforce the target behavior instead of any number of other behaviors that the subject might otherwise exhibit. For example, just because you are reinforcing a behavior does not mean that something may also be punishing it at the same time, resulting in ineffective training. As another example, just because you are training through one contingency does not mean that others are not competing with you for control over the subject's behavior (i.e., the subject is distracted).

A simple experimental arrangement involves a pigeon who may peck at one of two buttons, one of which generates reinforcement each time, and the other never generates reinforcement. The pigeon will spend time pecking the button that generates the reinforcement and not the other button. This is a form of ***behavior economics***—animals tend to behave efficiently to maximize their access to reinforcers. Indeed, this is implicit in the notion of reinforcement itself and we may consider it a law of behavior. We can expect animals (including humans) to exhibit the behavior that generates the most reinforcement relative to other behaviors. Arrangements that are more complex involve the availability of opportunities to peck one of two buttons that generate reinforcement for every sixth response or ninth response, respectively. In the initial acquisition stage, pigeons will spend time pecking both buttons, but once the responding stabilizes, the pigeon will settle on pecking at the button that generates the richest reinforcement at a higher rate.

The ***matching law*** states that the distribution of behaviors will match the frequency of available reinforcers. One of our goals as trainers is to arrange the environment such that the target behavior requires less response effort than other behaviors, and generates more reinforcers or more effective reinforcers than the various other behaviors that might otherwise occur at that time. There are many ways to make these kinds of manipulations. Using highly effective reinforcers for the target behaviors and mild deprivation beforehand will ensure that we are utilizing the most effective reinforcer. If the most effective reinforcer was previously available for some other behavior, that reinforcer can instead be made contingent on occurrence of the target behavior alone. We can minimize the ***response effort*** (i.e., the energy expenditure required to exhibit the behavior) for the target behavior, and increase the response effort, where possible, for competing behaviors.

We often simplify contingency analyses by only identifying the contingency exerting the most influence over the target behavior. However, sometimes a finer-grained accounting of the episode and the result makes consideration of concurrent contingencies appropriate. You might find when you are trying to train a behavior that the rate of progress is slower than you expected. This may evoke an analysis of the concurrent contingencies operative in the environment at that time. Often, these concurrent contingencies involve competing contingencies vying to evoke a different behavior (i.e., distraction). Other times, the competing contingency may involve a previously unidentified aversive stimulus related to the behavior you seek to train. In any given situation, even one that might seem simple for the most part, there may be many contingencies simultaneously exerting control over the behavior exhibited in that situation. Conflicts between concurrent contingencies might involve a conflict between two different behaviors or a conflict between the consequences generated by the same behavior. Indeed, both of these types of conflicts can exist at the same time. There might be other evocative stimuli present, all of which are exerting some control over competing behavior. In other situations, multiple sources of added and subtracted reinforcement are available, and perhaps multiple sources of punishment are available. The prevailing behavior will be the net result of the accumulated evocative capacity of all evocative stimuli present.

For instance, from a previous example of Jake lying down on his mat, the treat was the added reinforcer, and thus we featured the treat prominent by using that contingency alone to explain the behavioral episode. Let us assume that training is not as smooth and reliable as you would like or expect, and so you analyze the situation in more depth and work toward the goal of more effectively controlling the behavior. Perhaps Jake had been playing for only a short time (indicating potential deprivation) at the time we presented the cue and we subtractively punished going to the mat by requiring him to leave his toy. In addition, in all contingencies of reinforcement, response effort is inherently punitive although the reinforcers usually outweigh the punitive aspects of expending the energy to exhibit the behavior that contacts the reinforcers.

The point is that most behavioral episodes are complex mixtures of more than one contingency and often involve competing contingencies that might result in slow progress, vacillating behavior, or some other disruption in the training process. Careful analysis can often help you identify the variables involved in the resulting behavioral outcome. You can then adjust the environment and make the training more efficient. When undertaking such an analysis, make a list of each contingency and then rank them in the order that reflects what you believe are the relative strengths of each contingency (i.e., with the first being the strongest, and the last in the list being the weakest). This added analytical clarity could help resolve these kinds of problems.

Law of Cumulative Complexity

As you likely are beginning to see, behavior can be complex! For instance, a given behavioral episode can involve multiple concurrent contingencies (some promoting and some competing) and complex sets of stimuli can interact together in generating and maintaining behavior, including covert (private) verbal behavioral processes. Indeed, innate respondent contingencies can be co-mingling with operant contingencies. The *Law of Cumulative Complexity* addresses this issue nicely:

> The natural physical/chemical interactions of matter and energy sometimes result in more complex structures and functions that endure and naturally interact further, resulting in an accumulating complexity. (Ledoux, 2014, p. 20)

Behavior can become very complex, challenging our ability to fully explain and predict the most complex episodes of behavioral. Some might say that psychology or cognitive ethology provide a way to explain such complex behavior. Indeed, within psychology and cognitive ethology, there is no shortage of explanations for complex behavior patterns, many contradicting the others and even themselves, let alone the laws of nature. When a discipline is not constrained to studying and hypothesizing natural events in ways that are observable, measurable and confirmable in a parsimonious manner, it is free to postulate all sorts of notions to explain the phenomenon in question. This does not make that discipline superior in any way. The challenge in explaining, predicting and controlling complex behavior, does not justify abandoning the *most* reliable, effective and parsimonious means of doing so—that is, via a natural science approach. Abandoning natural science in favor of a psychology or cognitive ethology perspective will not yield greater explanatory and predictive power or an effective technology for controlling behavior. Indeed, these disciplines are notoriously ineffective in controlling behavior. Furthermore, behaviorologists have made much progress in expanding the behaviorological approach to analyze the most complex behaviors. See Fraley (2014) for good examples of analyzing complex episodes of behavior via a natural science approach. See Ledoux (2014, pp. 28–34) for a good introduction to some areas of behaviorology applied to extremely complex behaviors such as the ***recombination of repertoires*** and ***equivalence relations***, which psychologists and cognitive ethologists have previously explained with the notions of "insight" and "mental transformation."

Respondent Conditioning

Respondent behavior is behavior that is automatically elicited by an antecedent stimulus, cannot be prevented by supplemental antecedent stimuli, and is unaffected by consequences. Respondent contingencies involve a two-term contingency of the stimulus–response form. This is in contrast to the operant three-term contingency of the stimulus–response–stimulus form. The body is structured such that when the antecedent stimulus occurs, the stream of energy from the environment causes the body to mediate the specific response. These reactions are controlled by a neural pathway that activates spinal motor neurons without the delay associated with routing neurological signals through the brain, allowing the reaction to occur relatively quickly (and without "second guessing" so to speak). We refer to the stimulus that elicits this reaction as an ***unconditioned stimulus***, and the response it elicits is an ***unconditioned response***. In other words, no conditioning is required for that response to occur. Since these relations exist as a result of biological evolution acting on the population, across the span of its existence, they are related to imperative biological requirements. Examples include salivating in response to food in the mouth and blinking when something touches the eyeball. Unconditioned responses can also include the release of certain chemicals into the bloodstream when the animal is startled (e.g., during the so-called "flight or fight" response).

Respondent conditioning is the process whereby a ***neutral stimulus*** (NS), which does not elicit the response in question, comes to elicit a response after it has been paired with an ***unconditioned stimulus*** (US) (or an established conditioned stimulus). An example of a neutral stimulus is the sound of a clicker before we have conditioned a response to it. The neutral stimulus becomes a ***conditioned stimulus*** (CS) once it elicits the ***conditioned response*** (CR).[44] The conditioned response is usually similar to the unconditioned response (UR). The statement that the neutral stimulus and the unconditioned stimulus are paired, refers to the streams of energy from each of the stimuli impinge upon the subject's nervous system at approximately the same time. The nervous system is then changed in such a way that the subject will mediate the response when exposed to the neutral stimulus (now a conditioned stimulus), assuming that it is occasionally paired with the unconditioned stimulus. I illustrate these processes in Figure 9.

---

[44] Confusingly, the NS is sometimes called a CS right from the beginning of the conditioning procedure, the idea being that after even the very first pairing, some conditioning has occurred, even if it is not enough for the stimulus to reliably elicit the response.

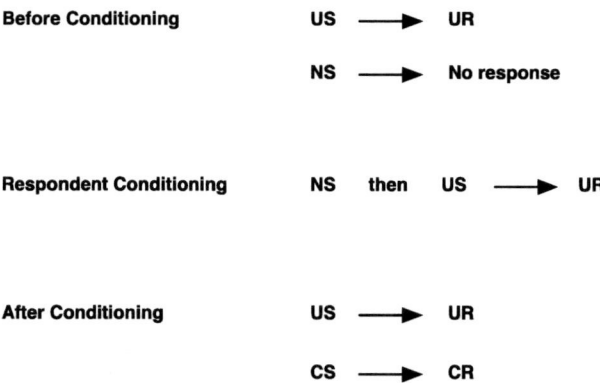

Figure 9. Diagram illustrating the process of respondent conditioning.

The two most effective respondent conditioning procedures are delayed conditioning and trace conditioning. In ***delayed conditioning***, we present the conditioned stimulus before the unconditioned stimulus and then ends after the unconditioned stimulus starts (but generally stops before the unconditioned stimulus stops; see figure 10). In ***trace conditioning***, we present and then remove the conditioned stimulus, followed shortly by the presentation of the unconditioned stimulus. Establishing a clicker as a conditioned added reinforcer uses trace conditioning. For effective conditioning, the unconditioned stimulus should follow the conditioned stimulus within a couple of seconds to achieve satisfactory contiguity. Simultaneous conditioning, in which we present the conditioned stimulus and unconditioned stimulus at the same time and backward conditioning, in which we present the unconditioned stimulus before the conditioned stimulus, are less effective and best avoided. Figure 10 illustrates these procedures.

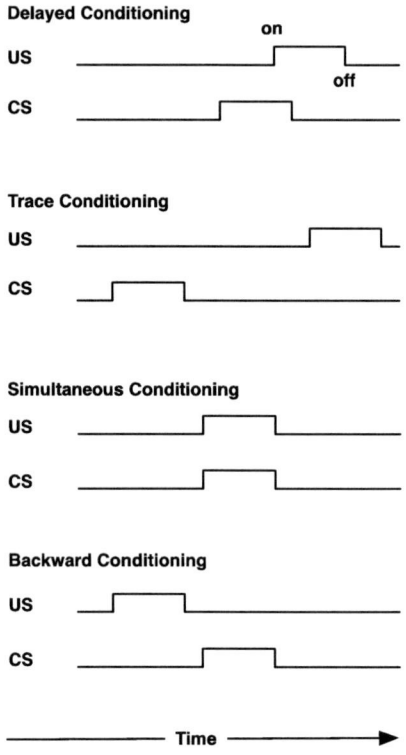

Figure 10. Respondent conditioning procedures.

A conditioned response can be maintained if the conditioned stimulus is at least occasionally paired with the unconditioned stimulus. Conditioned responses can be decreased or eliminated via **respondent extinction**. There are two respondent extinction procedures. The first involves presenting the conditioned stimulus without the unconditioned stimulus repeatedly until the conditioned stimulus again becomes a neutral stimulus. The second involves presenting the unconditioned stimulus randomly with respect to the conditioned stimulus (Moore, 2008, p. 126). Most commonly, when respondent extinction is called for, the former procedure is used.

Other processes can also affect the likelihood or magnitude of respondent behaviors. Repeatedly presenting an unconditioned stimulus generates a gradual and temporary reduction in the magnitude of the unconditioned response, a process referred to as **habituation** (Pierce & Cheney, 2008). In respondent extinction, the conditioned response is affected by repeatedly presenting the conditioned stimulus without the unconditioned stimulus, whereas, in habituation, the unconditioned response is affected merely by repeated presentation of the unconditioned stimulus. Habituation relates to unconditioned and not conditioned responses. Some sources suggest that habituation is applicable to conditioned responses but if repeated exposure is accompanied by the unconditioned stimulus, then the response will not decrease in likelihood or magnitude. If the response is not accompanied by the unconditioned stimulus, then this reduction in responding is respondent extinction. Habituation is temporary, whereas respondent extinction is much more durable, if not permanent. However, less conditioning is required to reestablish the conditioned response after extinction. If a stimulus elicits a particularly aversive response, repeated presentation may result in an increase, rather than decrease, in responding, and we refer to this process as ***potentiation*** (Fraley, 2008, p. 677; Catania, 1998). Finally, startling emotional arousal can briefly increase the magnitude of responding to *other* stimuli, a process called ***sensitization*** (Catania, 1998).[45]

Appreciating the basic respondent conditioning process is important for at least two reasons. First, respondent conditioning is how conditioned reinforcers are established and maintained. Second, ***emotional responses*** are respondents, involving the glandular secretion of various chemicals into the bloodstream. We refer to the after-effect, or neural awareness-related behaviors, of this emotional arousal as ***feelings***. Words such as fear, panic, anxiety, happiness, anger, joy, and so on, are simply the labels applied to describe feelings (i.e., the feelings come to evoke these verbal labels).[46] Emotional arousal can influence operant contingencies by temporarily structuring the body to make certain stimuli more or less likely to evoke operants, or to make certain reinforcers and punishers more effective. Emotional arousal can render operant behaviors exhibited during that aroused state more energetic. As such, emotional arousal is part of the antecedent condition in operant contingencies. Emotional arousal can even act postcedently as an endoreinforcer itself.

Measuring Behavior

It is not always necessary to track behavior quantitatively in animal training, particularly with simple basic training project objectives. However, in some more complex projects, it can be very useful. Indeed, a basic familiarity with formal methods of measuring behavior can benefit trainers, even when they are only informally tracking behaviors.

Informal versus Formal Measurement

A hallmark of a natural science approach is that it is precise, objective, verifiable and accountable. Most trainers exhibit a belief that their methods are successful but few can prove it. Measurement allows us to prove (or refute) just that. Measuring and tracking changes in behavior across time increases precision in describing the behavior and its changes, provides a finer scaled determination as to what the training is achieving, increases accountability, and avoids bias or self-deception behaviors. Assigning numbers and units to specific features of the target behavior allows us to

---

[45] The word sensitization is commonly used, mistakenly, to refer to potentiation. Sensitization relates to responsiveness to *other* stimuli.

[46] Many people refer to emotions as fear, panic, joy and so on but this is not accurate. We may characterize and label feelings, which are our awareness and experience of emotional behaviors and a distinct set of behaviors, but emotional behaviors per se are simply the release of chemicals into the bloodstream.

determine objectively "how much" of the behavior is occurring. By tracking the behavior across time, the trainer can determine how the strength of the behavior changes, particularly in response to training. This can provide data about whether you need to adjust the training plan, and eventually, whether or not you achieve the specific behavior objectives.

Tracking behavior quantitatively involves significant response effort, and under most basic training conditions, trainers respond effectively and sufficiently to training situations without using elaborate quantification methods. Trainers exhibit informal quantification of the features of behavior that control their training behaviors and usually do so to an adequate level in basic training. It is generally only when an unusually high degree of precision and accountability are required (such as when undertaking research, training extremely complex behaviors, or resolving complex problem behaviors) that quantification becomes sufficiently reinforcing. While behavior technologists working to resolve complex problem behaviors are commonly exposed to contingencies to measure and track behavior quantitatively, trainers engaged in bringing relatively simple behaviors under appropriate controls usually do not require formal measurement practices. Given that this book does not address the resolution of complex problem behaviors, the coverage of measuring behavior is described at an introductory level. This level of coverage provides an expanded repertoire adequate to improving informal quantification and a basis from which to continue studies on quantification of behavior, if the discussion evokes interest in doing so.

Behavior manifests with various properties or along various dimensions, and it is frequently useful to track a behavior by one or more of these specific properties. The following are the most common measures of behavior.

Measures of Behavior

Count

A *count* is the simplest quantification of behavior and simply involves counting the number of instances the behavior occurs. This alone is not usually very useful, but some computational measures based on counts can be quite useful.

Rate

The *rate* is the quotient resulting from the number of times the subject exhibits the behavior, divided by the number of time units across which trainer recorded the behaviors. For instance, if you measure rate of responding for 60 minutes and the subject exhibited the behavior 30 times, the rate is 0.5 (30 divided by 60). If there are multiple timeframes, you can add all of the times together and add all of the responses together to generate a single rate. Alternatively, an average rate can be determined by finding the mean average across timeframes.

Frequency

The *frequency*[47] is the quotient resulting from the number of fulfilled opportunities to respond divided by the total number of opportunities the subject had to respond. If you provide a subject with 10 opportunities to exhibit a behavior in question and they exhibit it in three of those instances, the frequency is 0.3 (3 divided by 10) or 30% (a derivative measure that can sometimes be useful).

Duration

In some instances, the *duration* that the behavior occurs is the most useful property of the behavior to measure. For example, if you measure a dog barking (or parrot screaming, or spouse nagging etc.) and they maintain the behavior, the duration of the behavior might be more useful or simply more accessibly tracked than other measures. For instance, if the dog barks or parrot screams when left alone, the duration might be a more useful measure of the barking or screaming behavior than the rate or frequency. Duration is most suitable for behaviors that the subject maintains in position such as sitting, lying down or maintaining eye contact. You may start a stopwatch when the subject commences the behavior and then stop it when they cease engaging in that behavior. Less formally (and less accurately), you can usually count off seconds (but do so covertly or else the behavior might come under control of the counting). When duration is measured, often it is a secondary measure, tracked along with rate or frequency. Alternatively, you can use it in small parts of the training, such as when you specifically work on increasing the duration criterion of the behavior.

Magnitude

Magnitude refers to the intensity of the behavior (i.e., how forceful it is). Let us say you measure barking, for instance. You could use a decibel meter to measure how loud the barking is. However, in most cases, this is not a realistic or informative measure.

Latency and Inter-response Time

It will not usually be your main concern but frequently, at some point within a training project, you will be exposed to contingencies to reduce the latency or inter-response time. *Latency* refers to the amount of time between presentation of the opportunity to exhibit the behavior and the occurrence of the behavior. *Inter-response time* refers to the time between consecutive responses (Fraley, 2008, pp. 249–250).

---

[47] Some behaviorology texts refer to frequency as "relative frequency" (e.g., Fraley, 2008). I will use the term frequency interchangeably with the term "relative frequency."

Topography and Extensity

Less common, although sometimes useful, are measures of topography, and extensity. **Topography** refers to measuring the form of a movement. **Extensity** refers to the distance over which the movement occurs, either linear or angular.

Choosing Among Measures of Behavior

Whether you choose to measure the rate, frequency, duration, magnitude and/or other properties of the behavior depends on which property is most relevant to the behavior in question (i.e., the measure that most reflects your objective for the behavior). There are some things to consider when deciding among these properties. What is the most appropriate/informative property of the behavior? What does the measure really indicate? When it changes, does it indicate how well you are doing with the training? Does it provide data regarding whether you are achieving or failing to achieve your objectives? If you identify exactly what you want to achieve, deciding on a measure will be easier and there will be more validity in the measure accurately reflecting what you portray it to measure.

Tracking the Target Behavior through Time

Whether you track behaviors through time precisely or not depends on your objectives. If you require more precision, as is common in some complex training projects or to help in resolving problem behaviors, this level of precision is appropriate. For simple behaviors trained mainly as "good manners" behaviors, we often estimate progress as training proceeds, but without graphing the results. That said, the behavior objective should always include quantitative criteria and you should always measure the behavior sufficiently, so that you will know when you have achieved the objectives. It is also a good idea to become familiar with quantitative tracking methods in case you are required to provide accountability that is more precise.

Under certain circumstances, such as when you take over a training project that is already well under way (e.g., when you adopt a new dog or a client brings their dog to you), you may begin by establishing a **baseline** for the target behavior. This simply means that you present the supposed cue through a few or several trials to determine the rate at which the subject currently exhibits the behavior before training. This can help you refine the behavior objective. In other cases, where there is no reason to believe there would be a rate above zero because there has been so previous conditioning, you will assume a rate of zero and begin training without a baseline measure.

Tracking the behavior quantitatively tells you the level, trend and variance of the behavior. The **level** is the measure discussed here; it tells you just "how much" of the behavior the subject exhibits. The **trend** is the angle of the line in the graph. It indicates whether the behavior is stable or is generally increasing or decreasing, as well as the magnitude of the increase or decrease. A horizontal or flat line on the graph represents a stable trend of the behavior. The steeper up the line is going, the more quickly the behavior is increasing; the steeper down the line is going, the greater the downward (decreasing) trend of the behavior. The *variance* refers to the "bounce" of the line, or how widely the strength of the behavior swings up and down (i.e., indicated by how jagged the line is). A high degree of variance usually indicates that you have not clearly/accurately established the evocative stimulus; when you cue the behavior you are sometimes evoking it and other times not.

Behavior measurement graphs have two dimensions:

- Horizontal axis

- Vertical axis

The horizontal axis usually represents time, either continuous or the number of the session or trial. We usually represent time in equal intervals, in units of minutes or seconds. If you use time, you might set the interval at 1 minute per unit (or 5, or 10, or whatever works best given the time frames involved). Start the intervals at zero and continue through, past where you will be plotting.

The vertical axis usually represents values for rate, frequency, duration, latency or magnitude. The intervals will usually be equal units and usually start at zero. Ensure that they continue high enough to allow for any measure possible or reasonably likely.

In order to record data on the graph, place a dot inside the graph plane where the time or trial number meets the measure. So, let us say you are recording the rate of a behavior every minute. In the first minute, if the behavior occurs three times, place the dot over the one–minute point on the horizontal axis and level with the three marking on the vertical axis. Then, in the next minute, record how many times the behavior occurred again, this time above the two–minute mark and at the level of the number of times that the behavior was exhibited. If the selected time measurement had been seconds, rather than minutes, the rate would have been 0.05, and the vertical units would have been indicated differently. However, for this type of rate, minutes seem more appropriate than the smaller unit of time does.

For contrast, this time, let us say you are recording the frequency (rather than rate) of a behavior through each session (rather than every minute). Your horizontal axis will be sessions rather than minutes in this case. If, in the first session, the behavior occurs five times through 10 opportunities to exhibit the behavior, then you place a dot above session one at 0.5 (remember, frequency would be five divided by 10 here) and continue on to session two. Notice that your vertical axis with a frequency measure will go from zero to one.

If, on the other hand, you measure the rate per session instead of per minute, as in the first example, then in your first session your dot will be above the first point on the horizontal axis at one, and at the level of the measure for that behavior in session one. Then, proceed to record the data through each time interval or session. Connect the dots as you enter the data, to form a line from dot to dot. Figure 11 provides a stylized graph showing the elements discussed.

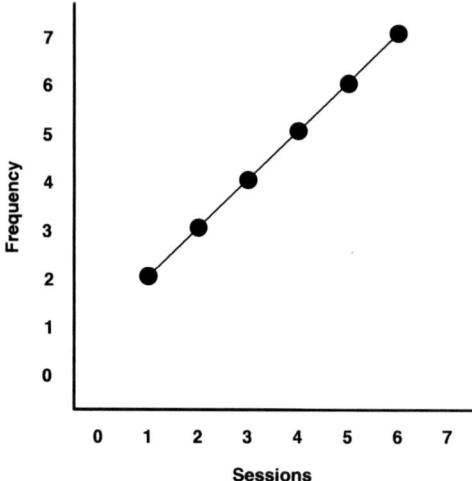

Figure 11. Stylized graph of behavior data, demonstrating the relationship between training sessions and the rate of the behavior.

If you established a baseline, draw a vertical line from the baseline to the time or session point where the change occurred, to clearly mark when the independent variable was changed. You will not need this line if you start training without a baseline. You should note any changes with the independent variable you make throughout data collection in the graph so that later you can see where the changes occurred and how the line on the graph illustrates the behavior changes. Include vertical lines with labels when you change any component of training that might influence the behavior (i.e., when any change to the independent variable is made). This can include changing a schedule of reinforcement, increasing or decreasing criteria requirements for duration, distance or distraction etc. If anything else changes in the environment that you did not control for, insert a vertical line for it as well. If the trajectory of the line recording the behavior changes, you will be able to ascertain what change in stimulus that change in behavior is likely related to. If it is in response to a change as indicated above then you can determine visually whether it is a brief adjustment blip or whether it is affecting the training in a more detrimental manner that requires changing your procedure or criteria. The line on the graph depicting the strength of the behavior represents your dependent variable and the vertical lines you make with labels or notes represent the independent variables. The combination of lines on the graph will provide precise feedback (consequences) regarding your training behaviors and hence bring your training behaviors under controls that are more effective.

There are different approaches to indicating the various components in graphs. I will describe one easy method below and provides a few important features in graphing techniques. Each training project requires its own graph format since there are different units of time and measures of the behavior appropriate to each project. For more information about graphical depiction of training, see Fraley (2008), Bailey and Burch (2002) or Cooper and colleagues (2007).

Examine the stylized graph in Figure 12 below. This graph depicts a 13 session training project. In each session, the dog is presented with 10 opportunities to exhibit the target behavior near the end of the session. Notice that the vertical axis units of the graph are pre-calculation units. In other words, the units go from one to 10 rather than from zero to one. This is just another way to depict the data. The first session was used to establish a baseline and in this case, the target behavior was exhibited zero times indicating it has not undergone any training thus far. After the baseline phase comes the training phase, and in this phase, we implement the training procedures. Notice how vertical lines are also used to indicate when training began after a baseline phase, when the continuous reinforcement schedule was changed to an intermittent schedule, and when distance and duration criteria were changed. As should be expected, the frequency of the behavior declines briefly in response to these changes, as indicated by the changes in the angle of the line.

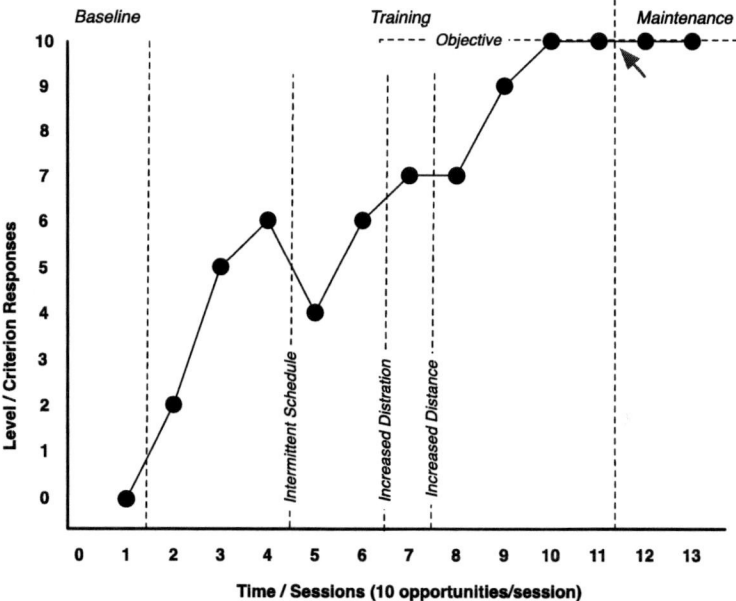

Figure 12. Stylized graph depicting a training project with 13 sessions in which the dog is provided 10 opportunities each to exhibit the criterion behavior. The graph depicts baseline, training and maintenance phases and the behavior objective. Milestone criteria changes are indicated by vertical dashed lines.

Continue the data collection process throughout the entire training project, including well into the maintenance phase, so that you can compare the baseline level (or the initial level when you started the training) with the maintenance level. This process will provide you with empirical feedback on exactly how the behavior is changing. It will indicate whether it is increasing or not and how the strength of the behavior is responding to your training (or not). This kind of objective accountability is a cornerstone of the behaviorological approach.

Continuing Education

Ledoux (2014) provides an excellent introduction to behaviorology and the principles and laws of behavior.

Fraley (2008) provides a massive and comprehensive treatment of a number of foundational and advanced areas of interest in behaviorology. Though not as current as Ledoux (2014), it is a tomb of a 1600 page book.

Both of the above books are worth their weight in gold and deserve a Nobel Prize, in my evaluation.

Courses provided through The Companion Animal Sciences Institute at www.CASInstitute.com provide a professional technologist level of education in behaviorology.

The Science and Technology of Animal Training

CHAPTER 3. AVERSIVE STIMULATION AND ITS PROBLEMATIC SIDE EFFECTS

Behavioral Objectives

The objective of this chapter is to measurably expand the reader's repertoire of behaviors in relation to describing and relating the principles of behavior. Upon successfully integrating the concepts outlined in this chapter, the reader, where exposed to contingencies to do so, will accurately:

- Describe the problematic features of using aversive stimulation

- Explain why punishment might work in a narrow sense, but not in the long-term

- Describe general alternatives to the use of aversive stimulation in training.

Introduction

In this chapter, I will describe aversive stimulation and the problematic side effects it generates, as well as explain why trainers and guardians use it.

Why is the Occurrence of Aversive Stimulation so Pervasive?

As is the case with all operant behavior, the use of aversive methods is maintained by reinforcement. Punishment administered with sufficient intensity, contingently (consistently), and contiguously (immediately), with regard to a target behavior will often result in a rapid suppression of the target behavior, assuming the net punitive effect overcomes the net reinforcing effect. Although aversive stimulation may result in robust and resilient long-term problems, the immediate effect is usually effective escape from the aversive stimulation that sets the occasion for the application of aversive stimulation on the part of the trainer. It is likely that most of those who use aversive procedures are not aware of (a) the long-term problematic effects they generate and (b) less problematic alternative methods.

Problematic Effects of Aversive Stimulation

An *aversive stimulus* is any event that functions (a) to evoke behavior that has reduced or terminated it in the past, (b) as a punisher if presented immediately following a behavior, or (c) as a reinforcer when withdrawn immediately after a behavior (Cooper, et al., 2007). Some authors use the term aversive synonymously with the term punisher (Miltenberger, 2008) or as a subtracted reinforcer (Vargas, 2013; Chance, 2009), but these are just more narrow applications of the term. Subtracted reinforcement, added punishment, and subtracted punishment all involve aversive stimulation. Extinction is also aversive in the sense that the subject behaves (the extinction burst and increased behavioral variability) to contact the now absent reinforcement, escaping the frustration (aversive emotional arousal) associated with blocked access to reinforcement. Added reinforcement is the only principle/procedure that does not involve aversive stimulation.

Although using aversive methods may generate effective escape from aversive conditions for the person using these methods, it is also extremely risky (Ledoux, 2014, p. 358; Sidman, 2001). Punitive methods tend to generate extremely robust longer-term problems. The risk of adverse side effects makes the use of aversive stimulation an unwise choice unless no better option is available and the objective is vital enough to justify the use of aversive procedures. It is important to note that expert application of aversive procedures can minimize the intensity of some of the problematic side effects, but side effects are not just an indication of improper application—they are an inherent result of salient aversive stimulation, and are common, even when specifically mitigated against. It is also important to note that some procedures are more intensely aversive than others are. Extinction and even subtracted punishment, for instance, are not as likely to generate as seriously problematic side effects as added punishment and subtracted reinforcement. Furthermore, some subtracted reinforcers and added punishers are more intensely aversive than others are. An irritating *"Psssst!"* sound is likely to be less intensely aversive than a shock to the neck. Some milder aversers are simply distractions.

Below is a brief overview of some of the problematic side effects commonly associated with intensely aversive stimulation. We can divide the problematic effects into respondent effects and operant effects, although these categories inevitably interact.

Respondent Side Effects: Problematic Emotional Arousal and Conditioning

Aversive stimulation can elicit emotional behaviors that are detected in ways referred to as "fear," "anxiety," or "panic." These aversive emotional behaviors function physiologically to energize and exaggerate escape behaviors (operants), as discussed in the next section. The stimuli present at the same time as the aversive stimulus tend to become conditioned aversive stimuli, which elicit the same problematic emotional behaviors as the unconditioned aversive stimulus. Furthermore, these responses can generalize, so that a wider range of stimuli elicits the emotional behavior. This can quickly become a robust, resilient and growing problem. The incipient preparation behavior associated with the initial conditioning can also come to elicit the problematic emotional behaviors, meaning, that preparing to exhibit a behavior that has resulted in generating aversive consequences will also generate the same emotional reactions.

Part of the problem with emotional behaviors is that they are robust, resistant and challenging to change. Eliminating problematic emotional behaviors is a very long and involved process. A few moments of punishment can generate extremely problematic emotional reactions that can take years to resolve, if they are ever resolved. An excellent source on this topic is Fraley (2008, pp. 909–921) as well as Sidman (2001).

Operant Side Effects: Escape Behavior

The escape and avoidance behavior generated by the application of aversive stimulation is often itself problematic, quite aside from the problematic behaviors that prompted the initial use of the punitive procedures. Escape behaviors are often aggressive in nature. A strong history of research has demonstrated that subjects exposed to aversive stimulation may counter-coerce and/or lash out and attack those inflicting the punishment or others present in the subject's environment or even others who merely share similar features with the person causing them pain. As mentioned above, aversive stimulation generates problematic emotional arousal that may generalize and also come to be elicited by other stimuli, which can energize and exaggerate these escape behavior. Other escape behaviors might involve flight from aversive stimuli or avoidance of them, which can also become quite problematic.

Escape behavior that some might characterize as "abnormal" is also possible. Behaviors characterized by some as "displacement" might seem out of place and strange. These behaviors often function to escape or avoid contact with certain stimuli as a form of self-distraction. Self-mutilation related behaviors are also common. They function to distract and can generate certain kinds of physiological processes that mask or counter aversive feelings. In dogs, paw licking or biting is common and in parrots, feather destructive behaviors are common. In humans, cutting one's self is common (as is drug use).

Aversive stimulation may also result in a reduction in creativity, resilience, and industriousness, as well as in general response depression, particularly if the aversive stimulation is inescapable and/or unpredictable or long-lasting.

Clarifying Punishment and its Role in Changing Behavior

This section provides clarity regarding common misunderstandings about the punishment of behavior.

Punishment, as a postcedent intervention on its own, superimposes an aversive contingency onto the original reinforcement contingency that was maintaining the behavior. The punishment procedure does not change the existing reinforcement contingency that was prevailing on that behavior on its own—that would be extinction. In that regard, we must consider punishment less efficient than extinction. Procedures involving punishment often result in post-punishment over-recovery of the problematic behavior, observed as a brief increase in the strength of the behavior (Fraley, 2008). At other times, punishment seems to suppress the behavior for an extended period of time after the punishment contingency is discontinued, but this is not always a result of punishment alone. Below, I will describe some relevant points of distinction between these results.

If a behavior occurs, it has a history of reinforcement; that contingency is in effect (assuming extinction has not been instated). When a punishment contingency is imposed on a behavior, that contingency is merely superimposed on the existing reinforcement contingency. If the reinforcer is stronger than the punisher is, the behavior will continue to occur, although perhaps at a reduced strength. If the punishment is stronger than the reinforcer, the behavior will be suppressed to some extent, depending on just how much stronger the averser is than the reinforcer. Indeed, the behavior may be suppressed to a rate of zero. But as soon as the punishment contingency is discontinued, the existing reinforcement contingency prevails again, and the behavior reequilibrates (i.e., stabilizes) to pre-punishment strength (Fraley, 2008).

If the behavior decreases dramatically, perhaps to a rate of zero, after the punishment contingency is discontinued, it is often a result of concurrent processes and not punishment alone. Whereas punishment suppresses behavior, extinction eliminates it, and often an extinction procedure is inadvertently instated simultaneously along with a punishment procedure; elimination of the behavior would then be a result of the extinction procedure and not punishment. Note that this can only occur if the punishment was ineffective and the behavior continued to occur. If the punishment procedure is very effective in suppressing the behavior, extinction cannot occur. In fact, an effective punishment procedure will protect the behavior against extinction because there is no opportunity for the behavior to occur and the usual reinforcement to be withheld. "Not only does extinction yield a more lasting method of behavior reduction than temporary suppression under punishment, an extinction procedure is also less likely than punishment to produce troublesome aversive emotional side-effects" (Fraley, 2008, p. 397). This is particularly true when extinction is used as part of a differential reinforcement procedure rather than alone. In his classic work, Estes (1944) elaborated on this point:

[A] response cannot be eliminated from the animal's repertoire more rapidly with the aid of punishment than without it. In fact, severe punishment may have precisely the opposite effect. A response can be permanently weakened only by a sufficient number of unreinforced [evocations] and this process of extinction cannot proceed while a response is suppressed as a result of punishment. The punished response continues to exist in the animal's repertoire with most of its original latent strength. While it is suppressed, the response is not only protected from extinction, but it may become a source of conflict. An emotional state, such as "anxiety" or "dread", which has become conditioned to the incipient movements of making the response, will be aroused by any stimuli which formerly acted as occasions for the occurrence of the response.

In other cases, punishment may create a behavioral void filled with some other behaviors, and these other behaviors may displace the punished behavior. Of course, this just means that if one is going to use punishment, one will need to also use other procedures in conjunction with it in order to achieve a lasting result. Punishment results in a decline in the rate of a specific behavior but it does not address the reinforcement contingency itself that the targeted behavior functioned to access. Punishment and reinforcement work in opposing directions (Azrin, 1956). If a particular behavior is punished and the reinforcer that was contacted is not made accessible by other behaviors, the subject may simply experiment with other behavior(s) in order to regain access to the reinforcer. If one behavior is punished, other behaviors will emerge for as long as the consequating stimulus maintains its reinforcing capacity. This *countercontrol* is often interpreted as the subject being "manipulative" or "dominant." While it is true that in some cases, these other behaviors might be acceptable, often any operant that functions to contact the reinforcer in question is a problem.

Another way in which a punishment procedure might result in long-term suppression of a behavior is if the emotional arousal generated by the stimulation is so intensely aversive that it causes an extreme aversion to the evocative stimulus that controlled the problem behavior. Some would call this level of aversion "phobic." The evocative stimulus becomes a conditioned punisher and this can generalize to other similar stimuli and even incipient components of the problem behavior itself. This kind of emotional reaction and escape/avoidance behavior might be so strongly aversive that the rate of the behavior decreases to zero and it takes an extended period of time for the behavior to return to the pre-punishment rate even if the punishment schedule is discontinued. This kind of suppression would be extremely unpleasant of course.

Research has shown that the use of aversive stimulation does not work any better than added reinforcement-emphasized methods and, in fact, the use of aversive stimulation is fraught with problematic secondary effects (see Sidman's classic work, 2001). In that sense, it is difficult to convincingly argue that aversive stimulation "works" in the real world in a long-term, putative sense. Many studies have shown that harsh aversive stimulation suppresses behavior (see Lerman & Vorndran, 2002). However, many of these studies considered the temporary suppression of a simple behavior, rather than the total effects of the procedure on the subject in the short-term or in the long-term. This failure to consider other variables beyond the rate of the discrete behavior in question may provide an inappropriate picture of the advisability of punitive methods.

In a particularly interesting study, Balaban, Rhodes and Neuringer (1990) hypothesized that as aversiveness of a punisher increases, its consequences generate what they call a "defensive response," which competes with the "orienting response" and subsequently diminishes the effect of a punishment contingency. Orienting responses are respondent reactions to attend to a stimulus, which increase sensitivity to external stimulation. A defensive reaction, on the other hand, stimulates the so called "fight-or-flight response," which reduces receptivity and limits the effects of external stimulation. Balaban and his colleagues hypothesized that the DR competes and interferes with the orienting response, and hence should negatively affect conditioning. To examine this, they assigned human subjects to two groups: the "informational punishment" (IP) group and the "aversive punishment" (AP) group. All subjects were given moderately challenging tests, which would ensure errors. When members of the IP group made an error, it was immediately followed by a brief tone sound. When the AP group members made an error, they were exposed to the same tone, but 20% of the time the tone was followed by a mild electrical shock. The researchers compared the two groups in terms of skin conductance responses, interbeat heart rate intervals, state-trait anxiety levels, skin temperature, and task exhibition. Surprisingly, they did not find a statistically significant difference between the two groups in global sympathetic arousal. What they did find was that the IP group did significantly better in their tests than the AP group! The tone in the IP group was found to elicit an orienting response. For the AP group, the tone became a conditioned stimulus eliciting a defensive response. The AP group did show higher skin conductance and cardiac acceleration than the IP group. Not only did the IP group behave more effectively, but also as the test continued, they continued to improve. These findings are consistent with the notion that the defensive response interferes with the orienting response, and the result is less effective conditioning. The activation of emotional arousal generated by aversive stimulation generates a set of physiological processes that is simply not conducive to conditioning.

A study carried out by Hiby, Rooney and Bradshaw (2004), comparing the behavior of dogs trained with punishment-based methods on the one hand, and added reinforcement-emphasized methods and "miscellaneous" methods (i.e., not obviously either punishment- or reinforcement-emphasized) on the other hand, found that dogs reported to be trained with added reinforcement-emphasized methods scored highest on obedience scores. Those who were trained using punishment ranked lowest, and those who were trained using both ranked in the middle. In none of the obedience tasks, were punishment-emphasized techniques most effective. Dogs reported to be trained with added reinforcement-emphasized methods were also found to have the fewest current behavior problems, whereas dogs reported to be trained with punishment, or punishment together with added reinforcement, were found to have the most current behavior problems. While this was a correlational study rather than an experimental one, and no causal relationships can be assumed nor confirmed, the results are consistent with many other studies on the topic, adding to replication in the field and increasing confidence in the results. What is also interesting about this study is its high degree of ecological validity—these were dogs living and trained in the real, dynamic, and complex world, rather than in a laboratory, and the dependent variables were real-world concerns.

Alternatives to Aversive Control of Behavior

This book should provide a decent treatment of alternatives to harsh punitive methods, so only a brief preview of the material follows. If the behavior objective calls for establishing a new behavior (resolving a behavioral deficit), the most powerful strategy for achieving it is to arrange the antecedent environment to make the behavior as likely as possible—prompt it (where possible) or shape it (if necessary)—and additively reinforce it. It may then be brought under the appropriate controls. If the behavior objective calls for reducing a behavioral excess, arrange the antecedent environment to make the behavior less likely and some replacement behavior more likely, reinforce the replacement behavior, and, if necessary, extinguish the problematic behavior. A major benefit of using added reinforcement-emphasized methods to resolve problematic behaviors—in particular, ones maintained by subtracted reinforcement—is that beneficial respondent conditioning occurs as a byproduct.

Continuing Education

Sidman (2001) is an excellent introduction to the harm that can be caused using aversive methods.

Courses provided through The Companion Animal Sciences Institute at www.CASInstitute.com address the risks associated with aversive methods and how to use added reinforcement-emphasized methods effectively.

The Science and Technology of Animal Training

CHAPTER 4. ADDED REINFORCEMENT-EMPHASIZED TRAINING STRATEGY

Behavioral Objectives

The objective of this chapter is to measurably expand the reader's repertoire of behaviors in relation to describing and relating the principles of behavior. Upon successfully integrating the concepts outlined in this chapter, the reader, where exposed to contingencies to do so, will accurately:

- Explain why decisions regarding the use of aversive stimulation are important and to whom it is important

- Identify ways of efficiently and effectively training without resorting to aversive stimulation

- Explain the assumptions and implications of models that propose a ratcheting up of aversive stimulation in reaction to difficulties in training

- Explain why finding the problem and resolving it is a more effective strategy than increasing aversiveness when facing challenges in meeting behavior objectives

Introduction

The previous chapter highlighted the importance of avoiding the use of aversive stimulation in training. In this chapter, I will provide guidance on how to plan for and implement training projects that emphasize and continue to emphasize added reinforcement-based methods. The strategy provided herein emphasizes diligence in finding added reinforcement-emphasized approaches to training. The focus is solely on training new behaviors, rather than objectives involving the elimination of problem behaviors. You can find a more comprehensive strategy that also applies to resolving problem behaviors in its most recent version online through the Association of Animal Behavior Professionals, linked to throughout the Professional Practices Guidelines.[48]

---

[48] www.associationofanimalbehaviorprofessionals.com/guidelines.html

Ethical Foundation

It is widely agreed among those from a wide variety of philosophical orientations that treating others in a coercive manner, where it is unnecessary to do so, is morally problematic. Ethical trainers recognize that the dignity and basic moral rights of others deserve respect. A cornerstone ethical principle in the helping professions is that we utilize the least aversive methods available.

The companion animals that trainers work with are vulnerable parties in the professional relationship established between the trainer, animal and the animal's guardian. The situation is similar to counseling relationships between a counselor, a child and the child's parents. Companion animals cannot provide informed consent regarding the training plans. Therefore, trainers must ensure that they consider the long-term interests of the companion animal and respect the animal's basic moral rights, by intervening in a minimally aversive manner (Association of Animal Behavior Professionals, 2008, principle 2.02; Behavior Analyst Certification Board, 2010, guideline 4.10).

Avoiding Extremism and Dogmatism

It is important to avoid exaggeration or excessive simplicity in this analysis. Accepting extreme arguments such as that "all forms of aversive stimulation are always sure to cause irreparable harm" or that "aversive stimulation is necessary in order to succeed in training" leads to dogmatic positions best avoided in favor of a careful consideration of the circumstances and dedication to utilize the least aversive methods possible.

Review of Aversive Stimulation

As indicated previously, an aversive stimulus is any event that functions (a) antecedently to evoke behavior that the stimulus has reduced or terminated in the past, (b) consequentially as a punisher if the stimulus is presented immediately following a behavior, or (c) consequentially as a reinforcer when the stimulus is withdrawn immediately after a behavior (Cooper et al., 2007). Added punishment, subtracted punishment, subtracted reinforcement and extinction, even in their mildest forms, all involve aversive stimulation to some degree. Only added reinforcement involves no aversive stimulation at all. If a stimulus meets the criteria listed above, then the stimulus *is* aversive. It is usually not difficult to predict which stimuli will function as aversive stimuli. Indeed, predicting which stimuli will be aversive is generally just as reliable as predicting which stimuli will function effectively as reinforcers. You can confirm these predictions once you have implemented the chosen procedure with the stimulus in question. Predicting *how* intensely aversive a stimulus will be, is usually quite reliable. Failing to reinforce non-criterion responses or withdrawing ongoing reinforcement is aversive, but it is generally not as problematically aversive as presenting harsh aversive stimuli or removing harsh aversive stimuli contingent only upon occurrence of the specific behaviors. Furthermore, noncontingent aversive stimulation is much more problematic than contingently applied aversive stimulation, and readily escapable and avoidable aversive stimulation is much less problematic than inescapable and unavoidable aversive stimulation.

Added Reinforcement-Emphasized Training *is* Maximally Effective Training

It is sometimes thought that aversive methods are more effective than added reinforcement-emphasized methods and that we "resort to" them where necessary but only as much as is necessary because they are unpleasant for the subject. Highly aversive methods, however, are *not* more effective than added reinforcement-emphasized methods. In certain narrow circumstances, highly aversive training methods might condition quick and effective aversions but if we consider the potential for disruption to training, side effects and problems with the long-term well being of the subject, this cannot be considered more effective than a well planned and implemented added reinforcement-emphasized plan—not if our goal is also to ensure long term behavioral well being. Trainers will have a better chance of success if they can identify exactly why they are having trouble with their added reinforcement-emphasized plan and fix it. Resorting to aversive methods tends to militate against resolving the original problems with the training plan.

Success with added reinforcement-emphasized methods requires proficiency in its application and a dedication to find solutions to problems when they are faced. Most trainers that resort to aversive methods are simply not adequately proficient in the application of added reinforcement-emphasized methods, including identifying what variables are causing the problem and fixing them. Some trainers exhibit a *belief* that they must "resort to" punitive methods when they fail to achieve success without them. Failure to achieve training objectives with added reinforcement-emphasized methods should prompt trainers to identify the problem causing the difficulty and make the necessary adjustments to the training plan to resolve them rather than "resorting to" more aversive methods that may hide, and compound, the difficulties.

Let it be understood right from the beginning of this discussion that intensely aversive methods are *not* more effective than well planned and implemented added reinforcement-emphasized methods.

Why Implement the Added Reinforcement-Emphasized Training Strategy?

I propose this strategy because of its careful attention to long-term effectiveness including the effects on the target behavior as well as the well being of the subject in general. What reinforcers are available to maintain behavior that comports with the strategy presented here? After all, it clearly requires a higher response effort and may indeed limit access to certain shortsighted, impulsively generated reinforcers. We sometimes call delaying an immediate impulsive payoff in favor of a much higher long-term payoff "wisdom" (Chance, 2009). Diligently working through the process of finding and solving the problems with one's training, as opposed to looking to more aversive methods, is a wise choice, partly because it avoids many of the disruptive problematic side effects associated with highly aversive methods which affect both the target behavior and the general behavioral well being of the subject. By avoiding the use of intensely aversive methods, you avoid the side effects associated with them that would otherwise be disruptive throughout and indeed after the training project, and you are free to focus on the real problem: finding and fixing what is wrong with the training.

Bringing behavior under the control of practices described in this strategy tend to generate pride-related feelings and thoughts as well. When professional behavior comports with this strategy, benefits accrue to the subject, guardian, and the individual trainer, as well as the field of animal training as a whole. The subject benefits from the standard by experiencing a higher degree of comfort and behavioral well being, being conditioned to exhibit acceptable adaptive behaviors that ultimately promote a more adaptive repertoire of social behaviors within the family. Furthermore, the subject will contact a greater number of added reinforcers. The guardian benefits from the standard by avoiding the necessity of dealing with the well-known side effects that commonly occur with the use of highly aversive methods, and their objectives will be achieved in an orderly manner. By providing effective and ethical training, the individual trainer benefits from stronger success rates, reduced risk of injury and liability exposure, increased business due to a good reputation, and the respect and trust of clients, colleagues, and allied professionals. The field as a whole benefits from the standard with market growth and increased respect from the public and allied professionals. Notice that these are the same reinforcers available for the adoption of all best practices and high-standard guidelines. In the long term, adopting a high standard of ethical behavior, including dedication to implementing this or similar strategy provides greater benefits to society than the failure to adopt such a strategy.

Key Features of the Added Reinforcement-Emphasized Training Strategy

The strategy I will present here proceeds under a radically different set of basic assumptions than many other models. There are many algorithms, flow charts and models available to provide guidance to trainers on how and when to implement more intensely aversive methods in their training plans.[49] They recommend minimally

---

[49] This includes models I have previously published (O'Heare, 2013). In fact, this chapter was originally destined to be another iteration of such a model and I rewrote it at the last minute. These aversion-ratcheting models often have the phrase "minimally aversive" or "minimally intrusive" in their title, alluding to aversion-ratcheting as a solution to failure. That is also why I am referring to the strategy provided here as "added reinforcement-emphasized"—to clearly put the focus on added reinforcers. The use of such aversion-ratcheting models is not accompanied by maliciousness and stupidity. These aversion-ratcheting models are the paradigm at present and have been for quite some time. The insidious implication they perpetuate is unstated and it simply rarely occurs to their authors that these implications exist. Indeed, the authors' "intent" is to help ensure that fellow trainers use as little aversive stimulation as possible, a laudable goal to be sure and a primary reason why the paradigm persists. I believe that stating these

aversive added reinforcement-emphasized methods to start and when the trainer faces failure, the algorithm justifies a small increase in the intensity of aversive stimulation in the training plan, followed again by the solution of greater levels of aversiveness if failure is reached again. The solution to failure justifies a ratcheting up of levels of aversiveness.

These models proceed under a false, usually unstated, set of assumptions that (a) aversive methods are more effective than added reinforcement-emphasized methods, (b) the problem when we face failures or challenges in training is one that increasing aversiveness can resolve. Indeed, the authors of such models may not be aware of the implication made by such models and that the model perpetuates these assumptions and the practice of solving challenges with increases in aversive stimulation.

However, (a) intensely aversive methods are *not* more effective than added reinforcement-emphasized methods, (b) a lack of more intensively aversive stimulation is *not* the problem causing the challenges in the training, and (c) increasing aversiveness is *not* the best solution to the challenges faced in the training. That is why the strategy I will propose here for handling challenges in training is radically different from the usual strategy.

If one fails to achieve their objectives with an added reinforcement-emphasized plan, there is a choice between identifying and fixing the problem versus increasing the aversiveness of the methods. Models that propose increasing aversiveness when such roadblocks emerge, perpetuate belief in a false choice between failure and increasing aversiveness. Again, the strategy I will present here rejects this assumption that aversive methods work as well or better than added reinforcement-emphasized methods and that the most productive solution for failure with nonaversive methods is to increase aversiveness; that we avoid the unpleasant aversive methods where possible, but where necessary, we resort to these unpleasant, though highly effective, methods. This strategy proposes that the most effective strategy is to identify and fix the problem with the training. This strategy does not assume that failure to reach training objectives can be solved with more intensely aversive methods. Instead, failure is more likely the result of a lack of proficiency in constructing and implementing added reinforcement-emphasized methods or when faced with a problem or challenge in executing the added reinforcement-emphasized plan, failing to identify and fix it. For example, if a training project is not going well, one might recognize that the dog is hyperactive and distracted and this is disrupting the training efforts. The most productive solution is not to increase aversiveness but rather to reduce the ambient distraction and generally ensure more exercise for the dog. Recognizing the problem is a skill as is finding a suitable solution. Expanding one's repertoire of such problem solving behaviors is more productive than increasing aversiveness.

The strategy I will present here emphasizes identifying and fixing problems with the training plan and/or its implementation. It also emphasizes objectivity and accountability through establishing precise quantitative behavior objectives and careful quantitative tracking of the behavior throughout the process. Failure to achieve the objectives prompts careful reevaluation of the behavior objective, the contingency analysis, application-related variables, and the choice of procedure. Failure to identify

---

assumptions and implications clearly and proposing a strategy based on a different kind of solution will help cause a paradigm shift in this regard.

and resolve the problem may prompt an increase in the motivative operations and ensuring that non-criterion behaviors are not reinforced. However, this does not involve the application of harsh coercive methods but rather failing to reinforce non-criterion behaviors and increasing the effectiveness of reinforcers for criterion behaviors. Remember, it is aversive to fail to provide reinforcers for a previously reinforced behavior and to subtract reinforcement that is present. This is quite different strategically, however, from coercing behavior with added punishers subtracted reinforcers and generally much less likely to generate problematic side effects.

A proficient trainer should be able to plan and implement added reinforcement-emphasized training and completely avoid harsh aversive stimulation. If the trainer faces difficulties, the best solution is not to increase aversiveness in the plan. The best solution is to identify the actual impediment and adjust the environment in such a way that resolves this difficulty. Where a trainer is frequently faced such difficulties, the best solution is not to resort to more aversive methods but rather to increase their own repertoire of effective planning and implementation of added reinforcement-emphasized training and to more effectively identifying and fixing problems when they face them. The solution is education, not aversion.

The flow chart in Figure 13 depicts this process.

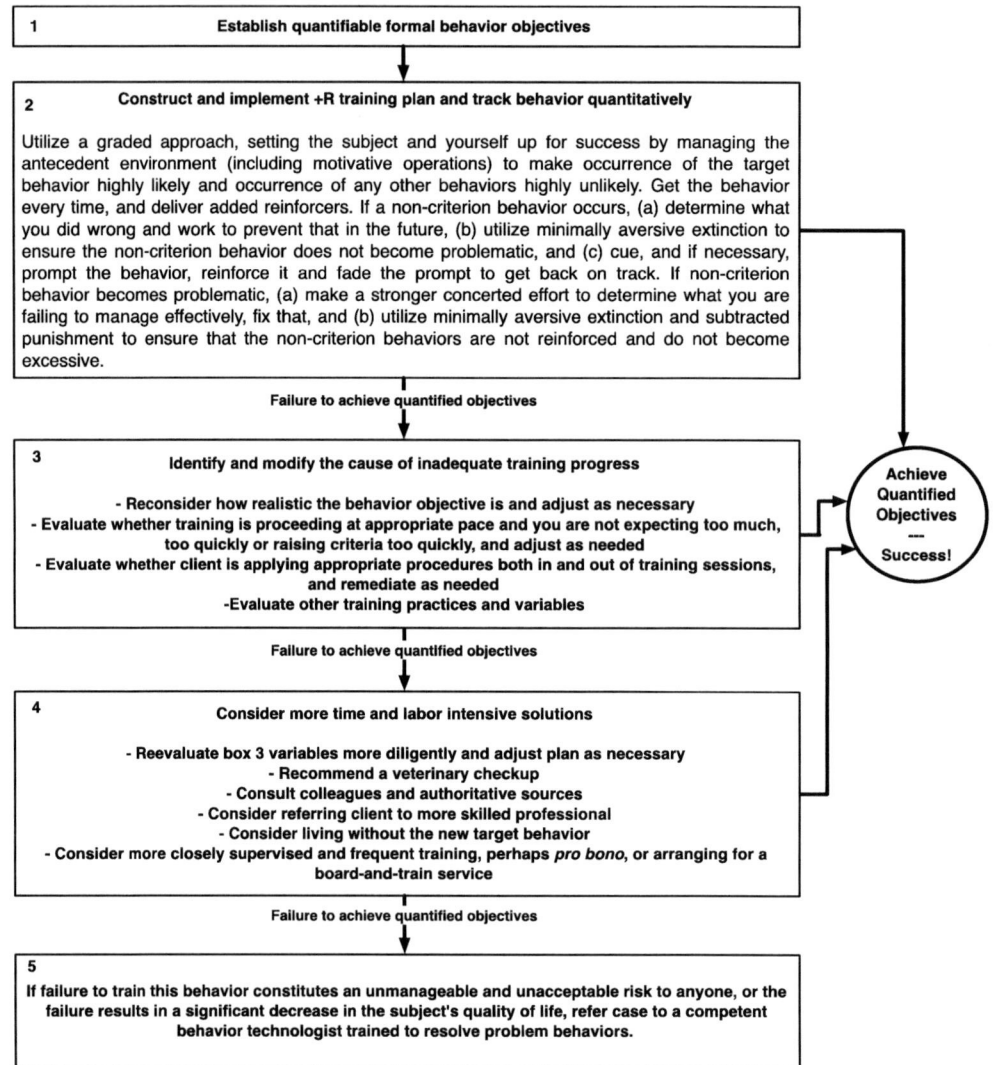

Figure 13. This algorithm provides guidance on how to identify problems in training plans and make adjustments to help achieve success in achieving behavior objectives.

Box 1. The first step in the strategy is to identify and operationalize specific target behaviors and quantifiable behavior objectives. Without clarity, specificity and objective accountability, success will be less likely. Operationalize the target behavior (i.e., described in a manner that is directly observable and quantifiable/measurable), not vague or speculative.

Box 2. In this phase of the project, you construct the training plan. The plan includes the objectives for the program, the basic strategy and procedures that you will implement, and the means to achieve the acquisition, fluency and maintenance of the new environment-behavior relation. The training plan is not a hodge-podge of anecdotally supported intuitions or "hit or miss" "tricks of the trade," the result of trying just another "tool" from a "tool box" full of different things that can be tried until one works. The training plan is an evidence-based application of strategies and procedures well supported in the natural science literature. Utilizing a natural science-based approach makes it far less likely that one will meet with failure and hence a supposed need to formulate a more aversive approach. Once you implement the systematically constructed training plan, you will track the target behavior in all but the simplest of cases.

At this stage, training plans emphasize added reinforcement. Utilize graded differential reinforcement. Set the subject and yourself up for success by manipulating the antecedent environment to make occurrence of the target behavior highly likely. Get the behavior every time, and reinforce it. The graded approach should minimize non-criterion responses. Where a rare non-criterion behavior occurs, (a) determine what you failed to manage effectively and work to prevent that from happening again, (b) utilize a minimally aversive extinction procedure, and (c) cue the target behavior, prompt if necessary to ensure it occurs and reinforce it, fading the prompt quickly to get back on track. As a minimally aversive extinction procedure, say "Oops" and fail to provide the reinforcer, pausing interaction for two seconds before proceeding. Take note of why the behavior did not occur in that trial. What variable did you expect too much of too quickly? Was there too much distraction? Control it. Was the subject becoming satiated? Take a break. Was something aversive occurring? Eliminate it. Then, get back on track. This is usually something you can handle on the fly as you train, as the reason for the non-criterion behavior is usually obvious.

Where you find that non-criterion behaviors are occurring more that "very rarely," make a concerted effort to identify what the impediment is. If you need to take a break to consider it, do so. Once you believe that you have identified the problem, arrange the environment to ensure that it is resolved and proceed. Where necessary, such as where a particular intractable specific non-criterion behavior is occurring regularly, ensure that the non-criterion behavior receives no reinforcement at all. This means instating a minimally aversive extinction and/or subtracted punishment procedure. Extinction ensures that the subject does not make contact with the treat reinforcer that is maintaining the target behavior you are training, that you do not deliver new reinforcers. Subtracted punishment ensures that no other reinforcers are allowed for a brief period of time after the non-criterion occurs either, that you withdraw ongoing reinforcers that might otherwise contribute to maintaining the non-criterion behavior. In this case, say "Oops," fail to provide the treat, and instead turn or look away, providing as little social contact as possible for as long as six seconds. Then, reengage the subject, cue and perhaps prompt the criterion behavior, managing the environment as best you can to prevent the non-criterion behavior, reinforce the target behavior when it occurs, fade the prompt and get back on track. Although withholding and removing reinforcers is aversive, it is usually minimally so, with little likelihood of generating problematic side effects and when utilized as a component of a graded differential reinforcement procedure, is highly effective. The emphasis however, should remain on antecedent control, a graded approach and the prevention of non-criterion behaviors from occurring. There is a reason that non-criterion behavior is occurring; identify why and manage it!

Box 3. A well-constructed and well-implemented training plan designed to achieve realistic goals ought to be successful, but there are so many variables involved in training, some of which occur outside of the presence of the trainer when the client carries out their "homework." Problems can occur and it is not always easy to identify and modify them. Is the failure due to unrealistic expectations? Are you requiring too much of a change in behavior too quickly? Are you jumping to new criteria levels before you have conditioned the previous levels fully? Are you failing to maintain minimal distraction, duration and distance to start or combining these variables too quickly? Is the problem just a matter of needing more time in order to ensure that you are moving at the subject's pace? Make the necessary adjustments to the plan, including controlling the variables that are causing difficulties. Set the subject up for success!

Make sure that if the client is engaging in any unsupervised training between consultations that they are carrying out the procedures appropriately. Have them demonstrate the training they have been implementing and remediate where necessary. Ensure that they are not engaging in appropriate training procedures during "training session" but then inadvertently counter-conditioning that training in "everyday life." This is common, for instance, in loose leash walking. You may need to also reevaluate what the client can effectively carry out by him or herself. Take whatever actions are necessary to ensure that the client is implementing the proper training at all times.

Finally, evaluate other training related practices and variables. This evaluation process is not a cursory "technicality" in which you recognize only obvious mistakes. If everything is being done right, then you *should* be achieving success. If you are not meeting your objectives, there is a problem with what you and/or the client has done so far. This is your opportunity to identify that problem and fix it, rather than resort to more aversive methods and tools. Aversive methods will not identify and fix the problem.

Consider the possibility that you may have misidentified the effectiveness of the reinforcers used for the target behavior or that there are competing contingencies interfering with training. Did you select procedures to address the target behavior appropriate for the situation? Have you adequately addressed the antecedent conditions? Many trainers focus on consequences and fail to appreciate the importance of antecedent conditions.

You need to look at all of the fine details, including: deliverability of the reinforcer; contingency and contiguity of delivery of the reinforcer; size of approximations; fluency of prerequisite skills; response effort and competing contingencies; and the schedule of reinforcement and the point at which the schedule is changed. Remember, competing reinforcers, are always available. Your goal is to ensure that you are controlling the reinforcers available for each behavior and that the relative effectiveness of each reinforcer is such that the subject will exhibit the target behavior rather than alternative behaviors.

Training can be complex in the real world, largely because of the dynamic nature of the environment and the variables that influence conditioning. When a well-constructed plan fails, this is largely where it does so. It can be a challenge to identify the application-related problems. If you have achieved some success, analyze why this success occurred. What differs in that situation as opposed to when the non-criterion behavior occurs? Often, video recording the training can help you analyze the problem and your approach. Consulting a colleague can also be helpful, as well as provide a fresh perspective on the training plan and its implementation.

Box 4. If you do not achieve your objectives, reconsider how diligent you were with previous steps, reevaluate the plan and be more creative. Refer to authoritative sources or consult a colleague with specific proficiencies that may help you succeed. Often, a fresh perspective is called for to identify problem areas and ways to circumvent them. Another option is to seek supervision for the case. This option has the added benefit of helping you develop your own formal proficiencies. It is also an excellent way to meet your training objectives, promote your professional development, and broaden your skill sets. If you have been making some progress but it is slow, consider accepting the fact that it will simply take longer to achieve your goals.

If these options are unavailable and you are otherwise still not able to identify the problem, you should consider referring the case to a professional with specific proficiencies related to the issues involved in the case. The Association of Animal Behavior Professionals[50] is a useful resource, particularly as certified members are behaviorologically oriented and specifically dedicated to using added reinforcement-emphasized methods. Referring the case to a certified member of the International Association of Animal Behavior Consultants[51] is another option. It is not a moral failing to lack proficiency in certain skill sets; recognizing and acknowledging a lack in specific proficiencies is laudable when you followed it up with a referral to a professional with the required skills.

Once more, increasing the aversiveness of the procedures will not identify and effectively manage the training errors that have resulted in failure. It is much more productive to focus on finding and correcting the problem, rather than adding aversive stimulation on top of your mistakes. Consistently finding it "necessary" to resort to increased aversiveness in training is an indication that the trainer is the one who needs more effective training.

If you have diligently reevaluated the case and researched authoritative sources; if consultation, supervision, or referral are ineffective or not viable options; and the plan is still not sufficiently effective, you should consider finding a different kind of solution (e.g., train a completely different behavior that still achieves what you need from it). Consider just how important the objective is. Perhaps it is worth simply living without this behavior.

If the behavior is vital to the subject's quality of life, it is time to consider escalating. However, the escalation is not in the aversiveness of the stimulation used in training but an escalation in the effort, time and resources expended to achieve success. Consider supervising the client's training behaviors more closely, perhaps being present for *all* training sessions and consider increasing the frequency of supervised training. This will help ensure that the client is well coached, more proficient and not sabotaging the training plan with their lack of experience. You might consider offering a discount, making this option more affordable in general, or providing extra training time on a *pro bono* basis. You might also consider arranging for a board-and-train service so that a professional can train the subject, and once you or a colleague has trained the subject, you or your colleague can coach the guardian on how to maintain the training. These options are more arduous for various reasons, but are worth considering if you have legitimately reached Box 4. It is extremely rare for a professional trainer to reach Box

---

[50] www.associationofanimalbehaviorprofessionals.com

[51] www.iaabc.org

4, let alone have to move to Box 5. This really only occurs when the trainer has mistaken a serious behavior problem as a situation requiring simple training as opposed to a serious behavior change program.

Box 5. Does failure to train this behavior constitute a significant risk to anyone or cause a dramatic hardship or reduction in quality of life for the subject? If so, it is time to recognize that the case does not require simple training but rather a full behavior change program constructed and implemented by a competent behavior technologist. Refer the client to a qualified professional!

No allowances are made for more aversive strategies that might include subtracted reinforcement of criterion behaviors or added punishment for non-criterion behaviors because these methods are simply not justified in simple training projects and would cause more harm than good. This might evoke surprise in some readers. However, there simply is no need for training components more aversive than described above. The problem is not a lack of highly aversive methods; it is the skill of the trainer involved. Aversive methods do not solve that problem.

Continuing Education

Sidman (2001) is an excellent introduction to the harm that can be caused using aversive methods.

Courses provided through The Companion Animal Sciences Institute at www.CASInstitute.com address the risks associated with aversive methods and how to use added reinforcement-emphasized methods effectively.

The most current version of this strategy that also includes resolving problem behaviors can be found on the Association of Animal Behavior Professionals web site at www.associationofanimalbehaviorprofessionals.com/guidelines.html.

The Science and Technology of Animal Training

CHAPTER 5. TRAINING STRATEGY AND PROJECT PLANNING

Behavioral Objectives

The objective of this chapter is to measurably expand the reader's repertoire of behaviors in relation to describing and relating the principles of behavior. Upon successfully integrating the concepts outlined in this chapter, the reader, where exposed to contingencies to do so, will accurately:

- Describe an efficient and effective general strategy for training animals based on nonaversive methods including how to establish specific quantifiable behavior objectives

- Describe how to plan a training project prior to training in order to make training the most effective and efficient it can be

- Describe how to arrange the environment to make the target behavior more likely, and how to administer reinforcement for the behavior to promote acquisition of the behavior

- Describe how to transition a training project from acquisition to training for fluency, undertaking generalization and discrimination training, bringing the behavior under stimulus control, and beginning to thin the schedule of reinforcement, as well as proofing behaviors against increasing distraction, distance, and duration

- Describe how to transition fluent behaviors to maintenance, transitioning to nontrainer-mediated reinforcers, finding a suitable long-term schedule of reinforcement and refreshing training when required

- Systematically prepare a training plan, ensuring that it includes all necessary information to proceed with implementation.

General Systematic Training Process

Training involves bringing specific behaviors under specific antecedent controls. This involves manipulating the antecedent environment in order to generate the target behavior, and manipulating the postcedent environment in order to reinforce the target behavior to the exclusion of other behaviors. More specifically, you prompt the target behavior in the presence of an antecedent stimulus and follow that with a reinforcer. Through repetition of this experience, the primary antecedent stimulus takes on stimulus control over the behavior (i.e., it becomes more likely to evoke the target behavior). Each new experience with the contingency increases the capacity of the antecedent stimulus to evoke the target behavior and the prompts are faded. Discrimination and stimulus generalization training fine-tune exactly what stimulus will evoke the behavior, and use setting generalization training to ensure that the behavior is reliable in appropriate settings.

There are several different basic strategies from which to choose in training animals. For example, it is possible to punish any behavior other than the desired one, until the subject reliably exhibits the target behavior. That is one strategy. The trainer could also arrange for aversive stimulation to be present until a target behavior occurs, and cease the aversive stimulation contingent upon exhibition of the desired behavior, thereby subtractively reinforcing it. That is another strategy. However, these strategies are fraught with problems, in terms of effectiveness and efficiency, as well as ethics.

A more efficient and effective strategy is to encourage occurrence of the target behavior, provide added reinforcement contingent upon it, gradually and incrementally increasing the level of difficulty of the task, helping to ensure success for the subject and trainer. One might use other, more eliminative principles of behavior, such as punishment, or extinction to discourage non-criterion behaviors, but it is best to avoid these where possible, in favor of a graded approach. A ***graded approach*** involves breaking tasks or goals down into smaller components or steps, or working on specific dimensions of a task one at a time, to ensure a greater likelihood of success and gradually integrating additional components or combining dimensions. For example, if the goal were to train a dog to sit under high levels of distraction, the trainer would arrange a hierarchy of distraction intensity, and start by training under the least distracting circumstances. Once the dog reliably sits in an environment with very few distractions, the trainer would introduce slightly more distracting elements, and repeat the training process through environments with successively higher levels of distractions until the dog exhibits the behavior reliably under highly distracting circumstances. In most cases, the training will not be successful if you attempt to train behavior in a highly distracting environment from the beginning. Another example includes bridling a horse. This actually involves a series of behaviors; the project is too big to complete in one step. Thus, the task is broken down into smaller, more manageable steps. First, the horse is trained to remain still, quiet and relaxed when they see the bridle. Then, the horse is trained to remain still, quiet and relaxed as someone approaches with the bridle. Then, you train the horse to lower his or her head when you present the bridle. You then train the horse to open his or her mouth to take the bit. Finally, you train the horse to remain still, quiet and relaxed with a lowered head while you fasten the bridle. After you have carried out all of these steps, it is much easier to combine the steps in the behavior chain of "bridling." I will discuss the graded approach in detail below as well.

I recommend that you utilize added reinforcement-emphasized methods, favoring a graded approach and a systematic series of steps be used in most, if not all, training projects. Becoming familiar with this general process facilitates the ability to design and implement training projects for novel behaviors with minimal or no need for aversive stimulation. Training should not be haphazard and mysterious. It should not be restricted to those who just seem to "get it." The approach here makes the process principle-based, empirically supported and systematic. Training is composed of behaviors, and as such, you can analyze and condition it.

Training new behaviors involves the following general steps:

- Phase 1. Preliminaries

  - Planning and Preparation

- Identifying and defining the target behavior
- Assessing the subject's current proficiency
- Preparing a formal behavior objective
- Identifying reinforcers and establishing the conditioned reinforcer
- Phase 2. Acquisition
  - Determine and implement antecedent tactics
  - Determine and implement postcedent tactics
- Phase 3. Fluency
  - Begin fading prompts
  - Begin thinning the schedule of reinforcement
  - Refine form, latency and speed parameters
  - Bring behavior under stimulus control
  - Proof against three D-parameters (distraction, distance, duration,)
  - Discrimination training
  - Introduce release stimulus
- Phase 4. Maintenance
  - Work toward maintenance of fluency

I will discuss each step below, along with some training tips.

Phase 1. Preliminaries

Planning and Preparation

Every training project should begin with a ***training plan***—a formal written presentation of the key elements that you will use to identify and achieve the objectives that the training requires. How elaborate the plan is will depend largely on the complexity of the project and the experience level of the trainer as well as the level of accountability required of the trainer. A very simple training project, such as for example, to train a dog to give a paw, might include only a statement of the behavior objective and perhaps a statement indicating the procedure that you will use. If the project involves complex procedures, a more elaborate written plan is required. If you are training someone else's companion animal, for or with them, you will likely want a more elaborate plan as well, in order to increase accountability and confidence in the training, which may be more than you need if you are training your own companion animal alone. I will describe all of the vital aspects of a formal training plan that might be called for in the most complex of plans. You can truncate the planning process for very simple training projects. I have provided a sample training plan at the end of this section.

Identifying and Defining the Target Behavior

The Target Behavior

An effective training project begins with identification of a **target behavior**, also known as the **behavior of concern**, though we often reserve this latter term for problem target behaviors. The definition of the target behavior should describe a single response class in functional and operational terms. Your ultimate goal might include more than one target behavior or a complex behavior that will require a graded approach and multiple steps. It may involve a behavior that the subject does not current exhibit and you cannot readily prompt in its final form, and it may involve a series of behaviors with one single cue that evokes the entire series of behaviors. Where complexity is involved, the plan should have a definition and formal behavior objective for each distinct behavior in the project. If your ultimate objective requires that you chain behaviors together or are otherwise implemented together, your final training project will involve the chaining or combining of the behaviors.

In resolving a **behavioral deficit** (simply training a new behavior), the primary target behavior is the behavior that you aim to train and bring under stimulus control. The "problem" behavior, if referred to as such, is simply the behavior that is evoked in the presence of the evocative stimulus or general training context before training occurs. Behavior is continuous and so there will always be some kind of behavior occurring. The new target behavior replaces this "problem" behavior. In most behavior deficit cases, the focus is simply on the target behavior you aim to train. If the current behavior in that situation is problematic in a practical sense, treat it as a **behavioral excess** case.

Clients commonly define the behavior under consideration poorly, reporting that the subject is "going crazy," being "dominant" or "stubborn," or that they want the animal to "be polite." Trainers need to operationalize these descriptions, as none of them constitutes actual behaviors, and they violate all of the requirements described herein. Your plan should not involve interpretations of the behavior, or assumptions or speculation about what it implies experientially, etc. However, the formal behavior objective that you establish should be clear, complete, specific, functional and operational; it should contain all of the necessary and sufficient conditions that will validly allow for accurate identification of when the behavior has, and has not, occurred. It should include any boundary identifiers (e.g., whether it must be exhibited within a specific amount of time, ruling out non-instances of the behavior, etc.). Vague definitions are sloppy, even for simple behaviors, and do not instill confidence by those with a stake in the project, such as guardians, shelter managers or certification boards looking at case studies. As a test of its reliability, you should be able to produce a written definition for your target behavior, and without further coaching or input, independent observers would all produce the same measures with regard to when that behavior was, and was not, exhibited.

For example, a slightly vague definition of "stand" for a dog might simply be to have all four feet, and only all four feet, on the ground. This is necessary but is it sufficient? A better operational definition rules out unintended behaviors (e.g., crouching) by requiring the extension of all four legs. In the first instance, you would click as soon as all four feet are on the ground and in the latter instance, you will require the subject to stand with all four legs fully extended before you click. Furthermore, let us say you go with this improved definition and provide it to five different observers who will separately observe a dog engaged in behaviors including standing. What if you cue the dog to stand and the dog finally does so after 30 seconds? Some observers may identify this as an instance of the behavior and others might decide it is not an instance of the behavior because it seemed unrelated to the cue. Leaving the training to guess on these matters will reduce reliability of the measure. The same thing could occur if, without being released, the dog barely stands before lying right back down. Most observers will identify the stand but some might decide that, because the trainer did not release the dog, it was not an instance of the target behavior. Do not assume that everyone will simply know (via "common sense") what does and does not constitute occurrence of the behavior. If you leave it unspecified, you have some chance of agreement between observers as to whether or not the behavior occurred, but it is much better to be accurate and precise. This attention to detail helps ensure a clear and unambiguous target behavior definition.

As another example, let us identify the target behavior as "sitting." Once you identify the behavior, you can identify factors such as the latency, speed, duration, specific forms of distraction that it must occur through, distance between the subject and the person delivering the cue, for instance. These specific criteria are not all necessary in all instances, but where appropriate, they provide a more precise behavior definition. For example, every trainer knows what an obvious instance of a dog sitting looks like and all would likely agree, reliably, when this behavior occurs in this obvious manner. However, what if the front feet are planted well forward of the dog for some reason, the elbows not touching the ground but still at an angle that might lead some observers to identify this as an instance of sitting and others not? Is this really what you want? What if the trainer delivered the cue and the dog does not begin moving for eight seconds, at which point, the dog slowly begins to sit? Is this really what you want? What if the dog favors one hip when sitting, as in the "lazy puppy sit?" Being precise with respect to what you want or what you mean when you tact the name of a behavior makes for clarity and contributes to greater efficiency and effectiveness in the behavior change project.

Obviously, there can be a point of diminishing returns. It is not necessary to include every possible topographic variation, going into excruciating detail. It is important, however, to consider likely variations or features of the occurrence of the behavior and stimulate concisely and precisely what is and is not to constitute instances of the target behavior. Certain contexts will require extremely detailed and careful definition, while others will not.

Preparing a Formal Behavior Objective

Assessing the Subject's Current Proficiency

Once you have defined your target behavior (or target behaviors if breaking the plan down into steps or preparing a chaining plan) with appropriate criteria, you should assess the subject's current proficiency with respect to the behavior(s). Can the subject currently exhibit the behavior in its final form? If they cannot, what approximation of the behavior does the animal currently exhibit? If they do exhibit the behavior, is it within the criterion latency? Are there other criteria that the behavior currently meets? Are there any other criteria that the behavior does not meet? Can you evoke the behavior with a different cue than the one you have planned for it? What is the stimulus or prompt that currently evokes the behavior? Appreciating the subject's current proficiency in mediating the behavior, and which criteria are currently met and not met, will inform your behavior objectives, particularly in terms of what time-frame it might require to train the behavior fully to meet all of the criteria. This will also inform what procedures you might require or where in the process you will need to begin. Prepare a statement identifying the proficiency of the subject in exhibiting the target behavior as part of the training plan.

Formal Behavior Objective Components

Now that you have defined the target behavior that you want to achieve with training and you have assessed the subject's current proficiency in exhibiting that behavior, you are ready to define the formal behavior objective. The **behavior objective** is your statement of exactly what you want to achieve from your training project. A formal behavior objective should include (a) the evocative stimulus, (b) the target behavior it is to evoke, in operational terms (including a suitable measure for it, and other criteria), (c) the reinforcer you plan to use to train the behavior, and (d) a tentative time point at which you plan to achieve the stated functional relation at the stated measure. Thus, the behavior objective takes the target behavior definition you prepared and stipulates the evocative stimulus and a time point at which you seek to have trained it.

Write down your formal behavior objective. You may want to also include it in the graph that you use to track the behavior (revisit Figure 12, if needed). A dotted line labeled "Objective" indicates the level of the behavior objective. Where the beginning of the maintenance phase (time), intersects the objective level lines (i.e., the grey arrow in Figure 12), you find the specific behavior objective point. The trend of the line during the training phase can be used to prompt continuing the training as is, or adjusting it in order to meet the objectives (or, if the trend cannot be maintained, reevaluating the feasibility of the objectives). In this case, you can see that the trainer achieved the behavior objective before the 12th session.

Graded Approach

The **graded approach**, as discussed above, involves breaking projects down into smaller steps. This makes complex training projects more manageable. It is best to incorporate these steps into the plan rather than "wing it." You have enough to react to while training; proper planning ahead of time takes off some of the pressure.

Kurland (2007, p. 32) provides a good example of a training plan that involves not only training a complex behavior but also resolves fear of stimuli involved in the training. When she sought to put a bridle on her horse, the horse would act disruptively at every opportunity. The primary goal was to have the horse lower his head and remain still and relaxed throughout the bridling process. As described above, this situation calls for a graded approach (i.e., breaking the behavioral episode into several distinct contingencies, each of which is addressed in turn). Breaking the training project down into steps makes the process much easier to address, as does breaking the target behavior down into component behaviors. In Kurland's example, sight of the bridle evoked head throwing and holding the head out of reach. In contrast to this, she wanted the bridle to evoke lowering the head within easy reach, and keeping it there. Instead of taking the bit, the horse exhibited jaw clamping. In contrast to this, she wanted approach of the bit to evoke mouth opening and acceptance of the bit. Head handling evoked flinging up the head, but she wanted head handling to evoke leaving the head lowered while you lift the headstall over the horses face. Next, the horse struggled to prevent the trainer from putting the bridle on, however, she wanted the horse to continue to leave her head lowered while the bridle went over her ears. Next, the horse struggled to prevent latching of the bridle at the throat. Instead of this, she wanted the horse to continue to stand quietly and relaxed to allow the trainer to fasten the latch.

A graded approach is appropriate for situations that do not involve problematic emotional arousal and escape behaviors as well, where distraction might be the only problem for instance. In the training chapters to follow, I will present orderly sequences of training steps for each training project. You will train features of the plan (e.g., distraction, distance, duration, etc.) sequentially, rather than concurrently. This is also an instance of a graded approach.

Functional Assessment

Sometimes, the behavior you aim to train occurs readily, just not in reaction to the specific stimulus (e.g., a vocal cue) that you want. In this case, a functional assessment is straightforward and the trainer merely identifies the contingencies they want put in place. When the chosen evocative stimulus currently evokes another behavior instead of the target behavior, we commonly refer to the other behavior as a "***problem behavior***."[52] At the lower end of the spectrum (i.e., where the behaviors are not causing any detriment), the behaviors are not particularly problematic. Problem behaviors increase the complexity of functional assessments. A *functional assessment* is a process by which the trainer determines the functional relations involved in the occurrence of a response class. This means that the trainer needs to determine the contingencies that currently exist, as well as identify the contingencies to they will put in place. These contingencies provide the information needed to replace the old behavior with the new behavior. It is important to determine what evocative stimulus is currently evoking the problem behavior, as well as the consequence that is currently maintaining that behavior. In this manner, trainers can control those reinforcers. If it is

---

[52] The emphasis in this book is not on replacing problem behaviors, as this is a rather complex topic, unnecessary for most basic training purposes. This brief introduction to replacing problem behaviors aids in addressing very simple problem behaviors that may arise when conducting basic training. For a more comprehensive coverage of behavior change programming, see *Changing Problem Behavior, 2nd edition.*

acceptable to utilize the same reinforcer that was maintaining the old behavior, do so. However, if not, you will need to identify a new reinforcer that will be as highly effective as possible. By taking control of the reinforcer for the current behavior, the trainer can prevent reinforcement. Instead, the replacement behavior can be prompted in the presence of the evocative stimulus and ensure that only the replacement behavior is reinforced. The trainer thereby brings the target behavior under the control of the appropriate evocative stimulus. We call this the *behavior replacement model*. For our purposes here, the "problem behavior" is usually the behaviors that might otherwise occur in a situation the trainer wants to train. Once you have carried out this assessment, you can simply apply the appropriate principles of behavior to promote the new behavior in place of the old behavior in that contingency. Behavior is continuous. Therefore, in training any behavior, the subject is always exhibiting some type of behavior, whether it is the "sit" that you want, or just standing and looking at you quizzically. The goal is to replace the current behavior (e.g., standing) with the sit.

In the case of our bridling example, each of the initial behaviors functioned to escape contact with the bridle, involving a straightforward analysis, as described in the previous section. It is important to remember that problematic emotional behaviors typically accompany behaviors maintained by subtracted reinforcement. The emotional reaction to the stimulus causes motivative conditions that make escape behavior more likely. Part of a comprehensive behavior replacement strategy is to not only control the reinforcers as described above, but also change the emotional reaction. By changing the emotional reaction, the evocative stimulus becomes irrelevant with respect to the escape behavior and the escape behavior unnecessary. By making heavy use of added reinforcers and a graded approach, the horse will become much more "comfortable" with the bridling process (i.e., the horse's reaction to the stimuli will be less emotionally problematic). The following replacement contingencies describe the training plan:

Sight of bridle $\rightarrow$ Remain still and relaxed $\rightarrow$ Treat$^{(+R)}$

Approach of bridle $\rightarrow$ Remain still and relaxed $\rightarrow$ Treat$^{(+R)}$

Present bridle $\rightarrow$ Lower head and remain still and relaxed $\rightarrow$ Treat$^{(+R)}$

Present bit $\rightarrow$ Open mouth and remain still and relaxed $\rightarrow$ Treat$^{(+R)}$

Putting bridle on $\rightarrow$ Keep head lowered and remain still and relaxed $\rightarrow$ Treat$^{(+R)}$

Latching bridle $\rightarrow$ Keep head lowered and remain still relaxed $\rightarrow$ Treat$^{(+R)}$

In this case, you have one grand objective of putting on the bridle but you are able to break the project down into several smaller, more manageable objectives. You can then start on working through the first one. Once one project is complete, you move seamlessly onto the next project until you achieve your grand objective.

To clearly establish your behavior objective, you should identify a tentative time point at which you expect to achieve the behavior at the specified strength. You will not always be exposed to contingencies to identify a precise quantification and time point, especially with simpler behaviors. But, it is a good idea to at least have a tentative idea of precisely the strength of the behavior you want and by what point in time, as this will help you determine whether you are on track, or not. It also helps you make the determination of whether or not you need to adjust your objectives, strategy or procedures. That said, "work at the subject's pace," so to speak.

I will provide an example of a formal behavior objective that contains all the required information. This is a simple training project. The verbal cue does not currently evoke a behavior (and there are no particularly important problem behaviors to replace), so the contingency analysis is straightforward. As you can see, this includes the evocative stimulus, the behavior, and the time point to be established. I will describe the behavior in detail. This is not usually necessary for very simple behavior, particularly if you are training your own dog. If however the behavior is more complex or you will be coaching a client to train their dog, this level of detail may be necessary to ensure precision and clarity.

Behavior Objective

S^{Ev}: Vocal "sit."

Behavior: The subject will contact their rear end to the ground (or as close as is physically possible) with front legs straight at ≈80–90 degree angle to the ground (back at ≈45 degree angle) and with front paws on ground.

Consequence: Treats and continued social contact.

Criteria: Latency: 2sec.; duration: until released, minimum. 1min.; frequency: 100% through 10 trials; distance: minimum. 20 feet; distraction: various; time point: 3 sessions at ≈5min. each.

Training new behaviors is all about bringing a target behavior under stimulus control. This formal behavior objective presents a definition for the target behavior that is likely to be highly reliable, and it provides adequate boundary data to ensure that you identify criterion, and non-criterion responses accurately, but it is also concise. The boundaries and the measures are clear; the frequency is established, and a time point goal is established. This allows you to devise a means for tracking the frequency of the behavior, if you are exposed to contingencies to track the progress quantitatively. You might do this with a simple chart or graph in this instance. The take-home message here is that if you include all of the necessary information concisely in a single place, you are well-prepared to settle on a strategy and plan for how you will achieve that objective, as well as to track progress so that your training is accountable and precise.

If you will be constructing a behavior chain, each component behavior will represent a training project of its own. If you are shaping a behavior, you do not need a separate training project for each approximation. The behavior objective will identify the target terminal behavior, but you will need to prepare a list of behavior objectives.

Final Word on Formal Behavior Objectives

Although the text provides a rather involved process of preparing a formal behavior objective, most experienced trainers are able to effectively truncate this process. They have trained hundreds, if not thousands, of sits (and other simple common behaviors) and they know all of the criteria for the behavior very well. This is not sloppy—it is just a matter of the trainer responding appropriately and proficiently to the contingencies without rewriting it each time. Trainers expand their repertoire of effective training behaviors by taking elaborate preparation measures. As their repertoire expands, they are able to begin training effectively with fewer and fewer elaborate stages. This does not mean that one should aim to avoid planning. Precision, clarity and accountability are always necessary. When working with a client and their companion animal, many experienced trainers will have handouts ready so they do not need to rewrite the formal behavior objective each time. With an early foundation of going through the process of preparing proper formal behavior objectives, one can safely truncate that process later on. Without this early foundation in precision and clarity, some experienced trainers end up just "flying by the seat of their pants." If you are a very experienced trainer, you might have been thinking that "this is just all too much" and it is when it has become unnecessary, but before that happens, the foundation in precision, clarity and accountability conditions one to exhibit the verbalization that "this is just all too much." Furthermore, experienced trainers will rather quickly find themselves in a position to truncate the simpler training projects, but establishing a formal behavior objective remains a useful skill when dealing with training objectives that are more complex, particularly when there are problem behaviors to resolve and complex behavior chains to train. Do not be too quick to forgo preparation of formal behavior objectives. A little extra preparation at first will go a long way in preparing you for truncating the process later when you have sufficient experience to do so. This extra preparation also provides the precision and clarity necessary for achieving success with highly complex training projects.

Identify Reinforcers and Establish the Conditioned Reinforcer

The next step is to identify the most effective *unconditioned reinforcers*. An informal approach of merely observing and noting what the subject currently expends significant energy contacting (ideally, in the specific context in which you want the target behavior exhibited) is sufficient; guardians are usually quite familiar with what works most effectively as well. You will probably want to rely on *generalized reinforcers* later in the process (i.e., using more than one unconditioned reinforcer and including fewer trainer-mediated reinforcers). Initially, an easily implemented unconditioned reinforcer is necessary. In most cases, a small treat works best, while in other cases, praise, and play, or contact with a specific toy will effectively strengthen behavior. With many parrots, the so called "drama reinforcer" works well.

Identify what kind of *conditioned reinforcer* you will use. I recommend clickers, as they produce a sharp, distinct and salient sound that will not be rendered ineffective by daily non-training exposure. For sensitive subjects, many of the button-style clickers are softer sounding than the older box shaped, hole-style clickers. You can wrap fabric around a standard clicker or keep it in your pocket to dampen the sound. Alternatively, you can click a ballpoint pen or press on the metal lid of a glass bottle (e.g., a glass Snapple® beverage bottle or a glass baby food jar) to make a soft "clicking" sound.

Some trainers find it challenging to master the physical skills/behaviors necessary to operate a clicker along with treats and leashes. In these cases, a tongue "*click*" can do instead. However, the clicker makes the exact same sound each time, making it more precise, and if your ambitions involve training at more advanced levels, you should consider expanding your clicker multitasking repertoire (or at least try to make the tongue click sound the same each time).

Conditioning a conditioned reinforcer is straightforward, and one of the first things you do in training the subject, if you have not previously done so. In order to establish the conditioned reinforcer, you simply repeat the sequence *Click!* → *Treat* several times. Follow the click with the treat within a second or two. Repeat this sequence several times with several seconds between each trial. Carry out several trials of this sequence, and ensure you avoid clicking after any particular behaviors many times and also that you do so in different locations to ensure that a single location does not become paired with the click. Avoid clicking without providing the treat, as a rule, to ensure a high degree of contingency, and a strong conditioned response.

Phase 2. Acquisition Stage Training

Once you have decided on a target behavior, have assessed the subject's current proficiency with it, and established a formal behavior objective, you can begin training. There are several decisions for you to make regarding how you will carry out the training. Although you must consider the whole contingency, and is integrally related, it is useful to consider antecedent and postcedent tactics separately for planning purposes. They will all come together as a unified strategy.

Antecedent Tactics

Behavior must occur in order to be reinforced and brought under stimulus control. Sometimes you can simply allow an uncontrived or unaugmented environment to evoke the behavior, and merely take the opportunity to reinforce it.[53] More often, it is most expedient for trainers to contrive circumstances to get the behavior to occur in its final form—this usually involves prompts. Once you have increased the rate of the behavior, you may fade the prompt. There are other operations that can contribute to generating behavior, quite distinct from prompting, including motivative operations that create a deprived or satiated body, which is more or less likely to exhibit the behavior when

---

[53] This is often referred to as "capturing" and is sometimes contrasted with reinforcing behavior that is evoked or prompted. Take note that behavior is *always* evoked. This distinction more appropriately refers to the evocative stimulus being mediated by the trainer versus something unrelated to the trainer's behavior.

exposed to the evocative stimulus. Alternatively, the current environment might be too weak to evoke the behavior in its final form, or it simply has no evocative capacity in that regard, but it might evoke an approximation of that behavior, in which case, you can simply reinforce that approximation and shape the final form, an advanced procedure that I will discuss below.

Prompting

I discussed *prompting* in the section on antecedent tactics in Chapter 2. Where you cannot readily cue a behavior, you can use prompts to help generate the behavior so that you can reinforce the behavior and strengthen other antecedent stimuli. The prompt is then faded, leaving only the now strengthened evocative stimulus. Aversive prompts are distracting and disruptive, so the prompts should be minimally aversive. Contrived trainer-mediated prompts might include pointing, luring, making sounds or even touching the subject. However, as physical contact can be distracting and disruptive, it is better to avoid such contact, if possible.

Motivative Operations

I discussed motivative operations resulting in satiation and deprivation in the section on antecedent tactics in Chapter 2. Establishing and abolishing operations act as function-altering conditions and change the function of existing antecedent stimuli. It is always useful to be aware of the current condition of the subject's body, with regard to satiation and deprivation of the reinforcers in question. When necessary, you may need to train during times of deprivation rather than satiation.

Postcedent Tactics

Reinforcement

This section provides a review of some principles of behavior related to reinforcement and clarifies a few common controversies or differing perspectives used in animal training.

In most cases, you will use both a conditioned reinforcer (e.g., clicker) as well as an unconditioned reinforcer (e.g., food treats) to increase or maintain the rate of a behavior throughout the acquisition phase of training and well into the fluency phase too. We use the conditioned reinforcer to administer reinforcement more precisely. This is most useful when training new components or parameters of behaviors. The clicker is used in the initial acquisition phase and also into the fluency stage when new features such as working on improving form, latency or speed of the behavior or introducing new levels of challenge with respect to distraction, distance or duration. Once there are no new features to introduce, you can discontinue the clicker and you can use just the unconditioned reinforcer, which will be on an intermittent schedule at that point. You should deliver the reinforcer either during or immediately following the behavior that you aim to reinforce. The unconditioned reinforcer should follow the conditioned reinforcer, ideally within a second or two but certainly within three seconds—the quicker the better. To maintain the strength of the conditioned reinforcer, it is best to follow every instance of the conditioned reinforcer with the unconditioned reinforcer. This is a general rule. You will still maintain the conditioned reinforcer if it is even just occasionally followed by the unconditioned reinforcer but to maintain the conditioning as strongly as possible, most trainers make it a rule to follow the click every single time with the treat.

The conditioned reinforcer is a postcedent term in the contingency of which it is a component. The behavior has occurred and the conditioned reinforcer and then unconditioned reinforcer represent the end of that trial through the contingency. This is why there is a general rule in animal training that "the click ends the behavior." Once you click, you do not require the subject to be engaging in any further ongoing behavior as part of *that* trial. If the subject was sitting, you click, and the subject gets up, that is okay. That trial is over. The consequence has occurred, you have reinforced the behavior and that is that.

That is the general rule in animal training. However, behavior is more complex than this. Let us explore the topic in a bit more advanced depth, as doing so will reveal further advice beyond the basic rule that may improve training. The interval between the presentation of the conditioned and then unconditioned reinforcer is important in some cases. If the subject regularly engages in a specific behavior during this interval, it can become chained to the previous behavior that was reinforced. For instance, if you train a dog to touch a target with their nose and after you click, you drop the treat to the ground, you may find, when you put the behavior on an intermittent schedule of reinforcement, that the dog touches the target and then dips their head down. This dip was chained to the target touch and the click has taken on stimulus control over it. So, placement of the treat, and more generally, any behaviors occurring between the conditioned and unconditioned reinforcer can actually be reinforced (or even evoked by the clicker), requiring care with respect to how treats are delivered, following the click.

Many trainers use the clicker both as a conditioned reinforcer and as an evocative stimulus to continue exhibiting the behavior. This is commonly referred to as the "keep going signal" although the word "signal" is inappropriate and better would be "***keep going stimulus***." Some argue that the clicker be used only as a conditioned reinforcer and that the strength of the conditioned reinforcer is maintained by following up with presenting the conditioned reinforcer again and in many cases, more times until ultimately the unconditioned reinforcer is presented. It is said that the clicker is not used as an evocative stimulus at all but rather a means by which the behavior can be reinforced without having to end the trial. It is said that each click is simply an added postcedent term in the contingency. The click functions as a conditioned reinforcer and is tied to other conditioned reinforcers in a chain, capped off with the unconditioned reinforcer at the end, when the trial is complete. According to this explanation, it is unclear why the conditioned reinforcer is required midway through the duration of a behavior. One might argue that trainers should extend duration gradually with a single conditioned reinforcer followed by a single unconditioned reinforcer at the end of the criterion duration. If extended gradually in this way, there is no need for repeated conditioned reinforcers (i.e., click, click, click, treat).

It is also often said that the keep going stimulus is used to "encourage" continued behaving, to keep the subject going. Trainers must be clear here. Encouragement is an antecedent process, not a postcedent process. Encouragement means the keep going stimulus is being used as a prompt or evocative stimulus. There is no reason why a stimulus cannot function as a conditioned reinforcer and as an evocative stimulus. For instance, when a series of discrete behaviors is chained together, completion of one behavior functions as a conditioned reinforcer for that behavior and also as the evocative stimulus for the next behavior in the chain. The conditioned reinforcers are maintained by the eventual pairing with the unconditioned reinforcer. Again, however, the need for encouragement (prompting and cuing continued behaving) is unclear when one may simply gradually increase the duration of the behavior without the use of a series of conditioned reinforcers or interjected cues or prompts. If a longer duration behavior is degrading, this does not necessarily argue for including "keep going" stimuli while the behavior is occurring in order to ensure the subject keeps going. Rather, it is an indication that the trainer is moving too quickly in extending the duration.

Identify the Required Procedure

Although flexibility is important, and the subject's progress should inform adjustments to the plan as appropriate, it is still important to plan your postcedent strategies for reinforcing the behavior.

Is the behavior a single discrete behavior that you can readily prompt in its final form? If so, you can simply differentially reinforce the behavior. If not, a more advanced procedure is required. If you cannot readily prompt the target behavior in its final form, you can use a *shaping* procedure. Is the target behavior a series of discrete behaviors that you aim to train with a single evocative stimulus at the beginning of the sequence?[54] If so, identify *chaining* as the procedure you will use and then treat each component behavior separately, which means deciding between differential reinforcement or shaping. Prepare a statement identifying the procedure, or procedures, that you will use to train the behaviors.

---

[54] Reference to "discrete behaviors" is convenient in discussing behavior chaining. However, please note that behavior is continuous, and the notion of discrete behaviors is a little misleading and arbitrary. Thus, we use the term here simply to distinguish between components in a chain of behaviors.

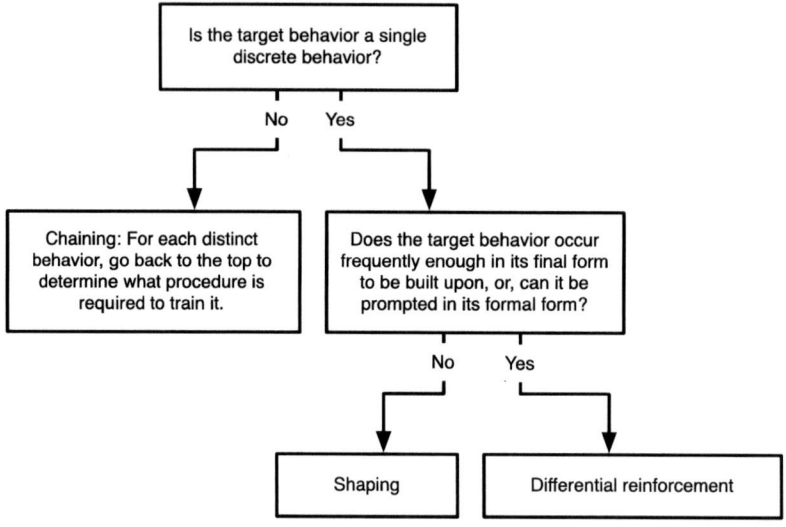

Figure 14. Flow chart for determining which procedure to use in training new behaviors.

How to Handle Non-criterion Responses

What should the trainer do if the subject exhibits a non-criterion behavior (i.e., any behavior that is not the target behavior, including seemingly no behavior at all) during training? The need to respond to a failed trial, or non-criterion behavior, indicates a failure to manage the environment well enough to ensure the subject exhibits only the target behavior. If the controls were present, the behavior would have occurred. The occurrence of non-criterion behaviors may mean that there was a failure to condition the target behavior, the trainer moved too quickly to the next level of challenge, or excess distraction was present.

Harsh behavior reduction procedures are best avoided in favor of a graded approach that improves the likelihood of setting the subject and trainer up for success, but occasionally non-criterion behaviors occur, even with careful attention to setting the subject up for success. When this happens, you have two tasks. First, avoid reinforcing the behavior. That means extinction, and where appropriate, subtracted punishment if there is an ongoing reinforcer present. Second, try to identify the source of the failure. What did you do, or fail to do, that resulted in inadequate environmental controls to generate the target behavior? If you are carrying out the first task, ensure that you give careful attention to this second task as well. Resolving the cause of the failure is more important than simply extinguishing or punishing a single trial in a training project, even though many trainers tend to be excessively focused just on the first of these two tasks.

As always, it is important to minimize the aversiveness of training, and to maintain the smooth pace of the training. Added punishment is unjustified and it is far too risky. Minimal use of extinction and/or subtracted punishment is effective, allowing you to get training back on track, but again, the occasion to consider these aversive procedures indicates your failure to manage the training project, and not the subject's failure, so at best, consider these damage control while you get back on track. You should always carry out an analysis of failures.

Which of the two principles of behavior reduction you use to prevent reinforcement of non-criterion behaviors, depends on whether the reinforcer involved is ongoing or not. Recall that, procedurally, extinction involves the failure of a reinforcer maintaining a behavior to follow that behavior, and that subtracted punishment involves the subtraction of an ongoing stimulus following a behavior. The distinction, in practice, is not always obvious. If you are training using treats, and cue a sit, but the dog lies down instead or looks over at a squirrel nearby, you would administer extinction, and withhold the treat that has been maintaining the sit behavior. If that treat was truly hidden, even to the dog's sense of smell, then it is indeed extinction, but as far as any awareness (smell, sight, etc.) that the treat is present goes, the same procedure might then be considered subtracted punishment if it ceases, contingent on the behavior. In some cases, your ongoing social contact is just as reinforcing as the treats, and in later stages, it can sometimes become even more prominent as a reinforcer. This is an ongoing reinforcer, and if you briefly subtract social attention contingent on occurrence of a non-criterion behavior, this would be an example of subtracted punishment (assuming that the non-criterion behavior is reduced as a result). In that case, you are really exhibiting both extinction and subtracted punishment, since you withhold the treat and subtract the ongoing social reinforcer. The basic distinction is preventing the reinforcer that maintains the behavior versus subtracting an ongoing added reinforcer. Another way of framing the distinction is that extinction withholds the reinforcer maintaining the behavior, while subtracted punishment subtracts some *other* source(s) of reinforcement. In any event, extinction is preferable, where available, because it manipulates the actual reinforcer that maintains the behavior. If a subject exhibits a non-criterion behavior and you withdraw your attention and the opportunity for reinforcement for several seconds and fail to provide the treat, which is a common procedure for non-criterion responses, you have actually applied both extinction and subtracted punishment.

You might start with a minimally aversive extinction procedure, saying "Oops," pausing for two seconds or so without delivering the treat when a non-criterion behavior occurs, followed by resetting the environment and providing another opportunity to exhibit the target behavior, perhaps this time with a prompt, less distraction or whatever other measures will help ensure the target behavior occurs. This leaves any other ongoing reinforcement present but this minimally intrusive approach is often quite effective. The vocalized "Oops" will become a **conditioned extinction stimulus** and **conditioned subtracted punisher**. If need be, you can include the prevention/elimination of *all* reinforcers when non-criterion behaviors occur. That means extinction and subtracted punishment. Assuming social attention is a reinforcer, this might look like this: when a non-criterion behavior occurs, say "Oops," turn away or look away without saying anything else, failing to deliver the treat and wait for five seconds, before turning back, resetting the environment and cuing the target behavior, perhaps this time with a prompt that you can fade afterward. It is worth trying the two–second extinction procedure before escalating to this version because it allows you to get back on track more quickly. If it is adequate, this is preferable.

Resetting the environment, as referred to above, simply refers to ensuring some kind of stimulation and behavior occurs after the non-criterion behavior in a way that differentiates the previous trial from the next trial. You always want to present the cue one time only, and never more than once. However, if you simply pause for a couple or few seconds and then recue the behavior, this might function as cueing the behavior multiple times. If you cue the behavior more than once, you will likely condition the subject to wait for further cues before responding—they will literally wait for the second, third or more cues before responding. Therefore, by including the conditioned extinction stimulus "Oops" and looking away or briefly stepping away and then returning, you ensure behavior occurs between instances of the cue and hence it is less likely that the subject will be conditioned to wait. Instead, the extinction and subtracted punishment are most likely to occur. In some cases, even more of a separation between trials is required and in those cases, you can briefly lure the dog into a slightly different position or cue another behavior, even just a short distance recall or sit, in order to reset the environment for a new trial through the contingency.

Always be on the lookout for escape behaviors during training. If the non-criterion behavior were actually an escape behavior functioning to forestall the interaction, then your attempt at subtracted punishment would actually end up being an instance of subtracted reinforcement of the non-criterion behavior. In other words, if the training is "unpleasant" for the subject and they act disruptively, and that results in your withdrawing for even a few seconds, you could actually reinforce that disruptive behavior. Always be aware of the rate of the behaviors involved. If disruptive behaviors are maintained or increasing in strength, this is an indication that you are actually reinforcing them, regardless of your intention to punish them. This is yet another pitfall of punishment.

Before moving forward, take note of the examples provided above. If the subject exhibits a behavior in response to some other aspect of the environment (e.g., looking at a squirrel), you have likely increased distraction levels too quickly. In this case, and in all others, this is not a "fault" of the subject; the subject simply reacts to the stimulation present and the history of conditioning they have been exposed to in the past. The "fault," lies in the planning and/or execution of the training project. But again, you too, are simply a "functioning physiology," your behavior is merely a reaction to the environment, and a product of conditioning, just as much so as the dog, cat, bird, horse, spouse, etc. So, go easy on yourself as well, and simply move forward with appropriate adjustments to your training behaviors. Avoid or minimize extinction and subtracted punishment trials, by utilizing a graded approach—put the subject (and yourself) in a position to succeed!

Phase 3. Fluency Stage Training

Training for fluency generally includes these practices and generally in the order presented:
1. Begin fading prompts
2. Begin thinning the schedule of added reinforcement

3. Refine form, latency and speed parameters
4. Bring behavior under stimulus control
5. Proof against three D-parameters (distraction, distance, duration)
6. Discrimination training
7. Introduce release stimulus

Planning Fluency Stage Tactics

Once you have trained the behavior through the acquisition stage in a minimally distracting environment, at a minimal distance and requiring minimal duration, you are ready to begin transitioning to training for fluency of the behavior. *Fluency* is a characteristic of operant behavior featuring stability and reliability, that is, the contingency is strong, and subject exhibits the behavior smoothly, and without hesitation.

Broadly, you will be fading prompts, thinning the schedule of reinforcement, refining the behavior features, proofing against distraction, distance and duration, establishing the evocative stimulus, and working on generalization and discrimination. It also involves introducing a release word to replace the conditioned reinforcer for duration-appropriate behaviors. We gear all of the refinement procedures toward bringing the project to a point of reliability in everyday life, including maintaining it on a sparse schedule of reinforcement.

Training to achieve fluency involves quite a bit of work refining the form, latency and speed of the behavior, and then the distraction, distance and duration, which trainers are so familiar with, as well as carrying out generalization and discrimination training, and thinning the schedule of reinforcement. The process of working through training to strengthen stimulus control, work on generalization and discrimination and improve distraction, distance and duration is usually referred to as *proofing* in the animal training field. We carry out generalization training to ensure that an appropriate range of settings evokes the behavior, and by suitable variation in the chosen evocative stimulus. We carry out discrimination training to ensure that the chosen evocative stimulus reliably evokes the behavior, while other stimuli do not. We manipulate schedules of reinforcement to make the behavior reliable and resilient against extinction, and to allow for variable responding, so that the form of the behavior can be fine-tuned.

Through the training process, you will be working on a number of specific features of the target behavior. You can work on some of these simultaneously, but you should work on many of them individually. When you raise the criteria for distraction (i.e., concurrent contingencies), the distance between you and the subject or the duration the subject maintains the behavior, briefly reinstate continuous reinforcement, and then re-thin the schedule. Thin the schedule gradually and seemingly randomly, but avoid thinning the schedule so quickly that ratio strain occurs, and the behavior actually extinguishes. Settle at a schedule that is as sparse as possible, while maintaining the behavior at an appropriate level.

The following sections are not broken into separate antecedent and postcedent sections because a chronological treatment is more effective here.

Begin Fading Prompts

Once you have the behavior occurring relatively smoothly with minimal distance, duration and distraction, you can begin fading prompts, in particular, treats used as lures. If you used the prompt as an unconditioned reinforcer (as with treat luring), fading should begin right away. Establish the evocative stimulus that you wish to use indefinitely, ideally after the subject exhibits the behavior reliably in its final form, and you have established all criteria. In some cases, you might be able to begin using the evocative stimulus right from the beginning. This would be the case for simple behaviors that you are at least 95% sure that the subject will exhibit in its final form when prompted (for example, this is often the case with sit and down). In other cases, you may need to shape the behavior in order to ensure that its form meets the criteria before you can start using the evocative stimulus.

Fading prompts that include the unconditioned reinforcer (treats in hand) is usually the first thing you do in the fluency stage. This often starts as soon as you have a few good solid trials prompted and reinforced. A prompt that you use both antecedently as a prompt, and postcedently as the unconditioned reinforcer, is unique. With other kinds of prompts, the prompt and the evocative stimulus both take on stimulus control, and once the evocative stimulus becomes strong enough on its own, the prompt can be faded, usually quite easily. It is important to quickly fade prompts that also function as the unconditioned reinforcer because these prompts are particularly salient—they quickly take on stimulus control over the behavior and will be more challenging to fade. Since the prompt is not the stimulus that you want to act as the evocative stimulus for the behavior, it is best to fade it before this occurs. However, this usually works well because the first part of the fading procedure involves completing the same luring motion without the food in your hands. By the time that you complete this part of the fading procedure, the evocative stimulus has become even stronger, and then it leaves you in the position of fading the luring motion (now, without the added distraction of food in your hands).

One of the most common amateur training mistakes is failure to fade prompts that include the presence of the unconditioned reinforcer. In this situation, the prompt becomes a function-altering stimulus for the behavior. This has led to misguided criticism of using unconditioned reinforcers in prompting (e.g., treats as prompts), and even in training with food altogether. Although it is a specious argument, it results in people abandoning added reinforcement-emphasized training (and they have now eliminated one of the most powerful and easily managed added reinforcers from consideration). The problem is simply a failure to fade this particular type of prompt.

To fade prompts that involve using the unconditioned reinforcer to lure a behavior, take advantage of the momentum in training. Lure two or three responses in rapid succession, and then lure again, but this time without the treat in your hand (your hand will still smell like treats, and this prompt is partly olfactory). The subject will almost certainly exhibit the behavior again—reinforce this. In most cases, you can simply continue luring without the treat in your hands from that point forward. If this is ineffective, lure with treats in hand, and out of hand, in a seemingly random manner, but continue to increase the percentage of prompts without treats in hand until you have switched completely over to not using food to lure the behavior.

Once you have the treats out of your hand during luring, begin fading the luring motion prompt by transitioning from a lure motion to a hand signal. You can use the hand signal as the evocative stimulus while the form and latency of the behavior are fine-tuned. It is best to leave installation of the permanent evocative stimulus until the subject exhibits the behavior meeting most or all of the objective criteria. I described these two kinds of fading procedures in Chapter 2, however I will summarize them again here. Gradually, make the lure motion look more and more like the hand signal. In fading the prompt, take advantage of generalization, which will promote occurrence of the behavior in these similar stimulus conditions. If the subject responded to the lure motion, they should respond to something very similar to it, and then once they respond to that reliably, they should again respond to something very similar, and on until your lure motion is actually the hand signal. This is the ***prompt fading*** procedure.

Begin Thinning the Schedule of Reinforcement aka Stretching the Ratio

In the acquisition stage of training, you use continuous reinforcement in order to achieve a steady and rapid increase in the rate of the behavior. As soon as the subject exhibits the behavior reliably, you enter the training for fluency stage, and it is time to start ***thinning*** the schedule of reinforcement. We also refer to this as ***stretching the ratio*** of added reinforcement. This is important because continuous reinforcement is challenging to maintain over the long term, and variable intermittent schedules produce behavior that is much more resistant to extinction. Variable intermittent schedules also facilitate refining the form of behavior via shaping. The thinning process is a kind of roller coaster ride through the training project. You may begin thinning it, but each time you begin working on a new level of distraction, distance or duration, you will return to continuous reinforcement and begin thinning the schedule again. Any time you increase the level of challenge, increase the density of reinforcement, and begin thinning it again. This will promote high levels of responding and effective conditioning. The thinning process will promote resilience.

Refine Form, Latency and Speed

Once you have established the hand signal as the "working cue," use it while you fine-tune the form, latency and speed. As usual, work these criteria separately. If you need to change the ***form*** of the behavior, that is, the way the body parts move, identify the exact change you need. You may be able to use a prompt to get that variation if it is not too different from the currently exhibited behavior. If that is not viable, you may need to use a series of approximations to shape the present form of the behavior toward the terminal form you want. When you are simply refining form, it usually does not require too many approximations in order to achieve success.

Reducing ***latency*** means reducing the time between the presentation of the evocative stimulus and the beginning of the behavior. To reduce the latency, identify the current average latency and set a new slightly shorter latency criterion that will ensure that most trials will meet the criterion. Reinforce only responses that occur in that interval. If the interval expires and the response has not been exhibited, administer extinction or subtracted punishment, and try again. Once responding within that criterion interval is stable, set a new one that is a little shorter, and repeat this process until you have reached the final target criterion latency.

The *speed* refers to how quickly the subject exhibits the behavior—the amount of time it takes to complete the behavior once it has begun. We condition this, too, gradually by resetting the criterion in successive trials to ensure we reinforce most occurrences at the new criterion, and building on that speed.

Work through each of the above criteria one at a time. Once you have good form, latency or speed, you should maintain those criterion levels. In other words, when you make progress on form, latency or speed, do not backtrack or relax that progress. When working one of these criteria, you should bring the schedule of reinforcement back to continuous reinforcement and thin it again (it will go much quicker on subsequent runs through thinning), and it is a good idea to relax the distraction, distance and duration parameters as needed (to be discussed below). However, as indicated above, once you have established the desired form, latency or speed, continue to maintain these without backtracking, but relax the distraction, distance and duration parameters and schedule of reinforcement.

To illustrate a problem of timing when training a new behavior, let us assume, for example, that you have a behavior that generally seems to take too long to complete. In practice, this likely means at least one of two things are involved. First, the behavior might be deficient if the subject begins exhibiting the behavior too long after you present the evocative stimulus, even if the behavior itself is exhibited in a quick motion once it begins (i.e., poor latency). Second, the subject might begin moving right away but exhibit the behavior itself quite slowly (i.e., poor speed). Whether sub-criterion latency or sub-criterion speed, you should work on it with a graded approach. Let us say the latency is deficient, and the body part currently begins moving only after three seconds, on average, after you present the evocative stimulus. In that case, keep the distraction, distance, and duration levels relaxed/minimal, and return to continuous reinforcement. Set the latency at ≤ 3 seconds, and begin running through trials until you get four reinforced trials in a row. That might happen right away, since the criterion was set at a point where the subject is already usually exhibiting it. Now set the criterion at ≤ 2.5 seconds. Reinforce only responses that occur within that interval, and extinguish non-criterion responses. Once you have achieved approximately four consecutive criterion responses, reset the criterion to ≤ 1.5 seconds and repeat the process at this reduced time. With this approach, gradually decrease the latency criterion until the subject is exhibiting the behavior to that criterion. If you are getting too many extinction trials, use smaller criterion jumps. For example, your first jump might be from ≤ 3 to ≤ 2.75. If the behavior was simply being exhibited too slowly, you would set a time unit for beginning to end of the behavior, and gradually, as described above, reduce it until the behavior meets the target criterion. Once you have achieved your goal for that criterion, put the behavior back on a thinning intermittent schedule for a few trials, and then work on another one, at first incrementally relaxing the distraction, distance and duration criteria.

Training is a constant balancing act between maintaining progress on the one hand and not pushing too hard on the other. If you set the criteria too low, you will make very little progress, and the subject will become bored and easily distracted. If you set the criteria too high, on the other hand, there will be too many extinction or subtracted punishment trials, and the subject will become frustrated. One result of such frustration is the subject becoming anxious or apprehensive, behaving excruciatingly slowly—the more intensely one demands a behavior, the slower it seems the subject moves. Finding just the right balance in the pace of training for the particular individual requires careful judgment. Keep training fun, upbeat, and reasonably achievable, yet challenging, but not overwhelmingly so. Indications of boredom and distraction or frustration are indications of a problem with the pace of your training. With experience, you will become conditioned to manage this pace effectively and smoothly during your training sessions.

Bring Behavior under Stimulus Control

At this point, the behavior is actually under stimulus control but usually only by the temporary evocative stimulus used through training. In most cases, this is the hand signal. At this stage, it is time to bring the behavior under stimulus control of the cue that that you want to establish as the permanent evocative stimulus for the behavior, usually a vocal stimulus.

Once the behavior is being exhibited in the appropriate form and with an appropriate latency (that is, you have the behavior the way you want it), this is a good time to transfer stimulus control from the hand signal to a vocal cue that you want to use for this behavior with a ***prompt delayed*** procedure. To do this, present the new stimulus (the vocal cue), followed immediately by the current stimulus (the hand signal), which will evoke the behavior and generate the reinforcement. Repeat this sequence (i.e., contingency) several times. The new stimulus should take on stimulus control and come to evoke the behavior without use of the previous stimulus, and you can then drop the previous stimulus from the sequence. Continue to reinforce the behavior throughout the process.

Proof against the Three D-Parameters (Distraction, Distance & Duration)

The ***three D-Parameters (or The Three D's)*** of training are distraction, distance and duration (Donaldson, 2005, p. 144). We keep all of these parameters at minimal levels during the acquisition stage of training in order to set the trainer and subject up for success. You will begin to improve these parameters one at a time in the fluency phase of training. To begin this process, it is vitally important to success that you work only with one parameter at a time, relaxing the others that you have worked on, when you introduce a new one. For example, if you started by working the duration of a sit at 30 seconds with minimal distance (standing right in front of the subject) and minimal distraction (no distractions around at all), and then you decide to work on the distance parameter, take the duration back to just one second and work the distance up gradually, say to 20 feet. Then, you have a choice. You can work both distance and duration up together, gradually, or you can work distraction first and then work on the combinations. Let us say you decide to work on distraction first. You will require only a couple of feet distance and a second or so duration and you will gradually work the distraction parameter up. Once you have all three worked up well individually, you will start combining them and work them up gradually together. You might be able to work all three up together but it is more likely that you will need to keep one relaxed and work two up together and work the combinations before working all three together. The main point here is that by relaxing other parameters and only working one at a time up gradually, as per a graded approach, you will achieve a high degree of success. It may feel like it takes a long time to work them all individually in this manner, but the high rate of success will actually help achieve your objective much quicker than the failure-rich tactic of doing them all at once.

Plan to proof behaviors against the three D-parameters to a higher criteria level than you expect to require in maintenance. Consider what levels of each of the parameters can be expected/required after training and proof the behavior against even higher levels to ensure that in maintenance, the behavior is less likely to deteriorate.

When working on a new D-parameter, reinstate continuous reinforcement and re-thin the schedule after it is smooth and reliable. Be sure to do this any time you introduce a new level of a D-parameter and each time you introduce a new D-parameter. We restrict the rule of working one D-parameter at a time, and when starting a new level or D-parameter, relaxing the other D-parameters as well as the schedule of reinforcement to these criteria. These parameters are worked in a similar manner as training to improve form, latency and speed, but remember that you should not backtrack or relax form, latency and speed, as is done with the three D-parameters. In both cases though, it is important to relax the schedule of reinforcement and re-thin it.

Distraction

Distraction refers to the imposition of competing contingencies into your training environment. In the acquisition stage, it is important to keep distraction minimal. However, in order to train a behavior to be reliable in everyday situations, gradually introduce distraction throughout the fluency stage, to ensure that you can reliably evoke the behavior, even when alternative competing contingencies are vying for control over the subject's behavior. When attempting to identify distractors, look for reinforcers (or punishers). Is there something else in the environment that the subject pays attention to or behaves to contact (or even escape)? The presence of other people or other animals can distract the subject from the contingency you are training. What these others are doing can contribute to the magnitude of the distraction. If the other person is sitting, faced away from the subject, and not engaging the subject in any way, this will be far less distracting than if the person is calling the subject or holding a toy or treat, etc. These are typical variables that you will manipulate as you incrementally increase distraction levels during your training sessions. In many cases, you can use access to the distractors as the reinforcer for exhibiting the behavior you are training. Always consider what other contingencies exist in the training environment, how much and what kinds of distraction you can have present and still achieve the training you need to achieve, as well as how you can increase the intensity of the distraction by very small increments to ensure you are successful. Throughout proofing against distraction, you will always be attempting to cover as many situations as you can and successfully modify the intensity of exposure.

Distance

Distance refers to the distance between yourself and the subject during training. In the acquisition stage of training, you keep distance at a minimal level and increase it later in the fluency training stage, where appropriate. In many instances, you will eventually want to be able to cue the behavior from further away than you did in your initial acquisition training. You achieve this gradually. Often, behaviors cued from a greater distance, without any specific distance training, will evoke coming to you in order to then exhibit the behavior right next to you. This is especially common with "sit," and it can be very useful to work distance into this behavior. For example, if there happens to be traffic between you and a dog, you need a way to ensure the dog will *not* come to you, but rather sit right where they are when the evocative stimulus is presented and remain there until released. In fact, you may wish to establish an alternative evocative stimulus that is visual rather than auditory for such occasions that extreme distance or noise prevents you from vocally evoking the behavior effectively; many trainers use raising their right hand high into the air to evoke this "emergency sit."

Duration

Duration relates to how long the subject maintains the behavior in position. As with the other parameters, maintain minimal duration in the acquisition stage, and increase it gradually in the fluency training stage. You will recall the duration schedule from Chapter 2. This feature relates to thinning that schedule in order to increase the duration of the behavior. Not all behaviors will call for increasing duration but some will. "Sit" and "down" are good examples of behaviors you should train the subject to continue holding beyond the point when they actually achieve the position.

Discrimination Training

The introduction and fading of prompts is a form of discrimination and generalization training in the sense that they increase and decrease the number of stimuli that evoke the target behavior. Further discrimination and generalization training is required though. Recall that discrimination is not something an individual "does" but rather a feature of a pattern of responding. Discrimination occurs—the subject does not carry it out or perform it.

One of the tasks at this stage involves inter-stimulus discrimination. Once you have at least three behaviors trained to this level, you will be able to evoke them in random order, so that you can extinguish merely changing position rather than exhibiting the behavior for which you present the evocative stimulus. For example, if a dog is seated and you cue "down" and there is a 50% chance the dog will "stand" and 50% chance the dog will "down," that means the evocative stimulus is only cueing a change in position rather than the behavior you want brought under stimulus control. Present cues for different behaviors in a random order, reinforce criterion responses, and extinguish non-criterion responses (i.e., responses that do not match the stimulus presented).

You should also narrow the range of evocative stimuli by extinguishing responding to similar yet unacceptable stimuli. Present the specific evocative stimulus, interspersed with similar stimuli. Differentially reinforce responses to the evocative stimulus and not to these other stimuli, thereby reducing the range of stimuli that will evoke the behavior. Use continuous reinforcement when working on discrimination.

Although some generalization will have occurred before this stage, there will be further generalization training to achieve. This usually involves presenting the evocative stimulus in different ways, in different places, or having other people present the evocative stimulus. Ensure that the subject exhibits the behavior reliably in various locations and under various conditions. For example, a good challenge in a group class for dog training that will drive this point home is to ask for the guardians to evoke a sit, which they will achieve with no problem. Then ask them to face completely away from the dog, or to lie down on their back, and present the evocative stimulus. Many of the dogs will fail to respond or be slow to respond. This demonstrates the importance of generalization.[55]

Together, discrimination and generalization training will promote occurrence of the behavior in the appropriate circumstances and not in others.

Introduce Release Stimulus

Once the subject exhibits the behavior fluently, it is time to discontinue using a clicker. We only use the clicker for training new features. For behaviors that involve maintaining a position—duration appropriate responses—you will need some way of releasing the subject. In order to transition from using the clicker to a release stimulus, simply begin using a release phrase such as "you're free," instead of the click. Follow this with reinforcement if the subject breaks position. If the subject does not break position when released, simply provide the unconditioned reinforcer in a way that requires the subject to break position. You can use the release stimulus to end a behavior, or you can simply present a new evocative stimulus for some other behavior. The release stimulus will become a conditioned reinforcer, especially if you follow-up with treats or other unconditioned reinforcers at least occasionally, effectively replacing the clicker.

---

[55] The novelty in such an arrangement may evoke approach and investigative behaviors, which would be distractions as well. This kind of training will address both, distraction in this way, as well as generalization.

Phase 4. Maintenance Stage Training

Working toward Maintenance of Fluency

Once the subject exhibits the behavior fluently, transition from active intensive training to daily *maintenance* of the conditioning. That is, once the evocative stimulus, under an adequate range of conditions, reliably evokes the behavior in its final form, speed and latency, even when distractions are present, and the schedule of reinforcement has been thinned appropriately, you are ready to maintain what you have achieved, so that it does not deteriorate.

Another component of maintenance training is transitioning from contrived trainer-mediated reinforcers to less contrived nontrainer-mediated reinforcers. This will help maintain a behavior on a very thin schedule. When you use added reinforcement to train a behavior, the occurrence of the behavior comes to elicit emotional responses that can help maintain the behavior. This endoreinforcement (i.e., reinforcement generated inside the body) contributes to maintaining the behavior, even when the trainer-mediated reinforcers are sparse. The behavior itself generates satisfactory reinforcement within the subject's body, and occasional trainer-mediated reinforcement helps maintain that contingency. Most professional dog trainers are familiar with the phenomenon of dogs rejecting treats, seemingly in order to get back to the training more quickly. This happens when occurrence of the behavior, and the changes that it generates in the environment, reinforce the behavior more effectively than treats.

A key task in this phase is to settle on a schedule of added reinforcement that will maintain the behavior adequately and is realistically achievable in the long run. If the behavior ever deteriorates, refresh the training. Also, ensure you refresh the training in different environments from time to time.

Constructing a Systematic Training Plan: Putting it All Together

We have addressed the components of a systematic training plan, as well as the type of training plan needed for a given training project (i.e., a short and simple plan versus an elaborate and complex one), dictated by the characteristics of the project. For training a simple behavior, identifying the target behavior objective and a few comments regarding how you will generate the behavior and reinforce it may be sufficient. When you will need instructions that are more detailed for a client, to increase confidence and accountability, or where the training will be complex, as in shaping or chaining projects, you should include more detail in the written plan. In this section, I will exemplify the components of a more elaborate training plan. I will use a simple behavior here for illustration purposes, but more details are included in order to present all of the features that we have explored.

Every training plan should contain the following information:

- Subject's name and breed

- Client's name

- Client contact information

- Formal behavior objective

- Assessment of proficiency

- Contingencies

- Conditioned and unconditioned reinforcers

  - Acquisition tactics

  - Antecedent tactics

- Postcedent tactics

- Fluency tactics

The first three items are simply to help keep track of the case, and this is only necessary if you are working with a client.

In the case of simple training projects, provide a formal behavior objective. In the case of a behavior chaining project, provide a formal behavior objective for each behavior in the chain. In the case of a shaping project, provide a formal behavior objective for the terminal behavior and a list of behavior approximations. I will discuss these topics in detail below. Where necessary, prepare any charts or graphs that will help you track the behavior over time.

Provide a statement regarding your assessment of the subject's current proficiency in exhibiting the target behaviors or components of it. What components of the training do you need to carry out and which ones do you not need to carry out?

Diagram the contingency you will condition. For instance:

$$\text{"Paw"} \rightarrow \text{Dog gives paw} \rightarrow \text{Treat}^{(+R)}$$

The evocative stimulus should be the final evocative stimulus you want to control the behavior. The behavior should be concise when diagrammed, as it will refer to the full behavior objective. The reinforcer should identify the unconditioned reinforcer you will use during the training process, through the acquisition and fluency stages. You may include the conditioned reinforcer as another term, or include it along with the single consequence term—most trainers simply forgo the conditioned reinforcer altogether when diagramming the training contingency. Diagram each behavior in a chaining project. If you are installing a replacement behavior to resolve a problem behavior, start with a functional assessment and the resulting contingency diagram for the behavior as it currently is and provide the replacement contingency you aim to train. If there are any other competing contingencies of note, identify those as well.

Present a statement identifying any antecedent tactics you will use to generate the behavior.

Present a statement identifying the postcedent tactics that you will use, including whether you will require added reinforcement only, graded differential reinforcement, shaping, or chaining. Include details on each behavior in a chain and approximation details for a shaping project.

Present a statement regarding what schedules of reinforcement you will utilize in the fluency stage and how you plan to proof against distraction, distance and duration. Identify those variables and how you will manipulate them to incrementally increase their intensity.

This does not usually need to be a very long, nor detailed, document in the case of simpler behaviors, but it is a good idea for initiating training projects to evoke preparation behaviors, even with simpler projects. The practice will become handy when you face a more complex training project that will require shaping approximations, and/or chaining techniques. The more extensive plans are most useful when you will be providing clients with training responsibilities that you cannot supervise. The clarity of a plan can be very helpful for them as they work with their animal.[56]

Below is a sample training plan for "give-a-paw," a simple behavior useful for illustration purposes. You will require much more elaborate information for more complex training projects, and for simple projects, you might elect to omit a few sections, particularly once you have extensive experience training the chosen behavior.

Training Plan
Subject's name, breed, age: Jake, Doberman Pinscher, 8 months
Target behavior objective

Behavior Objective

S^{FA}: Presenting hand, palm up in front of dog.

S^{Ev}: Vocal "Paw."

Behavior: Jake will raise his left paw off of the ground and place it on the person's palm.

Consequence: Treats (veggie burger bits) and continued social contact.

Criteria: Latency—2sec.; duration—1sec.; frequency—100% through 10 trials; distraction—various*.

* Distractions inside the house to include various people around, engaged in various activities including handling toys and talking, as well as with other dogs around, playing. Same thing but outside of house as well.

Assessment of proficiency
Jake currently lifts a paw slightly when people are near, and will usually place paw on hand if I present the hand right beside his paw, at a frequency of 80%. Not under stimulus control of a vocal cue, and hand needs to be lower to ground and closer to paw than acceptable.

Contingencies

---

[56] When having clients work on training on their own, ensure that you demonstrate the procedure, discuss how they are to progress, and provide them written instructions for their reference. Have clients demonstrate the procedure, and remediate any deficiencies before leaving them to train unsupervised. The chapter on training humans will provide detail on this topic.

"Paw" → Dog gives paw → Veggie burger bits[(+R)]

Conditioned and unconditioned reinforcers

Clicker as conditioned reinforcer and veggie burger bits as unconditioned reinforcer.

Acquisition tactics

Prompt with hand presented near paw, light touch to ankle if necessary. If unsuccessful, plan for shaping project. Fade later by gradually having hand further up and away from Jake.

Establishing operations: Train before meals and utilize most effective reinforcers only for training through acquisition and fluency.

Postcedent tactics

Graded differential reinforcement (shaping if necessary). Minimally aversive two–second extinction for non-criterion responses. Change to five–second extinction and subtracted punishment if necessary.

Fluency tactics

Fade food prompt right away.

Utilize continuous reinforcement for the first several responses, and once stable, an intermittent variable ratio schedule, gradually thinned based on progress.

Work latency if required. Work duration first up to one second, then work distraction gradually, including other people present, then other dogs, then outside of the home in yard, and then away from home. With each change, return to continuous reinforcement and re-thin.

10 Laws of Training

1. Plan your training ahead of time, including a target behavior objective, the strategy you plan to take and procedures you will need to utilize as well as any lists of approximations or task analyses. Break the project down into small manageable steps.
2. Keep training sessions short, so that participation remains reinforcing and the subject does not satiate with respect to the unconditioned reinforcers used or with respect to social contact.
3. End sessions with successfully reinforced trials, even if you have to relax distraction, distance and/or duration in order to generate a series of successful responses, and ensure the last few trials generate reinforcement. Ideally, end the session to engage in some other reinforcing activity.
4. At the beginning of each new session, review some of the progress made in the previous session—this will ensure smoother transitions in new sessions.
5. Present the evocative stimulus once only. If the target behavior does not occur, reset the environment with a brief lure to a different position or cue another preparatory behavior, identify why the target behavior did not occur, manage the environment to ensure occurrence of the behavior in the next trial and proceed to the next trial.
6. Participation in training should always remain highly reinforcing. Use tone of voice, energy level, pace of training, direction of attention, prevention of failure etc., to keep the training reinforcing.
7. Start with minimal distraction. Arrange the environment to make occurrence of the target behavior, rather than other behaviors, highly likely, putting the subject and yourself in a position succeed. You can gradually introduce greater distraction at a later stage in the project.
8. Before moving to a new phase in a training procedure or raising a criterion parameter, ensure the rate of the behavior has stabilized at the current phase or step to ensure the subject's nervous system—the thing being conditioned—can keep up with the pace.
9. When you begin working a new D-parameter, relax other D-parameters and the schedule of reinforcement. You can combine the D-parameters after you have worked each separately. However, when working form, latency and speed, maintain the progress made—do not relax these features when introducing a new challenge.
10. Once the behavior is occurring reliably, transition to generalize training during everyday life rather than specific contrived "training" sessions. Generalize the reinforcers as well.

Continuing Education

Courses provided through The Companion Animal Sciences Institute at www.CASInstitute.com address all of the topics covered in this chapter.

The Science and Technology of Animal Training

CHAPTER 6. ADVANCED TRAINING TECHNIQUES

Behavioral Objectives

The objective of this chapter is to measurably expand the reader's repertoire of behaviors in relation to describing and relating the principles of behavior. Upon successfully integrating the concepts outlined in this chapter, the reader, where exposed to contingencies to do so, will accurately:

- Define shaping and describe under what circumstances this procedure, as opposed to others, would be most appropriate

- Determine when prompting would be most appropriate

- Describe a systematic series of steps necessary in planning a shaping project

- Prepare a list of behavior approximations

- Define chaining, differentiate this from sequencing and describe under what circumstances this procedure, as opposed to others, would be most appropriate

- Differentiate between forward chaining and backward chaining, and determine which is most suitable for particular behavior objectives

- Conduct a task analysis to determine which behaviors will constitute the chain.

Introduction

Sometimes, the target behavior does not occur in the subject's behavior repertoire at all in its final form and you cannot readily prompt it. Other times, the objective calls for a series of behaviors, exhibited in sequence, with only one trainer-mediated cue at the beginning. Other times, the objective calls for several behaviors to be well trained such that the trainer can cue them in any order very quickly and the subject will move to the next behavior seamlessly upon completing the behavior before it. These are examples of situations in which more advanced training procedures are required, rather than simply prompting and reinforcing the target behavior. In this chapter, I will describe the three most powerful advanced training procedures used for such situations: shaping, behavior chaining, and sequencing.

Shaping

Definition and Appropriate Uses

Shaping is a procedure that involves the differential reinforcement of *successive approximations* of a target behavior. Shaping is a special type of differential reinforcement. All differential reinforcement procedures change the *rate* of a target behavior. Shaping changes the *form* of a terminal behavior, making it a unique type of differential reinforcement. Shaping involves a series of standard (i.e., rate-changing) differential reinforcement procedures carried out in succession. You increase the rate of an approximation of the terminal behavior through differential reinforcement, and then use the same process with another, closer, approximation to the terminal behavior. You repeat this process until the subject exhibits the terminal behavior. The *rate* of each approximation is increased one at a time, allowing for training the next approximation until you achieve the terminal behavior *form*.

Thus, not every response is precisely the same. This variability in the different properties of a response class from one occurrence to another facilitates selection and reinforcement of some variations over others. The initial approximation is set such that it will occur frequently enough for the trainer to reinforce it. The unaugmented environment may evoke the approximation, or the trainer may prompt it. Once that response class has a history of reinforcement, it will occur more frequently. Then, you put that response class on extinction in favor of the next approximation. A common byproduct of extinction is increased behavioral variability. Ensure that the second approximation is set such that it will fall within the range of variant response class forms resulting from the extinction process. In other words, the next approximation cannot be too different from the one you just finished establishing. You then target this new approximation for reinforcement and its rate is increased. The shaping process continues until the subject exhibits the terminal behavior fluently. We refer to this reiteration through cycles of differentially reinforcing approximations toward a terminal behavior form as shaping.

Shaping is an appropriate procedure for single discrete behaviors that do not occur frequently, and where you cannot readily prompt the behavior its final form. If you can prompt the behavior in its final form, then a single differential reinforcement procedure is preferable. If the behavior is actually a complex series of discrete behaviors that you want exhibited sequentially with a single trainer-mediated evocative stimulus at the beginning of the sequence, chaining is the most suitable procedure, although each discrete behavior in the chain must be trained and may involve shaping. We also use shaping in many simple training projects in order to fine-tune the form of the behavior as needed.

When considering whether a behavioral episode involves a series of approximations of a terminal behavior or a series of discrete requisite behaviors, determine whether the initial behaviors are actually approximations of the terminal behavior or simply requisite behaviors. For example, the trainer wants to train a subject to turn in a circle and the subject does not currently turn in a full circle at any time. The decision needs to be made whether a series of discrete behaviors is required to train the subject to turn in a circle (in which case, chaining is required) or whether there are approximations toward this terminal behavior (in which case shaping is required). The initial behaviors might involve turning the head, followed by turning the head to the point that the shoulders also turn, followed by turning the upper body sufficiently to turn the hips, followed by turning to the point that a step is required, followed by a turn of 120 degrees, followed by a turn of 180 degrees, followed by a turn of 270 degrees, followed finally by a full turn of 360 degrees. This is a clear example of a single terminal behavior for which approximations are required and rendering a shaping procedure necessary.

An alternative example involves putting on a baseball cap. The trainer may need to train the subject to walk to the table where the cap is sitting, followed by picking the hat up, followed by putting it on their head. This is a clear example of a chain of discrete behaviors. Walking to a table is not an approximation of putting a baseball cap on one's head; it is a prerequisite behavior.

As mentioned, one way to help ensure you are clear on whether you require a single shaping plan or a chaining plan is to consider whether the initial steps involve training a single dimension of the final behavior. Turning in a circle requires the subject to keep going in the circle whereas putting on a baseball cap that is not already in one's hand, requires completely different dimensions; walking to the table, picking up the cap, and then placing the cap on one's head. The trainer may still need to shape some or all of the requisite behaviors but they must first determine whether the shaping plan alone or a behavior chaining procedure is called for. The definitive test, however, for determining whether a behavior episode is made up of approximations or a series of discrete behaviors is whether the early segments are always required for the behavior to occur or not. Turning a bit to one side is always necessary to achieve turning all the way around in a circle. Walking to a table is not always required for putting on a cap.

Shaping is a postcedent procedure because it manipulates what comes after the behavior. That is, shaping involves manipulating consequences in order to change the likelihood of the behavior on subsequent occasions. Shaping, itself, does not specify any antecedent conditions; it does not specify whether a trainer-mediated stimulus evokes the behavior or not. The somewhat colloquial term *"free-shaping"* indicates a shaping program in which prompting is avoided. There is value in avoiding prompts in some shaping projects, particularly when the trainer is attempting to reinforce creativity[57] and persistence in general, as well as the specific behavior in question.

Trainers should consider some limitations to shaping. Cooper and colleagues (2007, p. 425) point out that shaping can be time consuming, progression through approximations is often erratic rather than linear and smooth, the trainer must be extremely attentive for indications of a need to change criteria, and it can be misapplied and promote problem behaviors. In addition, in order for shaping to executing well, the trainer requires extensive experience and proficiency in a number of requisite trainer behaviors. If one does not execute shaping well, the subject (and the trainer) can become frustrated, which can cause significant disruption to training and indeed other aspects of the subject's life.

---

[57] "Creativity" refers to novel and productive responding.

Planning Behavior Objectives and Lists of Behavior Approximations

Preparation for a shaping project generally requires more planning than for most other procedures. As in any training project, one must establish a behavior objective. First, define the evocative stimulus and an operational definition of the terminal behavior, including a suitable measure for it (e.g., specific rate, frequency, magnitude, and/or duration). Finally, set a time point at which you expect or need to achieve the terminal behavior. The time point should be tentative and flexible, but it is good practice to set a time point nonetheless. This might be after a specific number of trials, sessions, opportunities to exhibit the behavior, or number of minutes, hours, or days. Once you define the terminal behavior, prepare the list of behavior approximations.

Determining the list of behavior approximations requires careful judgment and flexibility. The best way to start is with observation of subjects who currently exhibit the behavior fluently. Find a subject who already exhibits the behavior, and either observe them exhibit the behavior, or better yet, video-record it and watch it repeatedly, perhaps in slow motion. You might be able to find such a video posted on one or more video sharing websites such as YouTube®. By watching the video of the behavior being exhibited repeatedly, you will be able to prepare a more accurate list of approximations than if your only option is to visualize the behavior occurring. With experience, you will become familiar with the approximations useful for some of the more common behaviors.

The first approximation must occur frequently enough to provide adequate opportunity to strengthen the behavior. You need to be able to easily prompt it or otherwise have it occur, so that you can reinforce it. Each approximation needs to make the next approximation likely to occur frequently enough to strengthen it also, whether it occurs within the normal range of variability when behaving the approximation before it or you prompt it. If the trainer is not able to quickly make the next approximation occur, it will likely be frustrating for both the trainer and the subject, and training will progress slowly. Remember, a high rate of reinforcement is required to maintain pace, high levels of responding, and prevent frustration. The approximations should be large enough that the subject is not likely rapidly skipping multiple listed approximations at a time, but small enough that progress remains smooth and efficient. Err on the side of smaller approximations because frustration is particularly disruptive to training. If you are not reinforcing at least every few seconds, your approximations are probably too large. Break them down into small enough steps such that that you can reinforce every few seconds to maintain a high rate of reinforcement. You should design the approximations to be as gradual as possible so that training moves smoothly from approximation to approximation. Ensure that each approximation is operational (i.e., describes specific body part movements or very specific function).

Next, for each approximation, it is good practice (though not always necessary) to break it down into a couple small sub-approximations, in case you need to quickly utilize them to prevent frustration. In your written training plan, you can include them as a subcategory that you can ignore unless you need them.[58]

At a more advanced level, look at each approximation for functional alternatives. For the particular behavior identified, is there an acceptable alternative response class form that still results in progress toward the terminal behavior? In other words, prepare yourself now for any step along the way where you might not immediately get the specific approximation you listed in your plan. It is possible that you will get a variation or response class form that is still acceptable, because it functions as a suitable approximation toward the terminal behavior. This may not be possible for all terminal behaviors and usually is more applicable to more complex terminal behaviors. You may need to continue a new branch of approximations from the alternative behavior to your terminal behavior. Otherwise, note where it connects back to your initial list of approximations. Perhaps you can put these alternative operants in a column to the right. This step is one that experienced professionals can usually achieve on the fly during training sessions. It is good practice to start planning with regard to these issues before you attempt to truncate the process.

Do not confuse the list of behavior approximations for a set of training steps you plan to use in training the behavior. The list of behavior approximations is a list of the subject's behavior approximations toward the terminal behavior you aim to train. This list starts with the first approximation and includes each successive approximation in order. This list tells you what to reinforce, and then once the behavior approximation is exhibited reliably, it is then targeted for extinction, and the next approximation toward your subject's terminal behavior is reinforced. The behavior approximation list describes only the *subject's* behavior and never the trainer's behavior. For instance, in training a horse to spin in a circle, you might set 'moving head to left' as the first approximation and 'turning head further to the left' as a second approximation and on through the horse's movements necessary to accomplish a full turn. A list that starts with something like "charging the clicker," or "I prompt…" cannot be a list of behavior approximations.

Implementation of the Shaping Plan

Shaping is an advanced training skill that requires practice in order to carry it out effectively and efficiently. A poorly executed shaping program can be frustrating for the subject as well as the trainer.

Managing smooth progression through the approximations involves several fluent trainer behaviors, including proficiency in making quick adjustments to both the plan and the techniques used at a given moment. For example, if you need to break a step down into smaller steps, or you need to add a prompt, you need to recognize these requirements quickly and consistently, and you need to make the adjustments quickly and effectively.

The list of approximations will be your guide to training, but remain flexible and prepared for off-list behaviors, be they acceptable ones or unacceptable ones. If the

---

[58] For example, in your outline, these sub-approximations can be on a separate, indented line under the main approximation.

training is going smoothly and you are experiencing very few extinction trials, there are no long periods of time in which the target behaviors fails to occur, and the subject is focused and participating, it is likely that you are right on track. If, on the other hand, the subject is losing interest, failing to exhibit criterion behaviors, or is exhibiting non-criterion behaviors, it is likely that you are moving at an inappropriate pace. If you move too slowly, the subject may become exposed to competing contingencies in the environment (i.e., lose interest). If you move too quickly, a greater number of frustrating extinction trials may occur. When beginning an approximation, there will necessarily be some extinction trials, but if you set the approximation and the pace appropriately, there should be very few. You should quickly progress to 95–100% reinforced trials.

If you reach a challenging part and the subject becomes frustrated, it can be useful to take a break. Toss a treat to change the focus temporarily away from the contingency at hand and reset the environment. Take the time to identify the variables that are causing the problem, so that you can fix them. Perhaps call the subject to you, request a simple previously well conditioned behavior, and provide an enthusiastic reinforcement package (treats, praise and energy). If need be, take a small break to do something different, be it fun or relaxing before proceeding again.[59] Occasionally, it seems as though a "frustration loop" develops and has the potential to deteriorate the training. Taking a brief break can be immensely helpful in this respect. If a subject responds to training with escape behavior, this bodes very poorly for effective training. If that occurs, you will need to take it very slowly, ensure success at first and ensure it is fun.[60]

It is important to remain at any particular approximation long enough to establish stability in responding, but not longer—mastery and maintenance are reserved for the terminal behavior. Moving to the next approximation before the behavior stabilizes will result in a greater number of extinction trials in the approximations to follow, and frustration will result. Remaining at an approximation too long will result in a conditioning history that is too strong and will require more extinction of that approximation in preparation for the next approximation. A good rule of thumb is that the training in any given approximation should proceed until the behavior has stabilized, but not longer, and that the training is progressing smoothly. Where it is not progressing smoothly, make necessary adjustments. Are you expecting too much or not enough? Are you moving too quickly or too slowly? Boredom and distraction are indicative of going too slowly, and frustration is indicative of moving too quickly.

If, when you move to the next approximation, the subject does not exhibit the behavior quickly, you need to be prepared. There is something to be said for allowing the subject to "figure it out" on their own—that is how they become conditioned to exhibit persistence and creativity. However, if it goes on too long, they can become frustrated, and this will definitely disrupt training. If the subject does not exhibit the behavior quickly, go back and work through several trials of the previous approximation, and then insert an intermediate approximation or two (these should

---

[59] Always be on the lookout for subtractively reinforced behaviors that will be disruptive to training. If a non-criterion behavior results in an end to a session that is aversive, you can expect that behavior to increase, which will be a behavior that is disruptive to training. The goal is for training to be reinforcing. If it is not, your area of focus should be solving this problem. Also, keep training fun so that escape behaviors are irrelevant!

[60] For rehabilitating this type of situation, you may find my book *Empowerment Training* helpful.

already be planned for in your list of approximations). You may have chosen an approximation that was too difficult. In this case, you can insert an easier approximation, which will more likely allow you to reach the target approximation. Then, you can continue moving forward with your plan. Having a plan will make it easier for you to quickly come up with intermediate approximations when needed. A well-placed prompt can also get you "over the hump" and back on track.

Chaining

Definition and Elements

A ***behavior chain*** is a sequence of discrete responses in close temporal succession in which each response produces a stimulus change that functions as a conditioned reinforcement for that response and as an evocative stimulus for the next response in the chain, and the entire chain of behaviors is maintained by a single unconditioned reinforcer delivered after the final response in the chain (Cooper et al., 2007, p. 436 & 690).

Notice that completion of each behavior serves a dual function; it functions as a conditioned reinforcer for the behavior the subject just exhibited, and it acts as the evocative stimulus for the next behavior in the chain. The opportunity to exhibit the next behavior in the chain reinforces the behavior, and this occurs for each link in the chain until the final behavior, which produces the trainer-mediated unconditioned reinforcer. This final reinforcement maintains the chain and the conditioned reinforcers that make it up. In other words, a behavior chain is a series of behaviors with one trainer-mediated evocative stimulus at the beginning, and unconditioned reinforcement after the last behavior—a series of behaviors is evoked.

To train a chain of behaviors, you break the project down into component behaviors (links) that you first train as separate projects, and then you link them together once the subject exhibits each individual behavior fluently. We refer to this linking process as chaining. For example, training a subject to retrieve involves training the subject to run to a thrown object, *and then* grab the object, *and then* return to you with the object, *and then* drop the object into your hand or at your feet. This complex sequence of behaviors becomes quite easy to train if you train each component behavior and then chain them together rather than trying to achieve all of the behaviors at once. Often, we use simple differential reinforcement to train each component behavior, but you may need to shape some component behaviors. Training each discrete behavior is a separate training project, requiring the use of procedures appropriate to that particular behavior. Behavior chaining is really about how to bring the whole sequence of behaviors, as a sequential set, under discrete stimulus control.

There are no trainer-mediated interjected cues (the evocative stimuli between each behavior are directly created by the completion of the behavior exhibited before it in the series of behaviors). Interjecting verbal or physical prompts or cues in the sequence of behaviors is not the same as chaining. The section on sequencing below elaborates on this distinction. I raise the issue here so that you may proceed without any misunderstanding. Prompts are not to be confused with interjected evocative stimuli. You may use prompts initially in training the component behaviors, or in linking them together, but they are faded as soon as possible.

Once you make the determination that chaining is the most suitable training method for the target behavior, you will need to perform the following four steps:

1. Construct and validate a task analysis.
2. Assess the level of proficiency of requisite behaviors.
3. Train the component behaviors.
4. Chain the component behaviors together.

Constructing and Validating a Task Analysis

The first undertaking in the planning stage of a chaining project is to construct and validate a *task analysis*. Constructing a task analysis involves breaking a complex behavior (i.e., multi-behavior chain) event into a sequential set of discrete component behaviors and describing each component behavior. The most useful way to construct and validate a task analysis is to observe a proficiently exhibited behavior chain. It is a good idea to observe multiple proficient occurrences of the sequence of behaviors to identify any useful variations that can inform the task analysis. Observing video is helpful because you can view the same behavioral episode multiple times. You can construct the task analysis from simple covert visualization of the sequence of behaviors, but this method is less reliable than planning based on actual occurrence of the sequence. In other words, if you "wing it," you are more likely to run into unforeseen problems during the training. You can also construct a tentative task analysis and run a quick pilot session, working through the sequence in order to refine the task analysis. From this careful observation, you can prepare a diagram that identifies each discrete behavior. The example below, involving a retrieve related sequence of behaviors, illustrates the product to be generated:

Given that a ball is thrown and the trainer says "git yer ball," the subject will:

- run to ball;

- take ball in mouth;

- run back to thrower with ball;

- drop ball at feet of thrower; and

- sit.

Below, I will provide a more detailed elaboration of the sequence of behaviors illustrating the actual changes to the environment that each response generates and acts as the conditioned reinforcer for the behavior it follows, and the evocative stimulus for the behavior to follow. It is not always necessary to provide such an elaboration and the format above is usually adequate.

- S^{FA} (throw ball) $\rightarrow$ S^{Ev}_1 ("git yer ball") $\rightarrow$ Behavior$_1$ (run to ball)

- S^{Ev}_2 (arrive at ball) $\rightarrow$ Behavior$_2$ (take ball in mouth)

- S^{Ev}_3 (getting ball in mouth) $\rightarrow$ Behavior$_3$ (run back to thrower with ball)

- S^{Ev}_4 (arriving at thrower with ball in mouth) $\rightarrow$ Behavior$_4$ (drop ball at feet)

- S^{Ev}_5 (dropping ball at thrower's feet) $\rightarrow$ Behavior$_5$ (sit) $\rightarrow$ Reinforcer

Assessing Level of Proficiency of Requisite Behaviors

Once you have constructed and validated a task analysis, your next undertaking is to determine which component behaviors the subject already exhibits and with what level of proficiency. There are a few ways to assess the subject's current proficiency level. First, identify which, if any, of the component behaviors the subject exhibits. Second, determine if you can evoke any segment of the behavior chain (i.e., more than one of the behaviors exhibited in the proper sequence without interjecting evocative stimuli or prompts). A third approach, useful in some cases, is the **multiple-opportunity method** wherein the trainer evokes the behavior chain and provides minimal prompts as needed to achieve occurrence of the entire behavior chain or as close to it as can be achieved (Cooper, et al., 2007, pp. 438–441). Avoid physical manipulation. Identifying where prompts are required and how salient the prompts need to be will inform you regarding proficiency level. All of these factors help you determine exactly what you will need to train and what you do not need to train.

Training the Component Behaviors

You must train each behavior in the chain before you can chain them together. The proficiency assessment will help you determine what exactly you need to train and what the starting points will be, but each discrete behavior is a training project of its own. Begin with your formal behavior objective and construct a training plan to achieve that objective. You may be able to simply differentially reinforce it or you may need to use shaping.

Behavior Chaining Methods

Once you have trained component behaviors, there are at least two ways to attach the links of the chain—carry out chaining a sequence of behaviors:

- Forward chaining
- Backward chaining

Forward Chaining

In *forward chaining*, you train the behaviors in the order that you want the subject to exhibit them once the chaining process is complete. There are multiple approaches commonly used in forward chaining, two of which I will describe.

Forward Chaining Method 1

For very simple and short chains, you may prompt or evoke each behavior in the sequence, reinforcing only at the end of the sequence. Prompt or evoke the first behavior, and then prompt or evoke the second behavior, and on, until you have worked through the entire sequence. Repeat this sequence several times, and you should then be able to begin fading the interjected prompts. Present the first stimulus, and then continue as before, but then either (a) gradually fade the prompts, or (b) delay the next prompt for a second or two to determine if the next behavior occurs, that is, if stimulus control is strong enough yet. If the subject exhibits the next behavior, reinforce it. See how far you can get through the sequence and reinforce for successful occurrence of the smaller sub-chains. You may then need to work on adding more behaviors to the end of that chain in the same way in order to get all of the behaviors chained together. If you cannot get far, you can try (a) several more trials, perhaps through a few sessions, or (b) fading the volume of the prompts, or (c) method 2 described below.

In the retrieve example just discussed, you would start by cueing running to the ball when you throw it, and present the vocal cue. Then you would evoke the picking up of the ball behavior. You can then fade the prompt you interject, and when you achieve this, you evoke bringing the ball back to you, and then fade the interjected prompt for that linkage as well. Now you have all three behaviors linked in a chain.

Forward Chaining Method 2

For slightly longer or more involved chains, or if you tried method one and it is not progressing well, you can reinforce after each component behavior, and gradually thin the reinforcers. The first method described above is a short cut that is often successful in simpler chaining projects, but in many cases, you will need to use this more elaborate procedure. Begin by cueing the first behavior, reinforcing it, and then evoking the second behavior and reinforcing the second behavior. Thin the schedule of reinforcement for the first behavior, but maintain the second behavior on continuous reinforcement. Once you are no longer reinforcing the first behavior, and you can maintain the chain adequately with the continuous reinforcement of the second behavior, begin evoking the third behavior after the second behavior. Maintain the third behavior on continuous reinforcement, and begin thinning the schedule on the second behavior as you did with the first behavior. Progress through the entire chain in this way, adding new behaviors to the end and thinning the schedule of reinforcement for the behavior before it until you have met the objectives.

We use forward chaining for relatively simple behavior chains, or when the first few behaviors are particularly easy to generate. The more proficient each component behavior is the more likely forward chaining will be effective.

Backward Chaining

In **backward chaining**, you train the sequence in backward order. Evoke the final behavior in the chain and reinforce it, and once you can reliably evoke that, prompt the second to last behavior of the chain, and then prompt the last behavior right away and reinforce. Fade the interjected prompt, and once this chain is exhibited reliably, start prompting the third to last behavior, followed by the previous chain, and then fade the interjected prompt so that all three behaviors are exhibited when the evocative stimulus is presented and reinforcement is provided only after the final behavior. Continue this process until you have trained the entire chain.

We often use backward chaining when the final behavior is critical in the sequence. In the retrieve example, you would start by evoking dropping the ball at your feet, and once that was reliable, you would evoke bringing the ball to you before dropping the ball at your feet, fading the prompts as you go, and on until you have worked through the entire chain.

You can use a limited hold extension when training behavior chains as well. In a behavior chain with limited hold, the subject must exhibit the behavior within a specified interval of time in order to contact the final reinforcer. This can be useful in instances where quick behavior is important. You can use the limited hold extension near the end of training in order to tighten up the speed.

Quality of the Chain

You must be concerned with the quality of the chain. Especially in relatively long chains, the subject may skip component behaviors, or behave sloppily in the initial part of the chain. This effect resembles the scalloping evident in certain fixed schedules of reinforcement. The subject responds poorly at first, and only responds adequately in the final steps of the chain because you provide reinforcement only at the end of the sequence. It is important to ensure that you train each component behavior to fluency, that you use highly effective reinforcers, and that you reinforce only full criterion occurrences of the behavior chain. Do not reinforce subpar chains. It can be tempting to reinforce any completion of the behavior chain, but what you reinforce is what you get.

Sequencing

Sequencing involves training a subject to exhibit a number of discrete behaviors and/or chains, and interjecting evocative stimuli after component behaviors or chains to initiate the next behavior or chain of behaviors. We refer to the entire behavioral episode as a sequence (Alexander, 2003) and we refer to the evocative stimuli used between component parts as **interjected cues**.

Sequencing can be extremely useful when a higher level of flexibility is required than a static behavior chain can provide. The example sometimes used involves a particularly long and complex agility run, in which the particular obstacles are arranged differently in each competition, and while each obstacle involves a chain of component behaviors that remain stable, the trainer must present evocative stimuli in order to get the subject to initiate the next appropriate behavior or chain of behaviors.

If your goal is to train a series of behaviors that will remain the same each time, and you will utilize a single cue to initiate that process, you will be training a behavior chain. If you simply evoke a number of behaviors sequentially (each with its own evocative stimulus), this is not a behavior chain. If the task requires flexibility (such as in an agility run), you may need to utilize sequencing. The process will involve training a number of behaviors and/or behavior chains, and using verbal or physical cues, "on the fly," to evoke the next behavior or chain.

Continuing Education

Courses provided through The Companion Animal Sciences Institute at www.CASInstitute.com address all of the topics covered in this chapter.

CHAPTER 7. TRAINING DOGS

Behavioral Objectives

The objective of this chapter is to measurably expand the reader's repertoire of behaviors in relation to describing and relating the principles of behavior. Upon successfully integrating the concepts outlined in this chapter, the reader, where exposed to contingencies to do so, will accurately:

- Implement a training plan for common good manners-related behaviors, applying the systematic strategies previously discussed based on a systematic plan previously discussed.

- Select appropriate equipment to aid in training

- Identify equipment that operates on principles of behavior that involve aversive stimulation versus equipment that does not.

Introduction

The behaviors discussed below are commonly trained in basic good manners classes. These are the everyday behaviors that help guardians manage their dogs, so that everyone can remain safe and the dogs can easily fit into human society. Training each behavior involves application of the systematic strategy outlined above. Think of the basic strategy outlined above as the formula and the information below as the specific details and extended examples. This chapter builds on the previous two chapters by providing more specific, application-related details.

Pay particular attention to the section on training "sit." This description includes many details that I will not repeat in the behaviors that follow it. These details are important, but it is too tedious to read the exact same material for every single behavior.

You might notice there is no "stay" behavior described in this chapter. Staying in position is simply the duration parameter of that behavior being maintained, and is therefore controlled by the evocative stimulus that evoked that behavior. For instance, the evocative stimulus for sitting colloquially means to "sit and remain in that position until another evocative stimulus is presented," rather than "sit and then you are immediately free to do anything else you want, except if stay is evoked." Thus, "stay" is redundant—it is more appropriate to treat it as a duration parameter. If the dog is already in the position you wish them to remain in for some duration, use "wait." "Wait" is similar to what one might refer to as "stay," except that the subject is already exhibiting the behavior and you did not initially evoke it. If the dog is already in a position you want them to remain in, cue "wait" until released, but if you evoke a behavior that involves duration, maintaining the position until released is simply the duration parameter of that behavior.

Training Dogs

Most dogs are quite receptive to training and do not quickly become satiated with social interaction or the reinforcers typically used. Training sessions of 15 minutes are not usually a problem for most dogs.

Equipment

Clicker and Treat Pouch

You will need a clicker. The i-Click™ Jewel clicker[61] is particularly comfortable, and if you plan to do a lot of training, you should invest in a good clicker (or two).

You can usually get stretchy wristbands available from various vendors that attach to your clicker, so that you can let go of it when necessary and it will remain readily accessible. You could also attach it a to a belt loop and click it right against your hip when needed. You should also get a treat pouch. Treat pouches look like the pouches used by rock climbers to hold their chalk. A good treat pouch holds itself open, easily facilitating quick treat retrieval. Most treat pouches clip onto a belt or waistband and many have edging to help prevent treats from bouncing out.

Restraints

One drawback of using a standard buckle neck collar attached to a leash is that if the leash tightens it may cause stress or even harm the dog in some circumstance. It may also elicit the dog's opposition reflex, so the dog pulls away from the leash. Furthermore, some dogs with small or tapered heads (e.g., Greyhounds) can slip out of buckle collars, even if they are tight on the dog's neck.

A standard, unrestrictive body harness can resolve the slipping problem, prevent the opposition reflex, and reduce stress. Most suitable harnesses have a ring positioned at the top of the dog's back, used to attach the leash to the harness. (This is not to be confused with harnesses that tighten to punish pulling on leash or act as subtracted reinforcers for walking closer to you.) Utilizing a standard body harness can often help resolve pulling problems. Do not fall victim to the "sled dog myth" that dogs will pull *more* on a harness than a buckle collar—they do not, and indeed, they generally pull less. It is also much easier on the neck. Again, it must not be too tight, as excessive tightness can prevent breathing, but should not be so loose that the dog can wiggle out of it if they move backwards away from the trainer. Some harnesses have extensive padding, while others do not. If the dog has very short fur, consider using the padded kind.

---

[61] www.clickertraining.com

There are harnesses designed to apply pressure, sometimes under the armpits and sometimes pinning the upper legs together, when the dog pulls, but in general, it is best to avoid any tool that applies aversive stimulation. Instead, focus on equipment that simply limits movement and facilitates the relevant training. There are several kinds of anti-pull harnesses, operating on various principles. Not all of them tighten to create pressure. Some simply attach at the front of the dog's chest rather than on their back so that the dog is forced to turn, rather than pull forward or is easier to guide. Some of these front attaching harnesses are the ones that pin the front legs together. Generally, a standard body harness is best but if you are evaluating "anti-pull" harnesses for special circumstances (e.g., a tiny handler/guardian and huge rambunctious dog combination) then identify the mechanism of operation and the principle it operates on and choose one that guides rather than tightens.

Use a standard 6-foot nylon leash.

Head halters (e.g., the Gentle Leader®, Snoot Loop®, and Halti® headcollars) attach around the dog's head rather than their neck or body and are similar to the head halters used with horses. A band typically goes around the back of the head and one goes over the muzzle. Commonly, the leash attaches under the chin although some attach at the back of the head, pulling the muzzle downward when tightened. One uses them to direct the orientation of the head, and therefore the direction the dog moves. Head halters can be uncomfortable and unpleasant (i.e., aversive) for dogs. While some dogs may become accustomed to them, it is generally advisable to avoid equipment that requires desensitization[62]—it will disrupt your training, at least at first. Many dogs do not become sufficiently comfortable with the head halters and continue to fight them. Under no circumstances should one use a head halter with an extendible leash (i.e., a "Flexi®-leash"). The risk of the dog hitting the end too quickly and injuring the neck is too great.

What Not to Use

Choke chains should *not* be used on dogs. *Period.* These are chains that form a noose and tighten around the dog's neck if they pull or if the handler pulls or yanks on the leash. These are used as added punishers for behavior they follow and as subtracted reinforcers for behaviors if loosened contingent on their occurrence. Because of the force generated on the dog's neck, the use of choke chains may cause damaged tracheas. Prong collars, with prongs or spikes that pinch or dig into the dog's neck, should also not be used, as they are based on aversive principles of behavior. Although the risk of tracheal damage is not as great with prong collars (due to the even distribution of force around the dog's neck), these collars cause more harm than good, as is common with all aversive methods and tools. Nor should you use collars that either shock or spray noxious substances into the dog's face when triggered by barking,

---

[62] By desensitizing, I mean training the subject to exhibit and maintain the target operant behavior without indications of stress or escape related behaviors. One achieves this by taking a very careful graded approach to exposing the subject to the stimulus and additively reinforcing relaxation-related behaviors occurring during that exposure as well as the target operant behaviors.

moving outside of a specific range, or by a button controlled by the handler. These tools simply cause more harm than good.

Name

A dog's name is nothing more than another evocative stimulus. In this case, the dog's name evokes attending/orienting to the person saying it, usually in preparation for another cue. You should not use it as a recall stimulus, as it is important to have an evocative stimulus that simply and immediately evokes increased attention to you without any other specific behaviors. Nor should you use the name to express disappointment or to evoke 'get off,' as lay people commonly use it. Having the name evoke attending to a person can be particularly useful in a multi-dog household, where you may want to evoke a behavior from an individual and not from any other dogs that are present. In that case, the name evokes attending behavior from the dog named and no other dog.

In most applications, you will conduct this training with very young puppies or else as a form of discrimination training if the dog currently approaches when someone calls their name and you seek to retrain them to attend to you but otherwise stay where they are. It is also used when dogs are newly adopted and the new guardian wishes to change the dog's name.

Phase 1. Preliminaries

Behavior Objective

S^{Ev}: Vocalized[63] "name."

Behavior: Orient to look at the vocalizer (without approaching).

Consequence: Treats and continued social contact.

Criteria: Latency: 1sec.; frequency: 100% through 10 trials; distance: minimum 20 feet; distraction: various; time point: 3 sessions at ≈5min. each.

Phase 2. Acquisition

The best approach for training attention and orientation to "name," especially if you are replacing approach behaviors with simple attending behavior is a graded approach

---

[63] The word "vocal" might evoke surprise in some readers expecting to see the word "verbal." Not all verbal behavior is vocal. In fact, most of it is not. Verbal behavior is behavior, the reinforcement consequating it being mediated by another individual that shares a history of conditioning the verbal behavior in a verbal community (Skinner, 1957/1992). When exposed to contingencies to vocalized verbal behavior, the word "vocal" is the appropriate term.

with special attention to the distance dimension. Ensure there is minimal distraction. Stand close to the dog. Say the dog's name, and chances are the dog will look at you— simply attending to the novelty of the vocalization. Click and treat. It is important to stand very close to the dog, so that they do not have the opportunity to approach you. The criterion for this behavior is simply to look at you. If the dog does not look at you right away, make a noise (prompt) that will ensure you get the dog's attention and then click and treat when you get it.

Phase 3. Fluency

Repeat a few more times, transitioning to an intermittent schedule of reinforcement and praise. Begin moving gradually further from the dog, at *very* small increments. If the dog moves toward you (other than to orient the body to better attend to you, administer a conditioned subtracted punishment, such as "Oops," turn away, and remain unresponsive for several seconds. Then turn back, go to the dog and try again, this time perhaps closer, so that the dog cannot move towards you, or with something between you, thus blocking the dog's access to you. This will rule out accidental reinforcement and maintain the sole behavior of looking at you. Try to avoid punitive trials though, as this is one of those "counter-intuitive" (remember 'task characteristics' from Chapter 1) contingencies that can be challenging to train, particularly if the name has previously been used as a recall stimulus. It is ideal to move at such a gradual pace that this does not happen, but even a very slow graded approach does not prevent occasional approach behaviors. Next, simply begin using the dog's name before cuing other behaviors, and avoid using it to as a synonym for "here" or "off" or any other specific behavior aside from paying attention to you. It will take on greater stimulus control as you use it in training.

Phase 4. Maintenance

Look for opportunities to occasionally evoke attention/orientation and reinforce it in order to maintain the fluency that you have achieved.

Sit

Phase 1. Preliminaries

Behavior Objective

S^{Ev}: Vocalized "Sit."

Behavior: Contact rear end to the ground (or as close as is physically possible) with front paws on ground, front legs straight and front paws within 5 inches of back paws.

Consequence: Treats and continued social contact.

Criteria: Latency: 2sec.; duration: until released, minimum 1min.; frequency: 100% through 10 trials; distance: minimum 20 feet; distraction: various; time point: 3 sessions at ≈5min. each.

The dog may achieve the position by sitting from a down position or sitting from a standing position, or even from lying on their back. It is important to address each of these common response class forms in training. You can train them at the same time simply by ensuring that you include the down and standing starting positions into your training sessions.

Some breeds cannot actually place their rear ends on the ground, so a *sit* for them will be for them to put it as close as they are capable (their upper rear thigh or rump area will touch their hocks as the joint articulates as far as it is capable). Eventually you will include the criteria that they exhibit the behavior within, say two seconds of it being evoked, and they will remain in the position until some other behavior is evoked, say a minute or two. As per a graded approach, begin with these criteria relaxed and build those up gradually.

The distraction criteria are vague in our behavior objective. That is partly because this is general, rather than specific to an individual and for conciseness purposes. If you allow a vague reference to distraction (or any other criterion) in your behavior objective, it is a good idea to include an addendum or footnote with some specific details. You might need the dog to sit close to traffic, in a crowded mall, with a bunch of other dogs or children running around, or when greeting all sorts of people. Identify the specific requirements and prepare a comprehensive list of the requirements you need for the dog to meet and the circumstances involved.

Identify the conditioned reinforcer and unconditioned reinforcer you will use. Use a clicker and small (approximately pea-sized) treats that do not take long for the dog to chew.

Increase the rate of responding by conducting training when the dog is not satiated with either your social presence or food. In order to ensure concentration and focus, train at a time when the dog is not too tired, but also not hyperactive. These establishing operations prevent other contingencies from taking control over the behavior. Ensure that you begin training in a low distraction environment—this means there are few competing contingencies that would control the dog's behavior. You can allow for increased distraction later in the process—for now, set yourself and the dog up for success with a graded approach.

Phase 2. Acquisition

In the case of "sit," prompting is usually a quick and easy antecedent strategy for getting the behavior to occur. To prompt sitting, place a small treat between your thumb and fingers. Ensure you have a good grip on the treat, so that the dog cannot grab it before you let it go. With palm up, allow the dog to sniff the treat. Move it around to ensure the dog is "targeting" it, that is the nose goes where the treat goes. Once the dog is targeting the treat, move the treat slowly over their head so that they crane their neck to continue targeting it. If they jump up to target the treat, it is likely that you are holding the treat too far above their head. In that case, quickly withdraw the treat and try again with the treat held a bit closer to their head. If they back up while you lift the treat over their head, withdraw the treat quickly and try again. If they do this again, you might want to perform the targeting with the dog's rear end close to a wall or corner so that they are unable to back up (just be careful that the dog does not "feel cornered" so to speak). As they crane their neck to target the treat, they should sit. You now have a criterion-meeting trial. Once the dog has exhibited the target behavior, immediately click the clicker, and then deliver the treat to the dog right away. We often refer to this protocol as merely "click and treat." The click ends the trial. Therefore, if the dog gets up after the click is sounded, that is not a problem. Carry out a few more trials. You should be able to exhibit the luring motion more quickly and effortlessly in each successive trial, as the prompt and other stimuli take on stronger control over the behavior.

Phase 3. Fluency

Begin Fading Prompts

After the first few trials, begin fading the food component of the prompt. Whenever you use the unconditioned reinforcer as a component of the prompt, you should fade this part of the prompt quickly, so that it does not become an establishing function-altering stimulus. This is the first thing you will do in the fluency phase of training. Start with a few rapid trials of the sequence with the treat in your hand. Then in the next trial, leave the treat in your treat pouch, and perform the prompt motion just as before. The momentum, the similarity of the trials and remaining odor of the treat in your hand will promote evocation of the behavior. Once exhibited, click and treat (i.e., reinforce). Through the next several trials, continue to reinforce on a continuous reinforcement schedule. In most cases, you can simply fade the treat-in-hand stimulus permanently this way.

If the dog seems "apprehensive" with this change, you can fade the prompt more gradually. In that case, carry out that first trial without the treat in your hand as above and reinforce. Then carry out another trial, this time with the food in your hand. Through the next several trials, alternate between having the treat in your hand and not, in a seemingly random manner, gradually increasing the ratio of food-out-of-hand to food-in-hand prompts until you can run through several trials without the treat in your hand, and the dog exhibits the behavior smoothly.

Always end your training sessions on a good note. If you believe that the dog is becoming satiated (i.e., bored) with the reinforcer, restless, or the training may soon slow down in terms of progress, end the session there. If things are not going as well as you would like, end the training session by cueing a behavior that the dog already exhibits fluently, reinforce the behavior and work on the new behavior later. If you believe the dog can continue without deterioration of the training, then continue, but always strive to end sessions before deterioration of any component begins.

To start your next session, briefly review the training from the previous session, to ensure a reliable start to training. There are a few new protocols to execute at this stage of training. You should now be presenting the prompt without food in your hand at all. The luring motion should be taking on stimulus control of the behavior at this point. Before this becomes too well established, begin fading the prompt, transferring stimulus control from the lure motion to a hand signal. The hand signal typically used for "sit" is a palm up motion from a straight arm to an articulated arm while you stand straight up. This resembles the lure motion, and so transferring stimulus control is an easy process. Taking advantage of stimulus generalization make the current stimulus (the lure motion) seem (in this case, look) increasingly like the new stimulus (the hand signal) during each successive trial over several trials. Do this incrementally and gradually, and the dog should continue to exhibit the behavior reliably through each trial. If not, you are probably moving too quickly through this process.

You should now be able to evoke the behavior with the hand signal alone every time. Use this temporary evocative stimulus until the behavior is fully formed the way you want it.

Begin Thinning the Schedule of Reinforcement

Until this point, the behavior has been on a continuous schedule of added reinforcement. You can now move to a gradually thinning, variable ratio schedule of added reinforcement. The goal here is to gradually thin the schedule of reinforcement in an indiscernible pattern (re-thinning it each time you introduce a new D-parameter or a new D-parameter level). Start by failing to reinforce a response, but quickly carry out another trial and reinforce that one. The extinction-generated aversive emotional arousal will not last long and you will begin conditioning persistence and resilience. Now, gradually increase the number of unreinforced trials to reinforced trials around a gradually increasing average. Randomize reinforced and unreinforced trials, to avoid producing discernible patterns. Go from a VR-2 toward, say, a VR-6 or VR-8. Move at a pace that continues to maintain the stability of the behavior. Beware of *ratio strain*, wherein the schedule is thinned too quickly, the dog becomes frustrated, and the behavior becomes unreliable or unstable and can actually extinguish. Be sure to always progress at the dog's pace.

Refine Form, Latency and Speed

Utilizing a graded approach, begin gradually working form, latency and speed, one at a time, as needed, until each satisfies the behavior objective requirements. Assess the features you need to change based on current proficiencies and the target criteria. This might include decreasing the latency or training a quicker motion from beginning to sit to being seated—speed. Work through one feature at a time. Refining the form means shaping. If any components of the form are inadequate, identify exactly what movement(s) needs to change and what they will look like when satisfactory. You may need a list of approximations if there is a significant difference between the current motion and terminal motion, but often, this kind of refinement simply requires slight shifts in form that can be achieved with one or two levels of approximations.

Bring Behavior under Stimulus Control

Once you have the behavior's form and latency to criteria, you should establish the permanent vocal cue. To transfer stimulus control from the hand signal to the vocal cue, simply repeat the sequence of new stimulus (vocal "sit"), followed by the old stimulus (hand signal), followed by occurrence of the behavior, followed by reinforcement. Achieve several trials through the contingency and the new stimulus should take on stimulus control of the behavior. Pause after saying, "sit," to determine whether the dog will sit in response to just the vocal cue. It might take an extra second or two, as the subject waits for the hand signal, but they will likely exhibit the behavior. If not, repeat several more trials and try again until the new vocal cue evokes the behavior on its own.

Proof against the Three D-Parameters

You now have the target behavior under stimulus control and it is time to begin "proofing" the behavior against dynamic real world challenges involving the three D-parameters. Remember to work only one at a time.

Start by introducing small distractions such as a motionless and quite person nearby looking away from the dog. Take the schedule briefly back to continuous reinforcement and thin it gradually with the new D-parameter in place. Introduce incrementally greater versions of the distraction, but again, do this at a pace that maintains the stability of the behavior. Once you are able to evoke the behavior under rather distracting circumstances, you might introduce another kind of distraction—evoking the behavior in other locations. For instance, instead of training in your living room, try several trials in the kitchen, remember to relax other D-parameters briefly, return to a continuous schedule of reinforcement and then re-thin the schedule. This can usually be done quickly, but always manage the level of frustration and stress, keeping these minimal and your training fun. Then you could attempt trials in a boring backyard (perhaps after already being out for a while to promote satiation with regard to the other reinforcers in the yard). Then you could run through the training with other distractors added in. Then you could run through the process away from the yard, on a sidewalk. Have you ever noticed cases in which a dog and guardian are in a pet supply shop and the guardian is very surprised when the dog does not sit in response to the "sit" stimulus? This is because they simply have not proofed the behavior to that point; it is probably a fast, reliable and sharp looking sit at home. Make sure to take opportunities to reinforce occurrence of the behavior in many different places and under many different circumstances. You do not need to get to this level of proofing before you can begin working the other D-parameters. Get a good start on distraction, but if you choose to work another D-parameter, relax distraction until you increase the other D-parameter. Then you can start combining them.

Distance will involve and require duration—therefore, it is best to work on duration before introducing distance. We train duration in the same way as distraction. Set a specific criterion and establish it reliably before increasing it again. Up to now, you have been reinforcing the sit immediately. So, begin to expect at least two seconds on maintaining the sit position before reinforcing. Move at the dog's pace to ensure that almost all of the trials meet the criterion. If you see the dog is just about to get up, resist the temptation to click and treat in order to avoid the non-criterion trial. You did *not* catch it in time. That will only result in reinforcing getting up. When the behavior fails to meet the duration criterion, administer extinction or subtracted punishment, identify your failure in generating the behavior (you likely waited too long or allowed excess distraction) and try again.

In the case of distance, you should be able to simply inch your way further from the dog through successive trials. Work up this D-parameter gradually as with the others and once it is well on its way, you can begin combining D-parameters. When you combine more than one D-parameter, remember to build them up gradually again.

Set the criteria in all of these parameters to ensure success in most trials, but with a degree of difficulty that maintains interest and progress. This judgment in maintaining smooth progress and minimal frustration is the trickiest set of trainer "chops" to teach trainers and the most challenging skill to acquire, primarily because it requires the generalization of many related behaviors to be exhibited reliably and quickly. Attend to expanding your repertoire in this regard; the appendix on trainer exercises will help. Attend to and recognize inefficiency and ineffectiveness in your training behaviors to differentially reinforce your own effective training practices. At the same time that you are training others (dog or a guardian), you are also training yourself.

Discrimination Training

Once you have two behaviors trained, you should work on inter-stimulus discrimination training. Run a series of trials, cueing one or the other in a random manner, reinforcing criterion behaviors and extinguishing non-criterion ones. This will be a good test of stimulus control, since many factors except for the evocative stimulus itself will be the same in both arrangements. Remember the differential outcome effect—use distinct reinforcers for each behavior and your training will be more efficient and effective. Once you have three behaviors trained, you will find it is even more challenging as you randomly evoke different behaviors. That is because with two behaviors the strategy of simply changing position can work, (assuming you don't evoke a behavior the subject is already in the position for). With three behaviors, this strategy will only be effective 50% of the time (rather than 100% with two behaviors).

Introduce Release Stimulus

Once you are ready to begin phasing out the conditioned reinforcer, which until now has been acting as the release, you can begin using a release stimulus in its place. The release stimulus will act as a conditioned reinforcer in place of the click. You can use "you're free" as the release phrase or pick something else but try to avoid phrases that are common in everyday discourse such as "okay." You can introduce the release stimulus as you train. Begin presenting the release stimulus right before treat delivery, instead of the click. If the dog does not break position when released, you can prompt it easily enough with open arms, backing up and praising the dog, or offering the treat at a slight distance, requiring the dog to break position in order to eat the treat.

Phase 4. Maintenance

Once you have achieved the final form, speed and latency criteria of the target behavior, it is under stimulus control, and reliably proofed through the three D-parameters, you can begin working toward maintenance. The line between fluency and maintenance phases is not always clear. You will likely want to continue to develop proficiency in new locations or with new distractors etc., but once you are well into the process, it is time to begin transitioning from the intensive training activities of the fluency development phase to less intensive maintenance of the fluency you have achieved.

Begin generalizing the reinforcers from just treats, for instance, to praise some times, and petting or perhaps a quick tug-of-war game, as long as these things are actually reinforcing. Remember also that the clicker is just for acquisition and training toward fluency. Start reinforcing with the unconditioned reinforcers and leave the clicker for new parameters. Begin using fewer trainer-mediated reinforcers, too. Use activity reinforcers (via the Premack principle) in order to help you maintain control over the behavior. For example, if eating is reinforcing, then take the opportunity to require a sit before allowing the dog access to their food. If going outside or having a leash put on acts as a reinforcer, require a sit while you open the door or prepare the leash. The same goes for sniffing fire hydrants or meeting other dogs, etc. The goal in this phase is to work toward simply maintaining what you have achieved through training with minimal contrived activity outside of the evocative stimulus. If at any point, any component of the training seems to be deteriorating, refresh the training by building that parameter back up.

Common Challenges

For some breeds of dog, the sitting position is seemingly uncomfortable and hence aversive. In these cases, you will be imposing an added reinforcement contingency over an existing punitive contingency. In practice, this usually means slow, awkward training and an unreliable final product. Although the sit behavior is useful, especially when you want a dog to wait without moving for more than a few seconds, or as a greeting behavior to prevent jumping up etc., it might be worth considering using a standing position. In this case, you can specifically work remaining in place for the stand as part of your proofing of this behavior.

Some dogs will tend to crane their necks and turn around, stand or jump up, rather than easily take a seated position to target a treat during prompting. If you are raising the treat too quickly or high over the dog's head, they will often turn or jump up instead of sitting. If this happens, experiment with moving more slowly and maintaining the treat closer to the dog. You do not have to keep the treat "out of reach" when luring; just do not let go of the treat until it is time to reinforce the behavior. For dogs that simply scuttle back, rather than crane their head to target the treat, consider training with the dog's back against a wall or corner, so that they physically cannot back up.

Down

Phase 1. Preliminaries

Behavior Objective

S^{Ev}: Vocal "Down."

Behavior: Contact both elbows, and either both hocks or either hip, with the ground.

Consequence: Treats and continued social contact.

Criteria: Latency: 2sec.; duration: until released, minimum 3min.; frequency: 100% through 10 trials; distance: minimum 20 feet; distraction: various; time point: 3 sessions at ≈5min. each.

As with "sit," "down" involves achieving a specific position that can be achieved in more than one way, requiring more than one kind of movement. For example, the dog could lie down from a seated or a standing position. Just as before, ensure that you train for both specific response class forms. The distraction criterion is vague here as well, because this is a general set of guidelines. Ensure it is specific to your requirements, or provide a brief addendum with some kind of specific criteria to refer to during your training.

Identify the conditioned reinforcer and unconditioned reinforcer you will use in training.

Phase 2. Acquisition

Utilize the same establishing operations as discussed for "sit."

In the case of "down," prompting is usually quick and easy. To prompt a "down," place a small treat between your thumb and fingers and lure the "down." Luring a "down" is a little trickier than "sit." You may need to kneel or sit on the ground to start. Once the dog is targeting the treat, move your hand slowly straight down to slightly forward, between the dog's front paws. Many dogs will bow their front ends rather than lower their whole body into a lying down position. Usually, if you wait a couple seconds, the rear end will also go down. If you cannot manage to generate the down position this way, you may need to sit with one leg bent, to form a triangle tunnel, and lure the dog to move through the tunnel. With large dogs, you may need to use a table or chair. In any case, form a tunnel with the dog on one side and you on the other, and lure the dog down and into the tunnel, so that they must lie down to continue targeting the treat. You would then need to fade the obstacle as well. Generating "down" can sometimes take finesse and creativity. As soon as the dog achieves the down position, click and treat, and repeat through several trials. You should observe that evoking the behavior becomes smoother and easier.

Phase 3. Fluency

Follow the same general steps to train for fluency of "down" as you did for "sit."

After the first few trials, begin fading the food-in-hand prompt as described for training "sit."

You should now be prompting without food in your hands at all. The luring motion should be taking on stimulus control of the behavior at this point. Before this becomes too well established, begin fading the prompt, transferring stimulus control from the lure motion to a hand signal. A typical hand signal for down is a palm down motion from an articulated arm to a straight arm, while standing straight up. This resembles the lure motion, and so transferring stimulus control is an easy process. Do this incrementally and gradually and the dog should continue to exhibit the behavior reliably through each trial. If not, you are probably moving too quickly through the process.

Up to this point, the behavior has been on a continuous reinforcement schedule. You can now move to an intermittent schedule of reinforcement and begin thinning that schedule. The schedule of choice for down and similar behaviors is a variable ratio schedule.

Now that you are reinforcing some and not all of the responses, and the schedule is not too sparse, you should take this opportunity to refine the form, latency and speed as needed. Work through one criterion at a time, and take the schedule back to continuous reinforcement, re-thinning it each time you modify the criterion.

Once you have the behavior in the form, and with a latency you are comfortable with, you are free to establish the vocal cue.

You now have the target behavior under stimulus control, and it is time to begin proofing the behavior against the three D-parameters. Remember to work only one at a time while relaxing the criteria for the previously conditioned parameters, and with each newly introduced parameter, re-thin the schedule from continuous reinforcement. Work distraction, distance and duration in the same way that you did for sit.

As explained under sit, you should promote discrimination between the evocative stimuli you have trained to this point. If you have only worked "sit" and "down," you can have the dog perform "pushups" by evoking each in turn. Remember to include some evocative stimuli for the position they are already in, to ensure that the evocative stimulus is not simply evoking change of position.

Phase 4. Maintenance

Once you have the final form and latency criteria of the target behavior, it is under stimulus control, and reliably proofed through the three D-parameters, you can begin working toward maintenance. Continue to develop proficiency in new locations or with new distractors etc., but once you are well into the process, it is time to begin transitioning from the intensive training activities of the fluency development phase to less intensive maintenance of the fluency you have achieved.

Begin generalizing the reinforcers as you did with "sit."

Common Challenges

As with "sit," lying down can be uncomfortable for some dogs. In these cases, as with sit, consider proofing a good solid stand and wait position instead. However, more commonly, we use the down position for longer duration maintenance of position than would be comfortable for a stand or even sit. In that case, consider specifically training a comfortable down behavior. First, ensure that where the dog is required to lie down is soft enough. This is particularly important for dogs with minimal fur and fat (such as Doberman Pinschers, Greyhounds, Boxers, and Dalmatians). Second, train the dog to lie down leaning to a side with one of their hips on the floor and both front elbows on the floor to ensure they will be comfortable. Find a position and solution that will work and if you cannot, consider using some other behavior to achieve your goals when a soft thick bedding material is not available.

For some dogs, lying down seems to be a more vulnerable position, particularly when they are around other dogs, and especially if unfamiliar dogs are present. Attend to the context in which you train and evoke "down." Consider using a "sit" or some other behavior that causes fewer vulnerability-related behaviors in the dog (i.e., reluctance, escape, emotionality). If remedial socialization and behavior change programming is required to reduce fear, consider doing so.

Small breed dogs in particular are averse to the down position. It can also be challenging to lure "down" and other position behaviors in some small breeds. Consider starting the training as you sit beside the dog for the initial acquisition stage.

Some dogs will lower their front end to target the treat, but never lower their rear end. In some cases, it takes a very long time before the entire body achieves the down position. As described above, consider sitting and creating a triangle tunnel with one bent leg to lure the dog through. Although physical manipulations are usually more disruptive, and even sometimes aversive in training, consider a very light touch to the top of the dog's back between the back hips. This is *not* a push or shove, but just a touch, which directs the dog's attention to the kinesthetic position of their rear quarters. If this works, it will work well and quickly, and you can fade that tactile prompt very quickly, but if it does not work quickly, it is best to abandon it. Please do *not* misunderstand these instructions. This is a mere touch to direct attention, rather than a pushing motion, and if it does not work right away, discontinue it. If the touch is strong enough, it will elicit the opposition reflex and fail, so it must be simply a brief light touch to elicit kinesthetic awareness behaviors. In cases where this might be appropriate, it works quickly, which in turn prevents a lot of frustration. A mere touch (discontinued if it does not work quickly) should not be aversive at all and is less aversive than the frustration of ineffective training. Use any physical manipulation with great caution and avoid such manipulation where possible. This criterion is speed. Remember that to modify speed, you gradually require just a slightly faster behavior until it is stable and then shift the criterion again and on like that until you have met the speed criterion objective. In some cases, this particular criterion can take a while to train. An alternative option is to gradually shape the down position.

Stand

Phase 1. Preliminaries

Behavior Objective

S^{Ev}: Vocal "Stand."

Behavior: Assume position with legs straight and all four paws on the ground with no other body parts touching the ground, at ≈90 degree angle with the floor.

Consequence: Treats and continued social contact.

Criteria: Latency: 2sec.; duration: until released, minimum 2min.; frequency: 100% through 10 trials; distance: minimum 20 feet; distraction: various; time point: 3 sessions at ≈5min. each.

The dog might stand from either a seated or a down position, so include both response class forms in your training. Training a stand can be useful for applications in grooming and veterinary exams. It is also useful in that it allows a third basic position (along with sit and down) for training inter-stimulus discrimination, which can improve training of "sit" and "down," the arguably more important of these three position behaviors. The distraction criteria is vague because this is a general example, so either ensure it is specific to your requirements, or provide an addendum with some more specific criteria.

Identify the conditioned reinforcer and unconditioned reinforcer you will use in training.

Phase 2. Acquisition

Utilize the same establishing operations as discussed for "sit."

In training "stand," prompting is usually quick and easy. To prompt, "stand" when the dog is seated or lying down, place a small treat between your thumb and fingers and get the dog to target it. Once the dog is targeting the treat, move your hand slowly up a bit and straight away from the dog toward you for a distance equal to about one of the dog's steps. The dog will stand to track the treat. Reinforce this the instant they are standing. Repeat through several trials until it is smooth and reliable.

Phase 3. Fluency

Follow the same general steps to train for fluency of "stand" as you did for "sit."

After the first few trials, begin fading the food-in-hand prompt as described under "sit."

You should now be carrying out the prompt without food in your hands at all. The luring motion should be taking on stimulus control of the behavior at this point. Before this becomes too well established, begin fading the prompt, transferring stimulus control from the lure motion to a hand signal. A typical hand signal for stand is to have your arm straight at your side with the palm toward the dog and then move it backward behind you (arm still straight). Use this hand signal as the evocative stimulus until the behavior meets criteria.

Now move to an intermittent schedule of reinforcement, and begin thinning that schedule.

Once you have the behavior in the form, latency and speed you are comfortable with, you are free to establish the vocal cue with the procedure previously outlined.

Begin proofing the behavior against the three D-parameters. Remember to work only one at a time, and use the advice previously outlined.

As explained for "sit," you should promote discrimination between the evocative stimuli you have trained to this point. At this point, you can use a three-way discrimination, which is particularly useful because the next evocative stimulus presented is not equivalent to a simple change in position; it would be one of at least two different positions. Once this is going well, include a test trial, cueing the behavior the dog is already in the position for to see if they remain in that position or change positions. Extinguish any changes in position and additively reinforce maintenance of that position.

Phase 4. Maintenance

Once you have achieved the final form, latency and speed criteria of the target behavior, and it is under stimulus control and reliably proofed through the three D-parameters, you can begin working toward maintenance. Continue to develop proficiency in new locations or with new distractors etc., but once you are well into the process, it is time to begin transitioning from the intensive training activities of the fluency development phase to less intensive maintenance of the fluency you have achieved.

Begin generalizing the reinforcers as you did with "sit" and "down."

Common Challenges

It can sometimes be challenging to train a maintained "stand," because the dog walks in the standing position (rather than the seated or down position). In these cases, it is simply a matter of working the duration component more gradually and using highly effective reinforcers, as well as managing concurrent contingencies (i.e., distraction) more closely.

Wait

Phase 1. Preliminaries

Behavior Objective

S^{Ev}: Vocal "Wait."

Behavior: Cease moving if in motion and remain motionless otherwise.

Consequence: Treats, food and continued social contact.

Criteria: Latency: ½sec.; duration: remain in place until released, minimum 2min.; frequency: 100% through 10 trials; distance: minimum 20 feet; distraction: various; time point: 6 sessions at ≈5min. each.

"Wait" is similar to "stay," except that you would not use it for duration maintenance for behaviors you have already evoked. In those cases, simply train the subject to exhibit the behavior until you release them—no secondary "stay" or "wait" cue is required. "Wait" can be useful at doors, when being fed, before grabbing a toy, or in any circumstance where you would like the dog to briefly halt their movement. Again, in actual cases, make distraction criteria specific in the definition, or provide an addendum with explicit and unambiguous distraction criteria.

Identify the conditioned and unconditioned reinforcer you will use.

Phase 2. Acquisition

Utilize the same establishing operations as discussed for "sit."

Training "wait" is all about gradually increasing duration. Begin with a simple wait-appropriate arrangement. Here, I will use waiting for you to present a food bowl as an example, but you can start with waiting at a doorway or other scenario. An advantage of starting with waiting for food is that it generates an effective reinforcer and the continuous schedule it necessitates is more suited to the beginning of a training program. In any case, you will promote generalization to these other situations as you proceed with the training. You can ensure more trials by presenting only an eighth to tenth of the dog's meal on each trial. You may adjust the specific procedure to allow for any differences in how you prefer to present meals. Make sure you see the challenges section below before deciding to start with this reinforcer.

Begin by approaching the dog's eating area with his or her bowl in your hand. Rather than requiring some specific behavior such as "sit," simply present a hand in front of the dog's face, palm facing them, and quickly place the bowl down (with the click in that hand), click, remove the hand and allow the dog access to the food. The palm hand signal will be in their way, between them and the food and should cause at least a momentary hesitation long enough for you to click, place the bowl down and remove the hand. You do need to carry out this sequence quickly at first to ensure success. If you need to eliminate the clicker from the sequence, you should be okay to do so as you will be providing the unconditioned reinforcer quickly. The dog will then begin eating the food, providing added reinforcement for the behavior. Remember that whether the dog is sitting or standing does not matter, and in fact, it will be more useful if you achieve a mix of these positions in order to achieve appropriate generalization and discrimination. Repeat this sequence through several trials until you have presented the entire meal. If the sequence is smooth and reliable at that point, you can move right to the fluency stage. Otherwise, repeat this for the next meal until the sequence is smooth and reliable with the minimal duration. If the dog barges through to the food before the click, simply say "Oops" while pulling the bowl away, bringing it back to the counter and trying again from there.

Phase 3. Fluency

Without prompts and especially prompts involving the conditioned reinforcer, fluency is a simpler process. Furthermore, because we use "wait" in instances where access to a reinforcer is simply delayed, there is no need to involve complex schedule thinning procedures, making the process simpler yet. Although training for form, latency and speed is not usually required, the three D-parameters are vital.

Once the sequence is smooth and reliable at this minimal duration, transfer stimulus control to a vocal cue. As before, say "wait," present the palm hand signal, and then release when the dog has ceased approaching. Repeat through several trials in order to condition the vocal cue, whereupon, you can eliminate the hand signal. Because the hand signal is so salient (it being between them and the food and preventing their approach), it may take a few more trials than usual in other training projects to get past this part of the training.

Once the vocal evocative stimulus is established, begin gradually increasing the duration criterion. If you were requiring the dog to wait for a half of a second, begin requiring a full second for several trials, and once this is smooth and reliable, set the duration criterion to two seconds. If the dog breaks the wait position before you release him or her, say "Oops," remove the bowl in as non-confrontational a manner as possible, and try again from the counter position. Move through the duration steps slowly, and avoid extinction and subtracted punishment trials as much as possible, as this will be frustrating for the dog and could lead to problematic side effects. Try to begin making the duration seem random around a gradually increasing mean average. For example, one second, then two, then a half, and then two, and then two, and then one, and then three, and then one, and then two, etc. Remember to adjust the criterion gradually enough to minimize extinction and subtracted punishment trials, and keep the progress smooth. Also, maintain minimal distraction while you work on duration.

Start utilizing a release word. To do this, say the release phrase, usually "you're free," right before you click. You can click as soon as the dog begins moving, but the use of a conditioned reinforcer is not usually necessary after a release.

It can be useful to begin working generalization with treats and toys, requiring a "wait" before accessing them when you place them near the dog. This allows you to administer more frequent trials, and makes it easier in different places. Start close to where you feed the dog, and as usual, when instating a new D-parameter, relax the others. For example, begin with minimal duration and work your way back up. This should go more quickly this time. Then, practice in different places, each time relaxing the duration and working it back up. Once this is going smoothly, begin using the "wait" cue for other situations, such as waiting before going through a doorway or the opportunity to play with another dog, etc. When you introduce a different item, reset the duration criterion and build it back up. Try to maintain minimal distraction through these trials. Once this is going smoothly, you can work in the distraction criteria. Begin using the wait for more highly effective reinforcers, that is, things that are tougher for the dog to wait for. In each new case of distraction, reset and rebuild the duration component.

Phase 4. Maintenance

Continue to generalize to different stimuli and settings. Also, continue to generalize the reinforcers, utilizing a wider range of nontrainer-mediated activity reinforcers (via the Premack principle). The task in the maintenance phase is simply to ensure that you continue to present trial opportunities to keep the training fresh. If, at any point, the training begins to deteriorate, refresh the training with more frequent sessions with initially relaxed duration and distraction, and build them back up.

Common Challenges

Some dogs will lunge or charge right through or around your hand for the food even before you put the bowl down. If you cannot get a "wait" long enough to put the bowl down, ensure that you are able to retract the bowl and prevent reinforcement. However, do *not* get into a wrestling match with the dog. Instead, try doing the exercise with a less reinforcing food item if need be and doing so after the dog has eaten a full meal. You may need to start with a less effective reinforcer. Find a reinforcer and motivative operation combination that will allow you to get through the trials successfully and build up gradually. The more effective the reinforcer you are cueing the dog to wait for, the slower you will have to go in terms of increasing the duration criterion and the more trials you will have to use. If you are getting too many extinction or subtracted punishment trials, this usually means that you are moving too quickly or not putting in an adequate number of trials before increasing the criterion.

Take it / Drop It

Phase 1. Preliminaries

Behavior Objective

S^{Ev}: Vocal "Take it."

Behavior: Take and hold an object in the mouth.

Consequence: Toy but eventually opportunity for next link.

Criteria: Latency: 2sec.; duration: maintain until released, minimum 10sec.; frequency: 100% through 10 trials; distraction: various.

Behavior Objective

SEv: Vocal "Drop it."

Behavior: Release from the mouth whatever is in it.

Consequence: Treats, food and continued social contact.

Criteria: Latency: 1sec.; frequency: 100% through 10 trials; distance: minimum 10 feet; distraction: various; time point: 6 sessions at ≈5min. each.

"Drop it" and "take it" are useful because they allow you to quickly get potentially dangerous things away from the dog. It is helpful to train "take it" and "drop it" together, since exhibiting one provides a perfect opportunity to exhibit the other and because "take it" is usually an effective reinforcer for "drop it." This also provides a good basis for training the dog to retrieve.

Phase 2. Acquisition

Utilize the same establishing operations as discussed for "sit."

Prepare a list of all the most effective reinforcer toys for the dog. Start with the least effective item, and if need be, ensure satiation. Hold the item, and encourage the dog to chew on the other end of it. Keep hold of the toy. Once the dog has it in their mouth, wait a few seconds, say "drop it," and then hold up a very effective treat. The dog will usually let go of the item to investigate the treat, especially if you started with a minimally effective reinforcer for them to hold onto and a relatively effective reinforcer to trade for it. As soon as they open their mouth releasing the toy, click, deliver the treat, and offer the toy back while you are still holding onto it. If you are confident that the dog will take the toy back, say "take it" after you present it to them, but before they actually take it. Repeat the sequence several more times, and then quickly attempt a trial without the treat in your hand. Behavioral momentum will make occurrence of the

183

behavior very likely. Repeat the sequence several more times with no treats in your fingers but maintain continuous reinforcement.

For training techniques like this one that are difficult to train with only two hands, it is usually most effective to hold the toy with one hand, and the treat and clicker in the other hand. Hold the treat between the index finger, middle finger and thumb. Keep the clicker in the palm of the treat hand, with the button available to the small or third finger. You could hold the toy and clicker in one hand, but the clicker will likely be too close to the dog's ear (the clicking sound might startle them).

Begin testing the effectiveness of the "take it" cue by waiting for a few seconds before presenting the cue. If they mouth at the toy before you present the evocative stimulus, say "Oops," and pull it away, reset the environment and repeat, offering the opportunity to wait for the toy again. Once the dog waits for the cue before attempting to take the toy, click and give the toy to the dog. Extinction and subtracted punishment are aversive, though mildly; try to minimize these trials with a graded approach. Start with just a few seconds and increase it gradually, ensuring you get very few punitive trials.

Next, run through this protocol with the next item on the list of most effective reinforcers. Work your way through each of the items on the list in this manner. Start to practice relinquishing constant contact with the item too, allowing the dog to simply drop the item, so that you can pick it up and offer it back to them. Occasionally, practice not offering the toy back to the dog. In that case, provide several treats that will take them a few seconds to find on the floor and eat while you place the item out of sight. It is important that you condition the dog such that the item sometimes will not come back. The distraction and the highly effective nature of the treats will help maintain the behavior under this condition.

Phase 3. Fluency

Once you have worked your way through most of the dog's favorite toys, begin practicing trials with other items outside of training sessions. Part of what controls the behavior is the trainer-mediated and controlled training session arrangement. Practice presenting the evocative stimulus with these items at seemingly random times, as well as when the dog is already in contact with the items, without you having encouraged them to take it first.

Begin proofing the behavior against the three D-parameters. In the case of "take it" and "drop it," all three are applicable—as always, work through one at a time and relax the others (as well as adjusting the schedule of reinforcement) when increasing a parameter level. Duration is a special case in training "drop it." Your goal is to administer the "drop it" cue, have the dog drop the item, and wait for an extended period of time, before "take it" is evoked or they are released. This allows you to present the cue from a distance and cover that distance while the dog continues to wait without taking the toy back. Practice in close proximity first, and gradually increase the time between the dog drops the item and you reinforce with the opportunity for the dog to take it again. Approaching the dog and item is likely to be a distraction, so work this aspect separately from duration training and then combine them afterward. After duration is covered, work distance gradually and incrementally. Distance, as usual, involves presenting the evocative stimulus from further and further away from the dog. You can work a sit into this sequence, if you like. This could be a behavior chain of dropping the item and then sitting.

Phase 4. Maintenance

Continue to generalize the behavior to various items, at various times, and in various places. If, at any point, any component of the training seems to be deteriorating, refresh the training, taking a run through building that parameter back up.

Common Challenges

If the dog has a history of aggressive behaviors evoked by others contacting toys that they are in possession of, you will need to be particularly careful and address that problematic behavior *before* proceeding.

One particular challenge you might face with some dogs is that they do not tend to take things in their mouths, particularly when you offer it to them by hand. In that case, break the process down into smaller steps. Begin with an item that the dog is more likely to take. Place it on the ground so that you are not holding it. Begin shaping interest in the item, and then closer and closer approaches to taking it. Initially, aim, for the dog to touch the item and then for taking the item in their mouth. Use highly effective reinforcers. Bring this under specific stimulus control. Once this is complete, you should find it much easier to train the dog to take it from your hands. To begin, hold the same toy close to their mouth, and reinforce tolerance and interest related behaviors through several trials. You may find that they begin to take it on their own, and if that occurs, reinforce it, and build on that. If they do not take it, touch the toy to their mouth to encourage them to take it. If none of this works, you may elect to work the "drop it" and "take it" behaviors without your actual contact with the toy. This will then likely involve shaping. Many dogs tend to retrieve with little or no dedicated training, and if this is the case, you might be able to take advantage of that to train the "take it" and "drop it" behaviors.

Here (Recall)

Phase 1. Preliminaries

Behavior Objective

S^{Ev}_1: Vocal "Here."

Behavior$_1$: Begin approaching the caller.

Consequence: Treats but eventually opportunity to exhibit next link.

Criteria: Latency: 2sec.; speed: at least a trotting gate; frequency: 100% through 10 trials; distance: minimum 40 feet; distraction: various.

S^{Ev}_2: Arrives at person.

Behavior₂: Sit in front of caller, facing caller, remaining still while collar is handled.

Consequence: Treats and release.

Criteria: Latency: 1sec.; duration: maintain until released, minimum 2min.; frequency: 100% through 10 trials; distraction: various; time point: 10 sessions at ≈5min. each.

This is a behavior chain since there are two distinct behaviors exhibited in sequence, with only one evocative stimulus at the beginning, and reinforcement provided by the trainer only after the last behavior. You can supplement the definition with the following task analysis:

Upon being called, the dog will:

- Approach caller

- Sit in front of caller, facing caller and remain still while collar is handled

If you have already trained "sit" to the maintenance phase, you will not have to train it as a part of this training project. If you have not trained it fully, you will need to do as part of this training project. I will describe a forward chaining procedure here, as the chain is short and simple.

Coming when called is extremely important. This behavior helps us manage where the dog is, it is useful in getting them away from potential problems, and it allows us to provide the dog with more freedom, stimulation and exercise. Recall training is often challenging, usually because guardians have managed to effectively counter-train it prior to being coached to train the behavior by a professional (and then often during and afterward training, as well). Three common mistakes are: (a) the vocal stimulus is used before the behavior is reliable, and hence the cue is weakened as an evocative stimulus; (b) guardians tend to inadvertently impose subtracted punishers, and in some cases, added punishers for coming when called, particularly outside of formal training sessions; and (c) guardians expect too much, too quickly, especially with regard to distance and distraction, which degrades the effectiveness of training. Avoid these three common mistakes, and the behavior will be much easier to train.

It is preliminary to training a recall to ensure that coming to people is always reinforcing and never punitive, or at least as close as possible to this ideal. This should be the guiding rule before, during and after training, both in training sessions and in everyday life. Reinforcing does not always mean providing treats or toys when dogs come to us. It means generally being a source of all sorts of reinforcers and the source of very few aversers. It means becoming a conditioned reinforcer, like a clicker. Avoid calling a dog in order to carry out tasks like nail trimming or getting the dog into a crate, if this will be aversive, and if it is, then work on desensitizing the dog to these things and otherwise reducing the aversiveness of them. Instead, work tirelessly to ensure that being near you, and coming to you, is as reinforcing as is possible!

A common pitfall scenario for training a solid recall occurs outside of training sessions, wherein the guardian has the dog outside and calls the dog to come inside. In one scenario, the guardian calls the dog and then brings them inside, which punishes coming to the guardian. In another scenario, the dog does not come to the guardian, because competing contingencies prevail, thereby weakening the recall stimulus. Thus far, there are two lose-lose scenarios.

It is best to avoid these scenarios to begin with, but if the guardian determines that it is unlikely that the dog will come to them in this situation, the guardian can either enforce the contingency by going to get the dog, or continue in their attempts to persuade/prompt the dog to come to them, or give up and wait for the dog to eventually come to them. Here again, there are lose-lose scenarios.

Guardians will frequently ask the trainer, "What do I do when I call the dog and they do not come to me?" In this case, remind clients not to attempt to evoke the behavior unless they are sure that the dog *will* come to them. In these early stages, a light long-line leash is an option that allows the guardian to gently control the dog. Another option is to wait for the contingencies to change (e.g., making coming inside more reinforcing than being outside). Feeding a meal or special treats upon coming inside can help in this regard, as can having favored toys only available inside.

You can of course manipulate the contingencies yourself, rather than wait for satiation of being outdoors and deprivation of being indoors. A squeaky toy might lure the dog in. Although not ideal, it does contribute to your reinforcing effect and minimizes certain lose-lose scenarios involving waiting or coercing the behavior. In any event, avoid forcibly retrieving the dog (often called "enforcing the command" by the trainers inclined to use this as a basic strategy in training the recall). If you go get the dog, this commonly results in one of two outcomes: either the dog runs away and you have now initiated the "keep-away game", or you do get the dog and the coercion of coming inside deteriorates your relationship (i.e., making social contact with you is punitive, rather than reinforcing). You do not want your approaching the dog to evoked running away, as this can be dangerous. Coercion is fraught with so many problems; it is best to focus on management and prevention, and if need be, even bribing[64] temporarily in emergencies. Arrange the environment so that you have effective reinforcers available to the dog if they come in when called, such as meals, special treats, or new toys etc.

This behavior chain involves the trainer handling the dog's collar. If this handling evokes escape behaviors, it is important to resolve that problematic behavior before proceeding to use it in training. In that case, carry out a few sessions with several trials each of delicately touching the collar and providing a treat for tolerant non-escape behaviors. Work your way up to quicker and more forceful grabs of the collar. You should do this at the dog's pace, in order to ensure that it does not evoke escape behavior at any point. The grabbing of the collar is not to be a very forceful form of handling but it is important that taking hold of the collar is not aversive.

---

[64] Bribing, in nontechnical terms refers to promising something to someone in order to get him or her to do something "wrong." The term is used more broadly here, as it currently is used in the animal training field to mean promising or showing the reinforcer before the behavior occurs in order to prompt the behavior.

Identify the conditioned and unconditioned reinforcer you will use.

Phase 2. Acquisition

Utilize the same establishing operations as discussed for "sit."

Because this behavior is frequently required in everyday life outside of training sessions, there are a few rules to which everyone should adhere. Avoid punishing either remaining near you or coming to you. For instance, avoid additively punishing coming to you by calling the dog to deliver a reprimand or to do something else that might be aversive such as clipping nails and so on. Avoid subtractively punishing coming to you by calling the dog away from reinforcing activities such as playing or interacting with another dog. Being near you and coming to you must always be reinforcing, ideally more reinforcing than any other contingency, which is a tall order so heed the rule.[65]

The "come-to-me" game is a great way to get started with recall training. You will require the dog and a helper. As always, start acquisition training in a very low distraction environment like a hallway or boring room. Each person should take turns calling the dog (without the "here" stimulus for now, unless you are *sure* the dog *will* come to you) away from the other person, clapping and encouraging the dog, in order to prompt the behavior.

If the subject exhibits the behavior eagerly and solidly, you should be able to take a short cut in the forward chaining procedure here. The short cut involves evoking "sit" as soon as the dog gets to you and then reinforcing the behavior chain. If this short cut works, proceed in this manner, cueing the recall and then when the dog gets to you, cueing "sit," and then clicking and treating once the dog is sitting. If this short cut is not viable, the full procedure involves clicking and treating when the dog arrives, and then evoking "sit" and reinforcing once the subject sits. The short cut is usually quite effective in this simple behavior chain and allows you to avoid transitioning from reinforcing both behaviors separately to reinforcing them together. You might find it useful to review the section on forward chaining.

Once this is well under way, you can begin calling the dog without a partner. Choose these times carefully! Ensure a high degree of likelihood that the dog will come to you, away from whatever they are doing.

Once these are fairly smooth and reliable, you can move on to training for fluency.

Phase 3. Fluency

Now, begin fading the prompt, and install a temporary evocative stimulus. You will still have some training to do, in order to hone certain features of the behavior (e.g., the speed with which they run to you, and the latency between evocative stimulus and initiation of the behavior), but it is useful to have a vocal cue to use in the meantime.

---

[65] One thing I personally do, because my dog loves in-shell peanuts, is drop one in the backyard and then, seemingly out of the blue, I call my dog and when he gets to me, I point to the peanut and he is allowed to pick it up. I do this at least two to three times per week. When he is in the back yard and I call him in the back yard, he eagerly comes running to me. This is just an example of one way to keep being near you and coming to you reinforcing. Find effective ways to keep the dog eager to run to you.

Some trainers like the phrase "come-on," but you can choose something else. (If the dog quickly comes when called at least 95% of the time, you can use the permanent cue at this point.) Begin fading the prompts by saying "come-on" in a happy tone, and follow that with the prompts. After a few trials, begin gradually reducing the prompts—the vocal cue, which should have taken on stimulus control, should evoke the behavior alone. If you need to provide gentle praise while the dog is on their way to you, you can use that but fade the praise prompts as well.

Now change to an intermittent schedule of reinforcement, and begin thinning that schedule as outlined previously. Unlike most other behaviors, it is usually advisable to maintain a recall on a rich schedule of reinforcement, to keep it highly reliable.

Begin refining the form, latency and speed of the behavior. As usual, work one at a time, and each time you introduce a new feature, bring the schedule back to continuous reinforcement. Remember, relax the three D-parameters and schedule of reinforcement, but do not relax form, latency and speed; once you achieve progress in these features of the behavior, maintain them.

Begin proofing the behavior against the three D-parameters. For this application, you will focus heavily on distraction. Begin incrementally increasing distraction at a pace that still allows you to achieve success in at least 95% of your trials. At first, it is usually best to arrange for highly contrived and controlled environments. Begin working in more real world distractions as you proceed to proof the behavior.

You might arrange for another kind of two-person recall game, wherein the person that the dog is near has toys or treats readily visible to the dog, providing a reinforcing competing contingency. The trainer calls the dog away from a short distance. If the stimulus fails to evoke the behavior, the distractor does not relinquish any of the reinforcers. Reset the environment and try again, adjusting the level of distraction as appropriate until you achieve success. Praise enthusiastically if the dog moves toward you in order to prompt or encourage coming to you, if needed. If the dog turns away, stop praising. Do not evoke the behavior again; present evocative stimuli only once per trial. Once the dog comes to you, evoke the "sit," gently take hold of the collar, click, provide a hidden treat, and allow immediate access to the distractor and the treats or toys in their possession. Repeat the procedure several times until the recall becomes reliable. Once it does, you can work on increasing the level and variety of distractions, including different locations. Evoking a recall away from a reinforcer is a big step—take it gradually.

This is probably a good time to perform the chaining procure. In fact, particularly if the short cut was effective, you might find that, by this point, the dog sits when he or she gets to you before you present the evocative stimulus for it. If that is the case, you can likely drop the interjected evocative stimuli and reinforce the complete chain. If not, test the stimulus control by waiting before presenting the "sit" cue to ascertain whether the dog will sit without the vocal cue. If they do, you should be able to drop the interjected vocal cue and reinforce the complete chain. If this does not work, either train through several more trials before testing the effectiveness of the stimulus control again, or begin gradually fading the interjected evocative stimuli. In this case, the evocative stimulus is vocal and so, on each trial, you would make the vocal cue quieter and quieter until there is no vocalization. Do not rush this fading process.

Once you have worked through increasing levels of distraction, several different types of distraction, and at a variety of locations, you can begin more spontaneous real-life trials. For example, you can call the dog away from increasing levels of distraction, ensuring that you reinforce the behavior. A good game to play at this stage is calling the dog away from a distraction of some kind only to immediately release and reinforce with the opportunity to reengage the distractor. This is useful, especially when the dog is playing with other dogs. In all of these cases, it is important to take a graded approach. Start by only calling the dog away from minimal distraction, then, increase the level of distraction gradually, only after the behavior is exhibited fluently.

Once you have the behavior's form, latency and meeting objective criteria, and you have worked distraction and distance up to appropriate levels, you are free to establish the final vocal cue. To transfer stimulus control from the temporary evocative stimulus to the permanent one, use this sequence: "here," followed by "come-on." This will evoke the behavior, which you then reinforce. Repeat through several trials, until you can drop the old temporary stimulus from the sequence.

Phase 4. Maintenance

Continue to generalize the reinforcers utilizing the Premack principle in order to help you maintain control over the behavior by providing the opportunity to exhibit other more effective reinforcer-contacting behaviors. If, at any point, any component of the training seems to be deteriorating, refresh the training by taking a run through building that parameter back up.

Common Challenges

By far, the most common challenge with the recall behavior is failure to achieve sufficient mastery at one level of difficulty before moving to the next. In other words, people usually expect too much, too soon in this training, and they fail to adequately manage the situation in order to ensure successful rehearsal. The recall faces competing contingencies almost by definition since the dog will usually be contacting some reinforcer when you call them. More than most behaviors (along with "off" and loose leash walking), distraction is a ubiquitous and powerful challenge. The reinforcement for approaching you must surpass the reinforcement available elsewhere.

Off

Phase 1. Preliminaries

Behavior Objective

S^{Ev}: Vocal "Off."

Behavior: Look away from whatever is being attending to but has not yet been taken in mouth.

Consequence: Treats and toy.

Criteria: Latency: 1sec.; frequency: 100% through 10 trials; distance: minimum 15 feet; distraction: various; time point: 5 sessions at ≈5min. each.

"Off" cue can be useful in helping direct dogs away from things that might harm them or to aid in other training endeavors. Although "off" is sometimes framed as "what *not* to do," the "off" behavior is best framed in a positive form. "Off" requires the dog to move their muzzle away from what they are attending to. We define the behavior objective as looking away although it can be useful to include stepping away from the stimulus as well, since proximity allows the stimulus to exert stronger control over the behavior. Proximity to the problem "thing" may become strong enough to exert control over the behavior again. It is a matter of defining the behavior. It can also be useful to evoke a "sit" immediately following the "off" stimulus, or you can chain the two behaviors, so that "off" evokes looking away followed by sitting and waiting. Alternatively, you can specifically train the dog to both look at you and sit. Again, define the behavior for your specific requirements.

"Off" can be a double-edged sword, and caution must be exercised when considering its use. It is common to inadvertently chain problem behaviors to "off." Any time that an evocative stimulus is presented during or immediately after a problem behavior, such as "off," to "redirect" a dog away from something, there is a possibility that the problem behavior will become chained to the "off" behavior. It is best to use "off" in the moments before the subject contacts something problematic (i.e., when the dog is merely attending to the stimulus from which you want to call their attention). In emergencies, you can use "drop it" to have the dog let go of something they have in their mouth. I considered leaving "off" out of the lineup in this chapter because of these pitfalls, but I personally have found it to be very useful and as long as one pays careful attention to avoiding the potential pitfalls, I believe that it could be useful for others as well.

The "off" cue, as presented here, evokes looking away from whatever the dog is paying attention to, but that which they have not yet taken in their mouth. This response class allows for various specific response class forms. It is good practice in this case to ensure that you include various specific response class forms in your training. In specific cases, it is also a good idea to provide an addendum to your definition of specific distractions. If you find that a specific response class form emerges in the training, you might also elect to adopt that narrower form as the response class.

Prepare a list of effective reinforcers, in rank order, that you can use for training. Start with a small treat as the first item, but ensure it is a minimally tasty treat to ensure success. This first one should be a small treat as it will be in your hands and used as the unconditioned reinforcer. The list should include as many items as possible, because it will be the list that you work off of all the way through to the maintenance stage. It should include things like garbage cans, other dogs, and fire hydrants—things that will attract your dog.

Identify the conditioned and unconditioned reinforcer you will use.

Phase 2. Acquisition

Utilize the standard training practices previously discussed regarding setting the dog and yourself up for success.

One way to train "off" is to shape it.[66] Begin by holding a treat in your hand far out to your side. Make sure the dog sees that there is a treat in your hand. Hold the treat out of the dog's reach and simply wait. Timing is vital here. Eventually, the dog will glance away from that hand. This is usually *very* brief the first time! The instant that they glance away from the treat, even very briefly, click and release the treat from your hand. Repeat the process several more times, requiring only a glance away from the treat. You should find that the dog more readily looks away from the hand as you repeat the process. Once you find this is proceeding smoothly, you can increase the difficulty by setting the criterion to a definite look away that lasts a full second. Repeat through several trials. Once it is smooth and reliable, set the criterion to two seconds, building it up gradually and incrementally in that manner until you have a look away for approximately five seconds.

Once these are fairly smooth and reliable, you can move on to training for fluency.

Phase 3. Fluency

There are no contrived prompts to fade, making the process a little simpler at this stage than it is for some other behaviors. Start by putting the behavior on a gradually thinning variable ratio schedule of reinforcement, and refining the form, latency and speed, as needed. At this point, the hand position functions to evoke the behavior. Once you have the behavior exhibited to criteria, establish the vocal cue for the behavior. Give the "off" stimulus, present the hand and the behavior should occur. Reinforce the behavior. Repeat through several trials until the vocal cue is established.

Once you have the vocal cue established, begin generalization training. Start by putting the hand holding the treats in different positions and directions. Then, once that is solid, begin applying the stimulus to treats that are not in your hand. Then, begin presenting the evocative stimulus for other things and in different places. Remember to relax the three D-parameters, and return to continuous reinforcement when introducing each new criterion feature. Each time you change the criterion, expect the dog to look away only briefly at first. You can gradually build the duration back up.

At this point, begin proofing the training against the three D-parameters. It is usually best to start with more controlled situations with trainer-mediated reinforcers; work with more intrinsically reinforcing real-world situations later. Start using more effective reinforcers such as better treats or toys, and then begin mild deprivation to further increase the rate of responding. It would be a good idea to plan some of your distraction items ahead of time. Disengaging from approaching or eye contact with another dog is often a high level distraction, as is approaching to sniff a fire hydrant or garbage can. Work through your list of increasingly strong reinforcers. Remember to

---

[66] This approach was conditioned in me in 1999 as I observed Carolyn Clarke's puppy classes at Carolark (http://www.carolark.com), instructed by Carolyn and Jo-Ann Steele and it has served me well ever since.

relax duration and distance when introducing a new feature. Similarly, you might work the duration up to 15 seconds. You may not need more than this for general use. Distance would include increasing how far you are from the dog and stimulus when you present the cue.

Once you have worked through increasing levels of controlled and contrived distraction, including several different kinds of distraction and a variety of locations, start more real-life trials, cuing the dog to disengage from increasing levels of distraction in the real world, ensuring that the behavior is reinforced, if not by contact with the stimulus they were called away from, then by other effective reinforcers.

Once the cue is effective in a wide variety of situations, begin thinning the schedule. It is usually a good idea to maintain a rather rich schedule of reinforcement for the "off" behavior.

Phase 4. Maintenance

Continue to generalize the reinforcers, utilizing the Premack principle in order to help you maintain control over the behavior, by providing the opportunity to exhibit other, more preferred, behaviors. Often the most effective reinforcers will be reestablishing contact with what you called them away from. However, this will not always be acceptable, so continue to include controlled and contrived trials for stimuli you can allow them to contact, and have highly effective reinforcers ready for times when you cannot allow continued contact with the stimulus in question. If at any point any component of the training seems to be deteriorating, refresh the training by taking a run through building that parameter back up.

Common Challenges

Some dogs, particularly active breeds, will jump up at the hand during the initial stages of your shaping program. You can try training at times when jumping up will be less likely such as after exercising. You can also use subtracted punishment or extinction trials for jumping up, if necessary. The best option may, however, be to lower yourself so the dog does not need to jump up to access your hand. You have to ensure you have a good grip on the treat and that you do not release it until you have clicked for a criterion response. The dog will mouth and nibble, but you will simply wait for the glance away. Typically, at first, the looks away are glances at you as if to say "Hey, what's up? Let go my Eggo." That is fine; reinforce that. This usually also involves significant patience waiting for the dog to calm down and look away from the treat.

Another common challenge is the potential for a behavior chain to develop. In some cases, you may reinforce attending to the object by providing the opportunity to respond appropriately to the "off" cue. In other words, the dog looks at the problem stimulus *because* that leads to the "off" stimulus and then a reinforcer. If moving toward and attending to something makes available the opportunity to move away from it in order to contact treats (as per the Premack principle), then you can expect the rate of contacting the problem stimulus to increase. Carefully observe for increased interest in problem stimuli after "off" training has started. It is best to manage the environment so that dogs cannot access problem stimuli and to use "off" only for situations where you did not anticipate contact with it. If a problem chain is conditioned, try to manage access to the stimulus as part of an antecedent control strategy instead of continuing to strengthen a behavior chain. Another tip is to use the "off" cue for a wide variety of items in training, but not extensively for any one item. If you find that you are using the "off" stimulus for one item a lot, you should find a way to preemptively manage access to that item to prevent use of the "off" cue.

Let's Go (Walking on a Loose Leash)

Phase 1. Preliminaries

Behavior Objective

S^{Ev}: Leash on.

Behavior: Walking adjustment behaviors such that the shoulders stay within 5 feet of the handler [Leash tightens at 6 feet indicating a non-criterion behavior].

Consequence: Treats and forward progress.

Criteria: Frequency: 100% through 10 trials; distance: maintain minimum 3 standard city blocks (approximately 1400 feet); distraction: various; time point: 14 sessions at ≈5min. each.

Work with a six–foot leash. Walking with a slack leash rather than a tight one is important and challenging to train. Some might say pulling is a "natural" behavior, but of course, all behavior is natural. It is simply that the general strategy of quickly approaching something reinforcing tends to be prepotent over approaching more slowly or walking away from it. Furthermore, most dogs move at a quicker pace than most people do. In addition, the opposition reflex can often encourage dogs to pull against a tight leash. Most dogs, even puppies, who pull on leash have a strong history of reinforcement for that behavior. Thus, it can be very challenging to avoid the problem behavior while you establish a preferable one.

Some trainers tend to frame this behavior in a negative form, stating what behavior to avoid, rather than the behavior to exhibit. However, we can frame the response class of walking on a loose leash positively as "walking with their head within five feet of the handler." The trainer may specify a side to walk on as part of their criteria (e.g., the trainer's left or right side). To train a formal "heel," the trainer may also specify a very close proximity level and specific position in relation to the handler's left leg. The reason most trainers frame loose-leash walking in negative terms is that a tightened leash is a clear and reliable indication that a mutually exclusive non-criterion behavior is occurring and hence the criterion cannot be occurring.

There are several ways to train loose-leash walking. I will provide one straightforward method here. Before beginning, prevent pulling on the leash as much as possible in non-training session times. Every instance of pulling on the leash will make progress through the training process more challenging, because you will have to counter-condition it. I recommend that you use a standard body harness, rather than a neck collar, as this will reduce the opposition reflex that will challenge your training. It is probable that there are times when the dog will be less likely to pull. Try to identify these times and begin training then. For instance, many dogs will pull less when they are walking inside rather than outside, or walking toward or away from home rather than the opposite, or after a good bout of exercise. Many dogs pull less in novel environments/neighborhoods. Work your way through a ranked list of increasingly challenging environments. This will help set both you and the dog up for success.

Identify the conditioned and unconditioned reinforcer you will use.

Phase 2. Acquisition

A graded approach works well for training loose-leash walking. Rather than providing the dog with the opportunity to either pull or not, and reacting differentially to each with reinforcement and punishment, set yourselves up for success by setting the criteria low, then gradually and incrementally increasing the level of difficulty, to keep the progress smooth and efficient, with a very high rate of success. Begin by holding the leash with your dog seated or standing still beside you, facing the same direction. Press the hand that is clasping the leash up against your belt line and keep it in this position. This ensures that the distance to the end of the leash is a stable six feet at all times, and the distance will not be modified by your reach. You can place your thumb into the waistband. This acts as a quick release so that you don't get pulled over if the dog bolts for some reason, your thumb will come out of your waistband, and you'll have a brief interval in which to prepare to maintain your hold on the leash and prevent a sudden jerk on the leash.

Although being on leash will set the occasion for keeping close to you, a vocal cue such as "let's go" can be useful since the dog may interact with the environment on walks, and you will want a cue to proceed with walking. When initiating your movement, say "let's go," take a step forward, and as the dog begins to walk, click and treat after their first step. Repeat through several trials. Once this is smooth and reliable, set the criterion to two steps. If the dog bolts forward at any point and the leash goes tight, say "Oops," stop, and do not move or engage the dog for a few seconds. Evaluate whether you are moving too quickly through the task increments, and reduce the number of criterion steps if needed, to ensure that you minimize the number of subtracted punishment trials. Proceed after the subtracted punishment interval and once the dog is not pulling on leash. In other words, count the time out interval and then when you have finished counting, you may begin walking again as soon as the leash goes loose if it was not already loose when the interval was up. Continue to work in this manner, varying the criterion in a seemingly random manner, until you can get several loose-leash steps and very few non-criterion behaviors.

Use direction changes to help get more steps per trial and avoid subtracted punishment trials. Walking on a loose leash is really made up of a long series of evocative stimuli and modifications to walking behavior; the dog attends to your trajectory and speed, and these stimuli evoke the changes in their trajectory and speed. When you change directions, you encourage the dog to pay close attention to your location, and it puts you further ahead of them, which allows for a greater number of steps before they pull on leash. It also increases interest. Once you have taken several steps and are getting to your current limit, try changing directions at a 45 to 90 degree angle. Consider using a greater angle relative to how close the dog is to the end of the leash. For example, if the dog is still walking close to you, a slight change in direction can spice up the walk a bit, but if the dog is getting close to the end of the leash and it is about to go tight, use a wider angle, maybe even 90 degrees or more. Avoid jerking the leash though. Gentle pressure on the leash will come to evoke attending to you for directional cues, which is easiest for the dog when they are closer to you so they can see you in their peripheral vision.

Novice trainers often use the leash during training to generate the behavior, rather than focusing on the training. Maintain a consistent leash length. Again, to keep the leash at a constant predictable length (not adding your arm length to it) and prevent inadvertent prompting with it, you can hold the leash with your hand pinned to your hip.

Phase 3. Fluency

Continue to work toward more steps between reinforcers, thinning the schedule as you proceed.

Next, begin proofing against the three D-parameters. You can use reinforcers found in the uncontrived environment as distractors. For example, if the dog indicates an interest (e.g., attention and approach behaviors) in sniffing a hydrant or meeting another dog, make access to these things contingent on walking close to you until you arrive at the thing. Reserve this exercise for when you are far enough along in the training to be confident that you will succeed. You can include a release phrase (e.g., "you're free") but one is not usually necessary. The "let's go" stimulus can be used to get back on track.

Duration would be how long you walk before reinforcing. Work on the duration gradually, and as usual, relax other D-parameters when you increase the duration.

Distance is not applicable to walking on a loose leash since it is an up-close behavior only.

Phase 4. Maintenance

Continue to generalize the reinforcers utilizing the Premack principle in order to help maintain control over the behavior with the opportunity to exhibit other more reinforcing behaviors. If at any point any component of the training seems to be deteriorating, refresh the training by building that parameter back up.

Common Challenges

Rehearsal of Pulling

By far, the most common challenge in training walking on a loose leash is that guardians allow too many reinforced occurrences of pulling behavior. If training is to be effective, the dog must *not* be allowed opportunities to rehearse the problem behavior, even though this is indeed an inconvenience. If walking on a loose leash is ever to be successful, trainers must extinguish pulling! Help clients brainstorm ways to prevent and manage situations where the dog might pull while on leash. Where necessary, use a distinct and salient restraint device when the dog will pull, to help protect the training associated with the usual harness (via discrimination training).

Diehard Puller

A strong history of reinforcement for pulling can be a major obstacle to training. In these cases, the same procedure discussed above can be used, but the progress will be much slower, and the training will be take longer, because pulling will need to be extinguished. In some cases, an anti-pull harness might be worth considering.

Trailing Behind

Some dogs hunker down like the proverbial "stubborn donkey" and refuse to move forward. They may or may not pull forward on leash at other times, but when a dog refuses to move forward, attempting to pull or otherwise force them forward, usually results in a longer battle. First, identify whether the dog may have medical problems, perhaps with arthritis or joint dysplasia, or perhaps the dog is heat intolerant. Determine whether the dog seems fearful of something specific, and if so, avoid that situation while trying to train loose-leash walking. Address the fear reaction with a proper behavior change program. If the dog trails behind in specific situations, then perhaps a previous experience with this situation was overwhelming or otherwise aversive. However, if none these factors appear to be an issue, then be patient. Wait a few moments with a slack leash, and perhaps then gently verbally coax (prompt) the dog forward. If the dog does not respond to this, discontinue coaxing, and begin a shaping program for moving forward. The benefit is derived from not only achieving reinforcement opportunities, but also that the added reinforcers will respondently counter-condition any problem emotional arousal occurring during the situation. First, be patient and have fun, *especially* at these times! Start by evoking a few other behaviors (e.g., down and sit or look) that you can reinforce and have fun with, perhaps including a game of tug-of-war. You could also try shaping the behavior. Observe for the slightest relaxation in stance or a forward lean, and click and treat, shaping forward motion. Repeat several times, and then adjust the criteria to require slightly more of a lean or movement forward. In this way, shape the walking forward behavior. Work your way up to places/situations that more readily evoke reticence.

Dog is Too Strong or Big Compared to Guardian

When the dog is exceptionally strong or large compared to the guardian, good training practices are even more important. Rather than manhandling the dog, identify and instate effective reinforcement. Carry out the training as described, but pay particular attention into ensuring successful trials, particularly with raising distraction levels. That said, this situation could be particularly challenging because if the dog does barge forward toward a reinforcer, the guardian may not be able to effectively ensure extinction. Consider using an anti-pull harness for these dogs for such situations. Avoid relying on anti-pull devices to make pulling aversive. Focus on controlling the dog's behavior with added reinforcers. However, for times when the guardian simply cannot effectively control the dog, the anti-pull harness may improve controllability. If the situation is extreme, then the guardian should consider hiring a capable dog walker.

Dog is "Crazy"

Although professional trainers, concerned with clarity, do not use such vague terms, trainers are quite familiar with the complaint that walking on leash is impossible, because the dog is "crazy." Determine specifically what actual behaviors the subject exhibits and what stimuli are controlling them. This encompasses various scenarios such as a dog that expends extensive energy engaging in many behaviors rapidly (e.g., bolting in different directions, chewing at the leash, bouncing or jumping). The key element is that the dog engages in numerous and various non-criterion behaviors rapidly. This is common with some puppies and dogs new to leashes, or dogs with bodies in an activity-deprived state. It can also be common with dogs that are fearful. These non-criterion behaviors function to self-distract or escape something.

If the dog is new to leashes, introduce the leash slowly, and shape tolerance-related behaviors with increasing contact with the collar and leash. If the dog is fearful, instate a behavior change program to change this, and come back to loose-leash walking when you can, or else find a time and place where you are able to bypass these behaviors. If it turns out that the problem is associated with excess stimulation, find a minimally arousing environment for the initial stages of walking on a loose leash. Work in increasingly arousing environments as part of proofing for distraction practices. Carry out the training after an extensive exercise or play session, or identify times when the problem is less likely to occur. Exercise is commonly very helpful in these situations. In any case, arrange to set the dog up for success as much as possible and maintain the basic strategies of training. Make the behavior you want more likely and other behaviors less likely, set the criteria low, and gradually and incrementally work your way through the levels of difficulty with a graded approach. Try to avoid participating in the "craziness" and instead be calm, contributing as little stimulation and reinforcement as possible to the situation. Consider cuing a few other behaviors and calmly reinforcing those if you believe you can get the behaviors under such distraction. Reinforce any instances of calmness in a calm way. In this way, you will shape calmness and participation in walking. Ensure the process is a calm type of fun for the dog.

Go to "Bed"

Phase 1. Preliminaries

Behavior Objective

S^{Ev}_1: Vocal "Go to bed."

Behavior$_1$: Proceed to a designated mat.

Consequence: Treats but eventually opportunity to exhibit next link.

Criteria: Latency: 2sec.; speed: average walking gate; frequency: 100% through 10 trials; distance: minimum 15 feet; distraction: various.

S^{Ev}_2: Arrives at mat.

Behavior$_2$: Lie down on the mat.

Consequence: Treats, comfortable bed and access to specific toy.

Criteria: Latency: 1sec.; duration: maintain until released, minimum 6min.; frequency: 100% through 10 trials; distraction: various; time point: 4 sessions at ≈5min. each.

This is a behavior chain comprised of two distinct response classes. In that case, you can supplement the behavior objective with the following task analysis:

Upon "go to bed" being vocalized, the dog will:

- Proceed to the designated mat
- Lie down on the mat.

Just as with recall training, there are two simple behaviors involved here, one of which is likely already trained to fluency. If the dog does not already exhibit "down" fluently, include training it here as a component project.

You can use this evocative stimulus to direct the dog to go to specific places. The actual stimulus you select is based on the location where you want the dog to go. You can use this to train a dog to go to a crate or bed, and this is the context outlined here. However, you can use it to direct the dog to other locations, or to find and go to specific people. The training process here is for the vocalized "go to bed" to evoke going to the mat or bed and then for the dog to down on the mat or bed.

Phase 2. Acquisition

Utilize the usual establishing operations.

The most effective and efficient way to train this behavior is with the use of a remote treat dispenser since it helps with administering treats from a distance. I will, however, outline the training here without such a device. Remember to begin in a minimally distracting environment.

Forward chaining is suitable for a simple chain of behaviors such as this. Assuming the "down" is fluent, begin by training the dog to go to the mat. Start by standing only about three feet from the bed with the dog beside you. Present the temporary cue of pointing to the mat, and immediately lure the dog with a treat to the bed. You face the same choice as with the recall as to whether you can use a short cut or carry out the longer chaining procedure. If the behavior of going to the mat is smooth and reliable, you may be able to simply evoke going to the mat, prompt it, and once it is exhibited, evoke the "down" and then reinforce that when the dog is lying on the mat. If this is not viable, evoke and reinforce going to the mat and then evoke and reinforce lying down on the mat. Repeat through a few trials. Begin fading the food-in-hand prompt to an empty handed prompt.

You should now have already trained both behaviors that compose the chain this point, though minimally and not yet linked together.

Phase 3. Fluency

Just as with recall training, you might find that the dog lies down on the mat before you present the cue. You may need to delay presenting the cue for "down" momentarily in order to determine whether "down" is now under stimulus control of arriving at the mat. If it is, reinforce this and repeat through several trials to further strengthen the stimulus control. If not, then you may need to fade the vocalized "down" cue, saying it quieter during each trial, until the stimulus is absent and the behavior occurs, evoked by arriving at the mat.

Put the reinforcement on an intermittent schedule, and gradually thin it in a variable fashion, as described previously for other behaviors.

Next, begin proofing against the three D-parameters. If you begin with duration, remember to keep distance and distraction minimal, and to briefly reinstate a continuous reinforcement schedule. Instead of clicking immediately upon the dog getting to the mat or lying on it, you should wait an extra second or two and then click and treat. Gradually work the duration up to a few minutes, and begin thinning an intermittent schedule of reinforcement again.

Once you have a reliable duration, begin working distance. Go back to continuous reinforcement and relax the duration feature as you start gradually increasing the dog's distance from the mat when you evoke the behavior. Ensure that you include the special feature of cueing the behavior, when the mat is not directly within the dog's sight. Again, once the distance criterion is reliable, put the behavior back on a gradually thinning variable ratio schedule.

Finally, you can proof the behavior against distraction. Start with minimally distracting stimuli present while relaxing the schedule of reinforcement, duration and distance criteria. Begin introducing incrementally more distracting stimuli into the environment, until the dog reliably exhibits the behavior, even under moderately to very distracting stimulation. Put the behavior back on a thinning schedule of reinforcement, and then begin working the features together.

Some dogs will run smoothly through this process, while others will require remedial work at certain stages. Be prepared to track your progress and adjust as needed to ensure success. You should now be able to evoke the behaviors from a greater distance, under distracting circumstances, and the dog will reliably go to the mat, lie down, and remain there until released. At this stage, replace the click with a release phrase.

Once the behavior chain is about where you want it in terms of the criteria, transfer stimulus control from the temporary cue of pointing, to a permanent vocal cue. Say "go to bed," and then point to the mat, at which point the dog will exhibit the behavior and you can reinforce it. Repeat through several trials until you can drop the old temporary cue from the sequence.

Phase 4. Maintenance

Continue to introduce new distractions as they become apparent, thin the schedule of reinforcement, and keep up regular practice to ensure the training is maintained. If at any point any component of the training seems to be deteriorating, refresh the training by building that parameter back up.

Common Challenges

One of the major challenges in training a dog to go to and lie down on a mat is to maintain the effectiveness of the reinforcer. Usually, lying on a mat, even with a treat coming their way after some period of time, cannot compete with concurrent contingencies. That is, the dog lies down, but then gets bored and gets up to do other things instead of staying on the mat. Use highly effective reinforcers. Another trick to addressing this is to consider providing some ongoing reinforcer at the mat location. You could tie a rope to a Kong, and attach that to the crate or an eyehook screwed into the baseboard near the mat. You could have treats in the Kong or peanut butter inside of it. This will keep the dog interested and make being at the mat reinforcing, since this is the only place they can access that particular reinforcer. You can also get remotely released treat dispensers that you can keep near the mat.

Training Challenges and Special Cases

Multi-Dog Training

Karen London and Patricia McConnell wrote a terrific book called *Feeling Outnumbered? How to Manage and Enjoy Your Multi-Dog Household*, in which they outline a basic strategy for working with dogs who reside together. Being among the first to abandon the outdated practice of "supporting the hierarchy," London and McConnell (2001) outlined an operant conditioning approach that has proven very successful for many trainers, myself included, over the years. I outline this general approach below.

Train Each Dog Individually

Guardians should arrange to have time alone with each dog for training. This can be incorporated into walks and play sessions, or exercise sessions as well as training. With multiple dogs providing a significant distraction for each other, it can be challenging to compete with, and soon, stimuli lose their evocative capacity. The guardian must train each dog individually to start. Whereas it might be omitted when training an individual dog in many cases, ensure that the dog's name precedes all other cues. This will be a cue for attention and a function-altering stimulus for exhibiting the primary behavior.

Train in Pairs

Once you are at a reliable point in individual training, you may start training the dogs in pairs, even if you have several dogs. Ensure you precede cues with the dogs' names. Carry out much the same training as you did with the dogs individually, but review it with the other dogs present to ensure appropriate discrimination training so that each dog exhibits the behavior that you present the cue for and other dogs do not. At first, other dogs may respond to the cues but discrimination will occur when you fail to reinforce those responses that were not preceded with their name.

You can also work on group cues, using a group name such as "Everyone," followed by the primary cue. Only those dogs who exhibit the behavior are given treats. This group cue will come in handy when the client wants all the dogs to respond similarly together. For instance, it can be useful to be able to evoke sit from all of the dogs.

Train in Groups

Once you have worked your way through training each pair combination to a reliable level of distraction, you can start working with combinations of three dogs. The training will proceed in much the same way as with pairs. Once significant progress has been made in these small groups, in each possible combination of three, you can start adding in any other dogs within the household, until you have your whole group working together. Clients should be encouraged to maintain the training by continuing to work with the dogs individually and as a group, and applying the training to everyday life when possible.

Training Toy Breeds

Toy breed dogs are frequently fearful, especially when close to people walking around, because they are at significant risk of being stepped on. Consider at least administering the acquisition stage of training while you are seated on the ground next to the dog, both to prevent emotional arousal from moving around near them, but also to get closer to the dog in order to achieve the maneuvers appropriately. This will need to be faded.

Another challenge with training some toy breed dogs, is not really a challenge with the dog him or herself, but rather with guardians. Because one can physically manipulate toy breeds so easily, people frequently do not train their toy breed dogs. Sometimes they simply fail to recognize that there is a reason to do so, and/or sometimes they simply find it challenging because of the dog's size. Lack of training is likely the major reason for the stereotype of the "snappy and yappy" toy breed dog. People fail to take these and other problem behaviors or lack of training as a serious problem. It is important to recognize the need for basic training and to emphasize this to guardians. Being "snappy and yappy" is *not* inevitable for toy breed dogs. A well trained toy breed dog is a joy to be around.

Social Contact not an Effective Reinforcer

For some individuals, social contact is not an effective reinforcer, and this can affect training quite dramatically—people become quite used to utilizing social reinforcement. Trainers often tend to identify a treat as the reinforcer for a behavior they aim to train, but reinforcement is frequently a package of stimuli including the social contact that comes with exhibiting social behaviors (e.g., during training) and receiving the reinforcer. Even if praise or touch are not used during training, and vocal and visual prompts are not used either (they frequently are), the social contact that otherwise accompanies training contributes to generating and reinforcing behaviors. With this major source of motivation reduced or eliminated, training can be more challenging.

Social reinforcement is composed of a combination of conditioned and unconditioned reinforcement. A history of aversive conditioning can overshadow the effectiveness of social reinforcement. Furthermore, a lack of effective early socialization can prevent social contact from becoming reinforcing. There might also be genetic influences on the effectiveness of social reinforcement. Careful history talking might provide clues as to the cause of the current social motivation. If the dog lacks a history of effective socialization, remedial socialization may help. If the dog has a history of aversive conditioning mediated by humans, a careful rehabilitation will be required. In either case (remedial socialization or rehabilitation), gradual graded added reinforcement can increase the capacity for social reinforcement. Generally, you can improve social motivation by reinforcement-rich social interaction. Identify and eliminating aversive social contingencies in the subject's daily life.

Satiation, with regards to contact with a person, can reduce the reinforcing effect of contact with that person. Normalizing the amount and intensity of contact with the person can reduce the satiation effect and increase motivation. This is not to promote social isolation and strong deprivation states! This is about fixing problem situations where too much social contact is causing a satiated body and hence aversive social relations. Next, increase the magnitude and rate of contact with reinforcing contingencies associated with the dog and trainer. Hand feeding of favored treats and food can help. Participate in games and other reinforcing activities. End sessions on a positive note before the dog becomes excessively satiated with the contact. Try to become a source of more reinforcement for the dog in general without overdoing it. As social motivation increases, training should become more fun and productive for both the dog and trainer.

Food not an Effective Reinforcer

For many dogs, food is simply not very effective as a reinforcer. In many cases, you can find better treats and deprive the dog of this most effective reinforcement to maximize its effectiveness. When this does not increase the effectiveness of the treat as a reinforcer, consider other reinforcers such as contact with a favored toy, perhaps involving a quick game of tug-of-war with the trainer. For some dogs, praise and certain specific kinds of physical contact can be an effective reinforcer. The trainer may need to experiment with different kinds of social contact to determine what is reinforcing; usually, gentle to moderate contact on a shoulder works well. Observe what the dog expends significant energy contacting on a daily basis for reinforcers, including activity reinforcers. Once you have a list of effective reinforcers, consider *mild* deprivation to increase their effectiveness.

Hyperactivity

Some dogs are so hyperactive that it becomes a significant disruption to training. This is common with certain breeds and with puppies/adolescents. The most common solution for this problem is exercise, manipulation of excess stimulation during training and shaping focus, attention and calmness. You must design exercise sessions carefully, ideally with the help of at least a veterinarian, if not a canine fitness consultant with specialized skills in working through aerobic exercise programs, and ideally implemented under supervision. However, in many cases, guardians alone can increase exercise, as long as they operate within the medical and biological condition of the dog, and gradually increase the dog's exercise level. The sessions should be scheduled regularly and ideally involve games to make it fun. Ensure warm ups and cool downs, and observe carefully for indications of fatigue. You might notice some benefits of moderate to intense appropriately implemented exercise programs after only a few sessions, but you will recognize most of the stable physiological benefits after several weeks of exercise. The trainer should also choose times of day that are more conducive to training.

On top of large muscle mass related physical conditioning, behavioral conditioning can also help. You can improve focus in general with training games that emphasize patience and focus, as well as creativity and persistence in general (see O'Heare, 2011).

General Sensitivity and Risk Averse

Dogs, who are generally sensitive and risk-averse, require a careful approach to training. Trainers need to attend to and recognize escape contingencies and sensitivities so that they can avoid these forms of stimulation while working through training. In some cases, these dogs are sensitive to social pressure and cower when someone towers over or approaches them. In these cases, train while sitting beside the dog, perhaps right on the ground. Face slightly away and avoid staring at the dog. Avoid sudden movements and be calm, but gently praising. Train in an environment that you are sure you can control. Take extra care to avoid aversive stimulation, such as techniques involving extinction or subtracted punishment trials. Ensure success by moving at an appropriate pace. Utilize a very careful graded approach, and begin introducing simple shaping exercises with minimal prompts. Generally, free-shaping can reinforce creativity, persistence and resilience in general (O'Heare, 2011). Severe cares may require a full behavior change program.

Easily Frustrated and Impulsive

Frustration refers to the blocked access to reinforcers and the emotional arousal elicited by extinction. *Impulsivity* refers to the tendency for a dog to seek out immediate smaller reinforcers when faced with concurrent contingencies with delayed access to a much greater source of reinforcement—they go for the quick fix, rather than putting in slightly more time and effort for a much greater gratification. One must handle dogs that are impulsive and easily frustrated carefully. They can be conditioned to delay gratification and experience less frustration if the contingencies are arranged so. The trick here is to take a graded approach, specifically with regard to response effort and duration features of training. Gradually train behaviors that require duration but move at a pace that allows the dog to succeed with contacting highly effective reinforcers that require a bit more effort and time. Similarly, use shaping exercises with minimal prompting to condition creativity and persistence that will allow the dog to easily work around frustrations. By emphasizing success generated by creativity and persistence, the dog will begin respond to frustration with other novel productive behaviors and strategies that will access the reinforcer, rather than strange non-criterion behaviors (O'Heare, 2011).

Continuing Education

Courses provided through The Companion Animal Sciences Institute at www.CASInstitute.com address all of the topics covered in this chapter.

O'Heare (2011) provides a detailed treatment of how to train for creativity, resilience and persistence as well as the rehabilitate companion animals suffering from disempowerment.

If you are looking for a good book to give to clients, consider these books:

- Dog-Friendly Dog Training by Andrea Arden

- The Culture Clash by Jean Donaldson

- Getting Started Clicker Training for Dogs by Karen Pryor

- The Power of Positive Dog Training by Pat Miller

- The Toolbox for Building a Great Family Dog by Terry Ryan

The Science and Technology of Animal Training

CHAPTER 8. TRAINING CATS

Behavioral Objectives

The objective of this chapter is to measurably expand the reader's repertoire of behaviors in relation to describing and relating the principles of behavior. Upon successfully integrating the concepts outlined in this chapter, the reader, where exposed to contingencies to do so, will accurately:

- Implement a training plan for common good manners-related behaviors, applying the systematic strategies previously discussed based on a systematic plan previously discussed.

The behaviors discussed below are the everyday behaviors that help guardians manage their cat, so that everyone can remain safe, and so the cat can easily fit in with human society.[67] Training each is the application of the systematic strategy outlined above, as with the behaviors discussed in the previous chapter on dogs.

Pay particular attention to the "targeting" and "sit" behaviors. These behaviors include many details that I will avoid repeating in the behaviors that follow them. The details are important, but it is simply too tedious to read the exact same material for every single behavior.

You might notice there is no "stay" behavior. Staying in the position resulting from the behavior is simply the duration parameter of that behavior, and is therefore, controlled by the cue that evoked the behavior. For instance, the cue to sit colloquially means to "sit and remain in that position until some other behavior is evoked," rather than "sit and then you are immediately free to do anything else you want, except if stay is evoked." Stay is redundant; it is simpler to treat it as the duration parameter. Where a cat is already in the position you wish them to remain in for some duration, use "wait." "Wait" is similar to "stay," except that the subject is already exhibiting the behavior and you did not evoke it. If the cat is already in a position you want them to remain in, cue "wait," but if you evoke a behavior that involves duration, maintaining the position until released is simply the duration parameter of that behavior.

Training Cats

People unfamiliar with cats sometimes believe them to be less socially motivated than dogs, however this is not accurate. The effectiveness of social contact as a reinforcer is simply different for cats than for dogs. Some (but not all) cats can satiate more quickly than some (but not all) dogs with regard to social contact reinforcement. This is not always true and generalizations are not hard and fast here, as you might guess from all the qualifying terms in this paragraph. Nevertheless, one can often get away with a long dedicated training session with dogs, even once they have become satiated with regard to the interaction but one can rarely get away with this with cats. It

---

[67] I would like to thank Jacqueline Munera, CCBC, CCBT, CAP2, for reviewing a draft of this chapter and providing feedback that made the chapter better.

is likely best in most cases to keep training sessions with cats shorter than with dogs; leave the cat "wanting more" as they say. The most important point about generalities is to remember that each subject is different; do not let generalities override consideration of each individual's particular requirements—that would take generalization to prejudice. And, if you are one of those who believe that "cats cannot really be trained," search the topic on public social video sharing web sites and you will be exposed to cat training events that will effectively recondition your opinion behaviors.

Equipment

For training cats, you will require a clicker, a treat pouch, and small treats that are consumable quickly. You will also need a targeting stick for some training projects. For training a cat to walk on a leash, you will need a leash and properly fitted body harness for them.

Targeting

Phase 1. Preliminaries

Behavior Objective

S^{Ev}: Vocal "touch"

Behavior: Nudge the target item with nose.

Consequence: Treats and play.

Criteria: Latency: 2sec.; frequency: 100% through 10 trials; distraction: various; time point: 3 session at ≈8min.

Target training is one of the most important and fundamental behaviors you can train because you can use touching a target to get a cat to go to any particular spot you want and even to remain at that spot as well as to engage in certain behaviors as they go to the target. For example, you may want to train a cat to go to a carrier or bed and you can use the target to train the cat to go there. As another example, you may want the cat to remain in a specific spot, such as on a grooming table or veterinary exam table, and you can use the target to train the cat to remain at the target until released. A final example might involve training a cat to spin in a circle; you can use the target to get the cat to follow it around in a circle and then the target can be faded. You can also use the target to train a cat to work through an agility course or to retrieve an object. It is an excellent starting point for training because it allows for conditioning the cat to take direction in general and to be patient.

You can use just about anything for a target. You may want to have a stationary target such as a very small traffic cone. It is also useful to have a handheld target stick.

You can buy targeting sticks with clickers built into them (e.g., the Clik Stik® clicker and target stick[68]).

Identify the most effective reinforcers for the cat. Carry out the training when the cat has not accessed the reinforcer recently to ensure they are not satiated and that it indeed will be an effective reinforcer. Increase rate of responding by conducting training when the cat is not satiated with regard to either your social contact[69] or food. Gather and prepare your equipment including the target, treats and a clicker.

Phase 2. Acquisition

Ensure that you begin training in a low distraction environment, so that there are few competing contingencies that could control the cat's behavior. You can allow for increased distraction later in the process—for now, set yourself and the cat up for success with a graded approach.

Start by presenting the target level with the cat's nose at a few inches from it. The novelty will likely prompt touching the target as the cat smells it. If this occurs, click and treat immediately while removing the target, which will reset for the next trial. If the cat does not touch the target, try rubbing a treat on the end of the target to prompt a touch. In some cases, you might need a little patience while you wait for the touch. If you think that this is taking an unreasonable amount of time, you may need to shape closer and closer approach approximations, until you achieve the terminal behavior of touching the target. Once the cat touches the target and you have reinforced it, the rate of the behavior will likely significantly increase. Keep the momentum going, ending the training session before the cat is satiated with regard to the reinforcer. In your next training session, review what you have achieved thus far with several touches before proceeding.

Phase 3. Fluency

Once the behavior is smooth and reliable under minimally demanding conditions, begin placing the target in various locations, so that the cat is required to move in different directions to nudge the target and even take a step to contact it, including lifting the front paws off of the ground to reach it.

Put the reinforcement on a gradually thinning variable ratio schedule of added reinforcement. You will take the schedule back to a continuous reinforcement schedule when introducing any new features or changes in criteria.

If you need to change the form, chances are that it will be a small adjustment. In this case, have a clear picture of exactly what form adjustment you need and shape that change by differentially reinforcing approximations toward that form. For example, the cat may be targeting with the side of their face near their whiskers and you want the touch to be right on the tip of their nose. If the latency is too long, gradually reset a

---

[68] http://store.clickertraining.com
[69] Social contact does not necessarily refer to touch. It can refer to simply being close by so that the cat can see and/or hear you.

shorter the latency criterion, reinforcing criterion responses and extinguishing non-criterion responses. Continue in this way until you achieve the appropriate latency. If the speed of the behavior is too slow, do the same thing, gradually resetting the speed criterion, requiring slightly quicker responses, until you achieve a suitable speed.

Once you have a reliable behavior trained with the appropriate form, latency and speed exhibited under minimal levels of distraction, you can transfer stimulus control to a vocal cue. As it stands, presentation of the target stick evokes the behavior. However, having the behavior under vocal stimulus control can allow for targeting other objects, which can be quite useful. To establish a vocal cue for targeting, simply present the vocal cue, such as "touch" and then present the target. The cat will exhibit the behavior and you can reinforce it. Repeat this sequence through several trials, reinforcing each occurrence of the behavior.

Once the behavior is reliable with an appropriate form, latency and speed, begin proofing against the three D-parameters. Duration is not generally applicable in this case unless you are also working on training the cat to remain at the target for an extended interval, in which case, you can gradually increase the duration. If you do work on duration, do not forget to use a release phrase, such as "you're free" when the duration is completed. Next, begin working distance by gradually requiring the cat to move further in order to nudge the target. By this stage, the cat should be moving at least a few steps, but vary the distance in a seemingly random manner, so that you gradually increase the distance to between five and 10 feet. Watch the cat for indications that you are moving too quickly. If the latency begins to increase, you will know that you are moving too quickly through the distance parameter. Once you have increased the distance to criterion, begin working on distraction. Have other people present and engaged in gradually increasing levels of distraction with them. Move at a pace that ensures success.

If you chose to bring the behavior under the stimulus control of a vocal cue, begin discrimination training. Reinstate continuous reinforcement for vocally cued responses and instate extinction for responses that you did not cue. Vocally cue targeting through several trials in succession, reinforcing each instance of the behavior's occurrence. Then, follow with presenting the target without having first evoked it vocally. Extinguish touching the target when you do not first present the cue. Repeat through several trials, vocally cueing the behavior or not in a seemingly random order and continuing to reinforce touching only when you present the vocal cue. Eventually, the behavior will only occur when you present the vocal cue. Begin re-thinning the schedule of reinforcement again.

Phase 4. Maintenance

Once you have the final form, speed and latency criteria of the target behavior, it is under stimulus control, and reliably proofed through the three D-parameters, you can begin working toward maintenance. There is not always necessarily a clear line between fluency and maintenance phases. You will likely want to continue to develop proficiency in new locations or with new distractors etc., but once you are well along the process, it is time to begin transitioning from the intensive training activities of the fluency development phase to less intensive maintenance of the fluency you have achieved.

Begin generalizing the reinforcers from just the treats you used to other treats and even to play (e.g., Cat Dancer® interactive cat toys, Neko Flies® cat toys,[70] or a small ball). Remember also that the clicker is just for acquisition and training toward fluency. Discontinue using the clicker; just use the unconditioned reinforcers or the release phrase and unconditioned reinforcers if you extend duration. In addition, begin using fewer trainer-mediated reinforcers. Use activity reinforcers via the Premack principle, in order to help you maintain control over the behavior. For example, if going out onto a screened porch or sitting on your lap is reinforcing, require a target trial or two before providing these highly reinforcing opportunities. The goal in this phase is to work toward simply maintaining what you have achieved in your training sessions, with minimal contrived activity outside of the vocal cue. If, at any point, any component of the training seems to be deteriorating, refresh the training by building that parameter back up.

In training targeting, you have actually trained several things. You have conditioned taking direction and attending to trainer-provided cues. Every time you train something, you are actually training a more collaborative/cooperative relationship between you and the cat. This will pay off when it comes time to train other behaviors.

Sit

Phase 1. Preliminaries

Behavior Objective

S^{Ev}: Vocal "Sit."

Behavior: Contact their rear end to the ground with front legs straight at as close to a 90-degree angle to the ground as possible.

Consequence: Treats and play.

Criteria: Latency: 2sec.; duration: until released, minimum 1min.; frequency: 100% through 10 trials; distance: minimum 20 feet; distraction: various; time point: 5 sessions at ≈5min. each.

There are at least two response class forms involved in this response class (sitting from a down position and sitting from a standing position). It is important to address these two common response class forms in training. You can train them both at the same time simply by ensuring that you include both response class forms into your training sessions.

Eventually, you will include the criteria that the cat exhibits the behavior within approximately two seconds of being cued, and the cat remains in the position until

---

released (e.g., one minute). As per a graded approach, begin with these criteria relaxed and gradually increase them.

The distraction criteria are vague in our behavior objective. That is partly because this is general, rather than specific to an individual and for conciseness purposes. If you allow a vague reference to distraction (or any other criterion) in your behavior objective, you should include an addendum or footnote with some specific details. You might need the cat to sit close to traffic, in a crowded mall, with a group of dogs, or children running around. Identify and list the specific requirements.

Identify the conditioned reinforcer and unconditioned reinforcer you will use (i.e., a clicker and small treats).

Phase 2. Acquisition

Increase the rate of responding by conducting training when the cat is not satiated with regard to either your social presence or food. Train at a time when the cat is not too tired, but also not hyperactive, to ensure that the cat's attention is directed to you and the training. These establishing operations prevent other contingencies from taking control over the behavior. Ensure that you begin training in a low distraction environment, so that there are few competing contingencies that could control the cat's behavior. You can allow for increased distraction later in the process—for now, set yourself and the cat up for success with a graded approach.

In the case of "sit," prompting is usually a quick and easy antecedent strategy for getting the behavior to occur. To prompt a "sit," place a small treat between your thumb and fingers. Ensure you have a good grip on the treat, so that the cat cannot grab it before you let go of it. With palm up, allow the cat to sniff the treat. Move it around to ensure the cat is "targeting" it (i.e., where the treat goes the nose follows). Once the cat is targeting the treat, move the treat slowly over their head so that they crane their neck to continue targeting it. If the cat jumps up to target the treat or swats at it with their paws, this likely means you are holding the treat too far above their head. In this case, withdraw the treat, reset the environment and try again, this time with the treat held closer to their head. If they back up while you lift the treat over their head, withdraw the treat, reset the environment and try again. If they do this again, you might want to train with the cat's rear end close to a wall or corner so that they are less likely to back up (just do not allow the cat to "feel cornered"). As they crane their neck to target the treat, they should sit. You now have a criterion response, an opportunity to reinforce. Once the cat has exhibited the target behavior, immediately click the clicker, and deliver the treat to the cat right away. The click ends the behavior so if the cat gets up after you click the clicker, that is not a problem. Carry out a few more trials in succession. You should notice that you are able to perform the luring motion more quickly and effortlessly in each successive trial as the prompt and other stimuli take on stronger control over the behavior.

Alternatively, you can use the target stick to prompt the behavior, if you prefer it. Carry this out similarly to luring with food except that you do not have to work to get the cat initially interested in the target; you have already trained target touching. Simply bring the target over the cat's head so that they crane their neck to touch it and have to sit by the time they touch the target. Otherwise, you would proceed just as described above.

Phase 3. Fluency

After the first few trials, begin fading the food component of the prompt. If you used the target stick, you will not need to do this. Whenever you use the unconditioned reinforcer as a component of the prompt, you should fade this part of the prompt quickly, so that it does not become a function-altering stimulus in that contingency and the behavior does not become "treat dependent." This is the first thing you will do in the fluency phase of training. Start with a few rapid trials of the sequence with the treat in your hand. Then in the next trial, leave the treat in your treat pouch, and execute the prompt motion just as before. The momentum, the similarity of the trials and remaining odor of the treat in your hand will promote evocation of the behavior. Once exhibited, "click and treat" (i.e., click the clicker, followed immediately by presentation of the treat). Through the next several trials, continue to reinforce on a continuous reinforcement schedule. In most cases, you can simply fade the treat-in-hand stimulus permanently this way.

If the cat seemed "apprehensive" with this change, you can fade the prompt more gradually. In this case, carry out that first trial without the treat in your hand as described above and reinforce. Then, carry out another trial, this time with the food in your hand. Through the next several trials, randomly vary having the treat in your hand and not, gradually increasing the number of trials with no treat in your hand, until you can run through several trials without the treat in your hand, and the cat exhibits the behavior reliably and smoothly.

Be sure to end the session on a good note. If there are indications that the cat is becoming satiated with regard to the reinforcer (i.e., bored), restless, or otherwise that the training progress may soon begin slowing down, end the session. If you believe the cat can continue without deterioration of the training, then continue, but always strive to end sessions before deterioration of any component begins.

To start your next session, briefly review the training from the previous session, to ensure a reliable start to training. There are a few new protocols to execute at this stage of training. You should now be presenting the prompt without food in your hands at all. The luring motion should be taking on stimulus control of the behavior at this point. Before this becomes too well established, begin fading the prompt, transferring stimulus control from the lure motion to a hand signal. The hand signal for sit is a palm facing forward while you stand straight up, arm held straight down beside the body, articulating the elbow so the palm sweeps upward. This resembles the lure motion, and so transferring stimulus control is an easy process. Taking advantage of stimulus generalization, on each successive trial through several trials, make the current stimulus (the lure motion) seem (in this case, look) increasingly like the new stimulus (the hand signal). Do this incrementally and gradually, and the cat should continue to exhibit the behavior reliably through each trial. If not, this likely means you are moving too quickly through the process. You should now be able to evoke the behavior with the hand signal alone every time. Use this visual cue until the behavior meets all form criteria.

If you used the target stick to prompt sitting, you will need to fade that as well. In this case, simply make the motion less and less prominent until you are barely presenting it at all. When you fade the target stick, replace it with the hand signal. In this case, use the hand signal, followed immediately by the prompt, which will evoke the behavior and you can reinforce. You can do this from the lure prompting motion with the target stick but it tends to transfer more easily with an already partially faded target stick motion. Repeat through several trials until the hand signal alone evokes the behavior.

Until this point, the behavior has been on a continuous schedule of added reinforcement. You can now move to a gradually thinning variable ratio schedule of reinforcement. The goal here is to gradually thin the schedule of reinforcement in a minimally discernible pattern (re-thinning it each time you introduce a new D-parameter or level of one parameter). Start by failing to reinforce a response, but quickly carry out another trial and reinforce that one. The "frustration" (i.e., emotional arousal elicited by extinction) will not have lasted long and you will begin conditioning resilience and persistence. Now, gradually increase the number of unreinforced trials to reinforced trials around a gradually increasing average. Randomly vary it to avoid discernible patterns. Go from a VR-2 toward, say, a VR-6 or VR-8. Move at a pace that continues to maintain the stability of the behavior. Beware of *ratio strain*, wherein the schedule is thinned too quickly, the cat becomes frustrated, and the behavior becomes unreliable or unstable, and may even extinguish. Move at the cat's pace. Dog trainers who begin training cats often thin the schedule of reinforcement too quickly with cats—thin the schedule carefully.

Utilizing a graded approach, begin working form, latency and speed, one at a time, gradually and as needed, until each satisfies the behavior objective criteria. Assess the features you need to change based on current proficiencies and the target criteria. This might include decreasing the latency or training a quicker motion from beginning to sit to being seated—speed. Work through one feature at a time. Gradually refining form means shaping. If any components of the form are inadequate, identify exactly what movements need to change and what they will look like when they meet the criterion. You may need a list of approximations if there is a significant difference between the current motion and terminal motion, but often, this kind of refinement simply requires slight shifts in form that can be achieved with one or two levels of approximations.

Once you have the behavior's form, latency and speed to criteria, you should establish the permanent vocal cue. To transfer stimulus control from the hand signal to the vocal cue, simply repeat the sequence of new evocative stimulus (vocal "sit"), followed by the old cue (hand signal), followed by occurrence of the behavior, followed by reinforcement. Achieve several trials through the contingency, and the new stimulus should take on stimulus control of the behavior. Pause after saying "sit," and wait for the sitting behavior. It might take an extra second or two, as the cat waits for the hand signal, but they will likely exhibit the behavior. If not, repeat several more trials, and try again until the new vocal cue evokes the behavior on its own.

You now have the target behavior under stimulus control, and it is time to begin "proofing" the behavior against dynamic real world challenges (i.e., the three D-parameters). Remember to work only one at a time.

Start with distraction. Start by introducing small distractions such as another person being nearby, but not exhibiting any specific or animated behavior. Take the schedule briefly back to continuous reinforcement and thin it gradually with the new D-parameter in place. Gradually introduce increasing levels of distraction, but again, do this at a pace that maintains the reliability of the behavior. Once you are able to evoke the behavior under more distracting circumstances, you might introduce another kind of distraction—evoking the behavior in other locations. For instance, instead of training in your living room, try several trials in the kitchen. Remember to relax other D-parameters briefly and re-thin the schedule. This can usually be done quickly, but always manage the level of frustration and stress, keeping these minimal and training fun. Make sure to take opportunities to reinforce occurrences of the behavior in many different places and under many different circumstances. However, you do not need to get to this level of proofing before you can begin working the other D-parameters. Get a good start on distraction, but if you choose to work another D-parameter, relax distraction until you work that other D-parameter up. Then, you can start combining them. Distance will involve and require duration—therefore, it is best to work duration up before introducing distance.

Train duration the same way as distraction. Set a specific criterion, and establish it reliably before increasing it again. Until this point, you were reinforcing immediately upon the cat exhibiting sitting behavior. Now, begin to require the cat to maintain the sitting position for two seconds before reinforcing. Move at the cat's pace to ensure almost all of the trials meet the criterion. When the cat fails to meet the criterion duration, administer extinction or subtracted punishment, identify your failure in getting the behavior, and try again.

In the case of distance, you need to increase the distance very gradually and carefully, so that the cat does not approach you and then sit. If the cat approaches you rather than sitting right away, where they are located, you have increased the distance too quickly. Ensure a solid sit "where-you-are" before increasing the distance. If the cat approaches you to sit, do not reinforce. Reset the environment and try again, this time ensuring you do not increase the distance too quickly. Once the behavior is reliable at the desired distance, you can begin combining D-parameters. When you combine more than one D-parameter, remember to build them up gradually again.

Set the criteria in all of these parameters to ensure success in most trials, but with a degree of difficulty that maintains progress. This judgment in maintaining smooth progress and minimal frustration is the trickiest set of trainer "chops" to teach trainers and the most challenging skill to exhibit fluently, primarily because it requires the generalization of many related behaviors to be exhibited reliably and quickly. Attend to expanding your repertoire in this regard; the appendix on trainer exercises will help. Attend to and recognize inefficiency and ineffectiveness in your training behaviors to differentially reinforce your own effective training practices. At the same time that you are training others, be they a cat or a guardian, you are also training yourself.

Depending on whether you have trained this behavior first, or if you have other behaviors trained, you should work on inter-stimulus discrimination training. Once you have two behaviors trained, you should run a series of trials, cueing one or the other in a random manner, reinforcing criterion behaviors and extinguishing non-criterion ones. This will be a good test of stimulus control, since many things except for the evocative stimulus itself will be the same in both arrangements. Remember the differential outcome effect—use distinct reinforcers for each behavior and your training will be more efficient and effective. Once you have three behaviors trained, you will find it is even more challenging as you randomly evoke different behaviors. That is because with two behaviors, the strategy of simply changing position will meet the criterion, assuming you don't evoke a behavior the subject is already in the position for—with three behaviors, this strategy will only be effective 50% of the time (rather than 100% with two behaviors).

Once you are ready to begin phasing out the conditioned reinforcer, which until now has been acting as the release, you can begin using a release stimulus in its place. The release stimulus will act as a conditioned reinforcer in place of the click. You can use "you're free" as the release phrase or choose another phrase. Begin presenting the release stimulus right before treat delivery, instead of the click. Ensure the food is far enough away from the cat that they will have to break position to eat it. If you are just using treats, then deliver the treats on the floor or from your hand at various orientations, but far enough from the cat that they will need to break position to access the treats. If you do not at least occasionally ensure that the cat breaks the position to get the treats, remaining in position might become a superstitious behavior.

Phase 4. Maintenance

Once you have the final form, speed and latency criteria of the target behavior, it is under stimulus control, and reliably proofed through the three D-parameters, you can begin working toward maintenance. There is not always necessarily a clear line between fluency and maintenance phases. You will likely want to continue to develop proficiency in new locations or with new distractors etc., but once you are well along with training, it is time to begin transitioning from the intensive training activities of the fluency development phase to less intensive maintenance of the fluency you have achieved.

Begin generalizing the reinforcers from just the treats, for instance, to praise some times, and a petting and perhaps a quick string-chasing game, as long as these things are actually reinforcing. Remember also that the clicker is just for acquisition and training toward fluency. Start reinforcing with the unconditioned reinforcers alone and leave the clicker for new parameters. Begin using fewer trainer-mediated reinforcers too. Use activity reinforcers (i.e., the Premack principle), in order to help you maintain control over the behavior. For example, if eating is reinforcing, then take the opportunity to require a sit before allowing the cat access to their food. If having a leash put on acts as a reinforcer, require a sit while you open the door or prepare for going for a walk on leash. If being picked up or being allowed to sit in your lap is reinforcing, require a sit before being allowed to engage in these behaviors. The goal in this phase is to work toward simply maintaining what you have achieved in your training sessions with minimal contrived activity outside of the cues. If at any point any component of the training seems to be deteriorating, refresh the training by building that parameter back up.

Wait

Phase 1. Preliminaries

Behavior Objective

S^{Ev}: Vocal "Wait."

Behavior: Cease moving if in motion and remain in position otherwise. *

Consequence: Treats and food.

Criteria: Latency: 1sec.; duration: remain in place until released, minimum 30sec.; frequency: 100% through 10 trials; distance: minimum 4 feet; distraction: various; time point: 6 sessions at ≈5min. each.

* Not moving and staying in position refers to the cat not walking, running, leaping, or otherwise progressing toward something. However, they may change positions (e.g., sitting rather than standing or the like). You may change this criterion as required.

This may seem like the absence of a behavior, but stopping and maintaining position is indeed a behavior itself. In fact, it is a response class requiring several behaviors involved in coming to a stop and balance related behaviors of maintaining position.

"Wait" is similar to "stay," except that it is not used for duration maintenance for behaviors you have already evoked. In these cases, simply train the cat to maintain the behavior until you release them—no secondary vocal "stay" or "wait" is required. "Wait" can be useful at doors, at feeding time, or in any circumstance where you want like the cat to briefly halt their advance. Again, in actual cases, make distraction criteria specific in the definition, or provide an addendum with explicit and unambiguous distraction criteria.

Identify the conditioned and unconditioned reinforcer you will use.

Phase 2. Acquisition

Increase the rate of responding by conducting training when the cat is not satiated with regard to either your social presence or food. Train at a time when the cat is not too tired, but also not hyperactive or playful, to ensure that the cat's attention is directed to you and the training. These establishing operations prevent other contingencies from taking control over the behavior. Ensure that you begin training in a low distraction environment, so that there are minimal competing contingencies that could control the cat's behavior. You can allow for increased distraction later in the process—for now, set yourself and the cat up for success with a graded approach.

Training "wait" is all about increasing duration gradually. Begin with a simple "wait"-appropriate arrangement. As an example, waiting for a food bowl to be presented is described here, but you can start with waiting at a doorway or another scenario. An advantage of starting with waiting for food is that it generates an effective reinforcer and the continuous schedule it necessitates is more suited to the beginning of a training program. In any case, you will promote generalization to other situations as you proceed with the training. You can ensure more trials by presenting only a sixth to an eighth of the cat's meal on each trial. You may adjust the specific procedure to allow for any differences in how you prefer to present meals. Make sure you see the section on challenges below before deciding to start with this reinforcer.

Begin by approaching the cat's bowl with kibble. Rather than requiring some specific behavior such as "sit," simply present a hand in front of the cat's face, palm facing them, and quickly place the bowl on the floor. The hand will not be maintained in front of the cat through the entire time they are required to remain in place, but for now, since the hand signal takes a second or two and the trainer only expects a second or two for the wait, it may seem that way. The hand signal just happens to block access to the food, giving you the time that you need to immediately click and allow access to the food (the clicker is usually held in the same hand as the pouring cup). The cat will begin eating the food, providing added reinforcement for the behavior. Remember that whether the cat is sitting or standing is irrelevant, and in fact, it can be useful to achieve a mix of these positions, in order to promote appropriate generalization and discrimination. Repeat this sequence through several trials until you have presented the whole meal. If the sequence is smooth and reliable at that point, you can move right to the fluency stage. Otherwise, repeat this sequence for the next meal until the sequence is smooth and reliable with the minimal duration. If the cat goes for the food very quickly after you present it, say "Oops" and pull the bowl away. Reset the environment and try again, perhaps a bit faster if you can, in order to get a criterion behavior that you can reinforce. If this is proving difficult, you may want to feed the cat half of their meal and then training for the second half to increase satiation a little bit.

Phase 3. Fluency

Without prompts and especially prompts involving the conditioned reinforcer, fluency is a simpler process. Furthermore, because you use "wait" in instances where access to a reinforcer is simply delayed, there is no need to involve complex schedule thinning procedures, making the process even simpler. Although training for form, latency and speed is not usually required, the three D-parameters are vital.

Once the sequence is smooth and reliable at this minimal duration, transfer stimulus control to a vocal cue. As before, say "wait," present the palm hand signal, and then release when the cat has ceased approaching. Repeat through several trials in order to condition the vocal cue, whereupon, you can drop the hand signal. This can be a challenging component of training because the removal of your hand may act to release the cat at first. Prevent access to the reinforcer if the cat proceeds before you release them but try to set the criterion such that you will get mostly criterion-meeting trials.

At this point, you will need to start utilizing a release phrase. To do this, say the release phrase, such as "you're free," right before you click. The unconditioned reinforcer will be access to what the cat was waiting for; you will not need to use treats. Right now, this will seem like a very quick sequence of events. However, the next step will be to increase the duration of the wait.

Once the vocal cue and a release stimulus are established and you no longer require the hand signal, begin gradually increasing the duration criterion. If you were requiring the cat to wait for a half of a second, begin requiring a full second for several trials, and once this is smooth and reliable, set the duration criterion to two seconds. As always, if the cat breaks wait before you release them, say "Oops," prevent access to the food in as non-confrontational a manner as you can arrange, reset the environment and try again. Move through the duration steps slowly, and avoid extinction and subtracted punishment trials as much as possible, as this will be frustrating for the cat and could lead to problem side effects. Try to begin making the duration seem random around a gradually increasing mean average. For example, one second, then two, then a half, and then two, and then two, and then one, and then three, and then one, and then two, etc. Remember to adjust the criterion gradually, enough to minimize extinction and subtracted punishment trials, and keep the progress smooth. Also, maintain minimal distraction while you work on duration.

It can be useful to begin working generalization with treats and toys (e.g., a Cat Dancer® interactive cat toy[71]), requiring the cat to wait before accessing the treat or toy when you place it near the cat. This allows for more frequent trials, and makes it easier to train in different locations. Start close to where you feed the cat, and as usual, when instating a new D-parameter, relax the others. For example, begin with minimal duration, and work your way back up to the criterion length of time. This should go more quickly this time. Then, practice in different locations, each time relaxing the duration and building it back up.

Once this is going smoothly, begin using the vocal cue "wait" for other situations, such as waiting before going through a doorway or the opportunity to play with someone etc. When you introduce a different item, reset the duration criterion and build it back up, and go back to continuous reinforcement and re-thin the schedule. Try to maintain minimal distraction levels through these trials. Once this is going smoothly, you can work in the distraction criteria. Begin using the wait for more highly effective reinforcers (i.e., things that are tougher for the cat to wait for). In each new case of distraction training, reset and rebuild the duration component.

Phase 4. Maintenance

Continue to generalize to different stimuli and settings. Also, continue to generalize the reinforcers, utilizing more nontrainer-mediated activity reinforcers via the Premack principle versus more trainer-mediated extrinsic reinforcers. The task in the maintenance phase is simply to ensure that you continue to present trial opportunities to keep the training fresh. If, at any point, the training begins to deteriorate, refresh the training with more frequent sessions with initially relaxed duration and distraction, and build them back up.

---

[71] http://catdancer.com

Common Challenges

Some cats will charge for the food even before you finish placing it on the floor. If you cannot get a wait long enough to place the bowl and click, do not get into a physical confrontation with the cat (i.e., holding them back with the hand signal hand). Instead, try doing the exercise with a less reinforcing type of food and/or training after the cat has eaten a full meal. You may need to start with another reinforcer that can be controlled more easily and/or is not quite as effective as a reinforcer to help ensure that you can get the wait long enough to build upon.

The more effective the reinforcer you are cueing the cat to wait for, the slower you will have to go in terms of increasing the duration criterion and the more trials you will have to run. If you are getting too many extinction or subtracted punishment trials, this usually indicates that you are moving too quickly or not putting in an adequate number of trials before increasing the criterion.

Here (Recall)

Phase 1. Preliminaries

Behavior Objective

S^{Ev}: Vocal "Here."

Behavior: Comes to the caller within 24 inches of caller's feet.

Consequence: Treats and release.

Criteria: Latency: 2sec.; speed: at least a trotting gate; frequency: 100% through 10 trials; distance: minimum 40 feet; distraction: various; time point: 10 sessions at ≈3min. each.

Coming when called is extremely important. It helps us manage where the cat is and helps us get them away from potential problems. Recall training is often challenging, usually just because guardians have managed to effectively counter-train it prior to being coached to train it by a professional (and then often during and afterward, too). Three common mistakes are: (1) the vocal cue is used before the behavior is reliable, and hence the stimulus is weakened as an evocative stimulus; (2) guardians tend to inadvertently impose subtracted punishers and added punishers for coming when called, particularly outside of formal training sessions; and (3) guardians expect too much, too quickly, and raise criteria too quickly without establishing a sufficiently strong foundation, especially with regard to distance and distraction, which degrades the training effectiveness. Avoid these three common mistakes, so that the behavior is much easier to train.

Identify the conditioned and unconditioned reinforcer you will use.

Phase 2. Acquisition

Utilize the same establishing operations as discussed for "sit."

Because this behavior is frequently required in everyday life outside of training sessions, you should adhere to a few rules. To reiterate, first, avoid punishing either remaining near you or coming to you. Avoid, for instance, calling the cat to you to deliver a reprimand or when you need to do something that might be aversive (e.g., clipping nails, being placed in a carrier, etc.). Also, try to avoid calling the cat when coming will end reinforcing activities, such as playing, unless the reinforcer will be effective enough to overcome the end of the activity. Being near you and coming to you must always be reinforcing, ideally more reinforcing than any other contingency, which is a tall order, so it is important to heed these rules. This should be the guiding rule before, during, and after training, both in training sessions and in everyday life. Reinforcing does not always mean providing treats or toys when cats come to us. It means generally being a source of all sorts of reinforcers and the source of very few aversers. It means becoming a conditioned reinforcer, like a clicker. As for all those aversive conditions such as nailing trimming and being in a carrier, work on desensitizing the cat to these stimuli separately.

Guardians will frequently ask the trainer, "What do I do when I call the cat and he or she does not come to me?" In this case, remind clients not to present the cue unless they are sure that the cat *will* come to them.

Chances are, the cat already comes to you under certain circumstances. These "circumstances" are evocative; take note of them and use them in training to prompt the behavior. You may then transfer stimulus control to a vocal cue of your selection.[72] Remember the sequence: Present the vocal cue "here," prompt the behavior, and then reinforce the behavior once it has occurred. So, for example, call the cat, shake the treat bag, when the cat comes running, click and provide the treat, and repeat. Then, test the evocative capacity of the vocal cue on its own and fade the can or bag shake. You can eliminate it if the vocal cue works well or you can move more gradually and incrementally, by reducing the volume and duration of the shaking sound through repeated trials. Choose these times carefully! Ensure a high degree of likelihood that the cat will come to you. Keep training sessions short—three or four successful trials are a good start. You may find you can only carry out one trial at a time because the cat is interested in remaining near you. This is fine and indeed several single trial sessions throughout the day is an effective way to carry out the training.

Since you are building on a behavior that already occurs and strengthening it, while establishing a permanent cue, the acquisition phase is usually quick. Once these are fairly smooth and reliable, you can move on to training for fluency.

---

[72] Of course, the stimulus does not need to be vocal. It can be a whistle or hand clapping. Auditory stimuli are good choices because the cat does not need to be within reaching distance or looking at you to experience it, but you do not technically have to use auditory stimuli either.

Phase 3. Fluency

Continue fading the prompts, and installing the permanent vocal cue. Continue fading the prompts by saying "here," and follow with the prompts. After a few trials, begin gradually fading the prompts—the vocal cue, which should have taken on stimulus control, should evoke the behavior alone.

Transition to a gradually thinning variable ratio schedule of added reinforcement. Ratio strain can occur easily in recall training, due to the competing contingencies and the high response effort involved in recall. Avoid thinning the schedule too far and too quickly.

Begin refining the form, latency and speed of the behavior. As usual, work one at a time, and each time you introduce a new feature, bring the schedule back to continuous reinforcement. Remember, relax the three D-parameters and schedule of reinforcement, but do not relax form, latency and speed; once you achieve progress in these features of the behavior, maintain them.

Begin proofing the behavior against the three D-parameters. For this application, you will focus heavily on distraction. Begin incrementally increasing distraction at a pace that still allows you to achieve success in at the very least 95% of your trials. At first, it is usually best to arrange for highly contrived and controlled training environments. Begin working in more real world distractions as you proceed to proof the behavior.

Once you have worked through increasing levels of distraction and several different kinds of distraction as well as a variety of locations, you can begin more spontaneous trials, calling the cat away from increasing levels of distraction, ensuring that the behavior is reinforced. Another good game to play at this stage is calling the cat away from a distraction of some kind (e.g., playing with a toy) only to immediately release and reinforce with the opportunity to reengage with the distractor. Start by only expecting to call the cat away from minimal distraction. Then, once the subject exhibits the behavior fluently, gradually increase that level of distraction. Avoid too many of these trials in a row, as they will become frustrating and aversive. Many cats are less tolerant of frustration than dogs (in general). This is not necessarily because they are cats, but because they have not undergone the kind of training that dogs typically do.

Phase 4. Maintenance

Continue to generalize the reinforcers utilizing the Premack principle in order to help you maintain control over the behavior by providing the opportunity to exhibit other more likely behaviors. If, at any point, any component of the training seems to be deteriorating, refresh the training by taking a run through building that parameter back up.

Common Challenges

By far, the most common challenge with recall is failure to achieve sufficient fluency at one level of difficulty before moving to the next. In other words, people usually expect too much, too soon, in this training, and they fail to manage the situations adequately in order to ensure successful rehearsal. The recall faces competing contingencies, almost by definition, since the cat will usually be contacting some reinforcer when you call them. More than most behaviors, distraction is a ubiquitous and powerful challenge. The reinforcement for approaching you must surpass the reinforcement available elsewhere.

Go to Mat

Phase 1. Preliminaries

Behavior Objective

S^{Ev}_1: Vocal "Mat."

Behavior$_1$: Go to mat.

Consequence: Treats but eventually opportunity to exhibit next link.

Criteria: Latency: 1sec.; speed: average walking gate; frequency: 100% through 10 trials; distance: minimum 15 feet; distraction: various.

S^{Ev}_2: Arrives at mat.

Behavior$_2$: Remain on the mat.

Consequence: Treats.

Criteria: Latency: 1sec.; duration: maintain until released, minimum 2min.; frequency: 100% through 10 trials; distraction: various; time point: 5 sessions at ≈4min. each.

This is a behavior chain since there are two distinct response class's exhibited in sequence, with just one evocative stimulus at the beginning and reinforcement provided by the trainer only after the last behavior has been exhibited. In that case, you can supplement the behavior objective with the following task analysis:

Upon "mat" being vocalized, the cat will:

- Proceed to the designated mat

- Remain on the mat

You might be wondering if remaining on the mat is really a behavior at all or if it is just a duration component of being at the mat. However, duration of "going to a mat" is not the same thing as remaining on a mat. It is a distinct response class, as it may involve various response class forms but they all result in remaining on the mat and none of them involve going to a mat. In other words, going to the mat is not a response class form of remaining at the mat and so they are indeed different response classes. Remaining at the mat may involve sitting, lying down, or even moving about, as long as the cat remains on the mat. Because there are multiple response class forms but clearly a specific response class, they should be identified as distinct response classes rather than treating remaining on the mat as a duration component of going to the mat.

Promote response generalization so that the cat will exhibit whatever behaviors make remaining on the mat more likely. Allowing any of various response class forms makes the response effort lower and helps ensure that the cat will remain on the mat. Here, two simple response classes are involved, each of which is likely already trained to fluency.

You can use this cue to direct the cat to go to specific places. The actual cue you select is based on the location you want the cat to go to. We often use it to train a cat to go to a crate or bed, but you can also use it to direct the cat to other locations, or to find and go to specific people.

Phase 2. Acquisition

Utilize the usual establishing operations and begin training in a minimally distracting environment.

Forward chaining is suitable for a simple chain of behaviors such as this. Begin by training the cat to go to the mat. Start by standing with the mat between you and the cat. Lure the cat to the mat with a treat in your hand or with the target stick. Once the cat reaches the mat, reinforce. Repeat through several trials, until the cat will readily go to the mat. Remaining at the mat is a simple matter. Since the cat is already on the mat, when you click for going to the mat, you are also really including remaining on the mat with a minimal duration (even if it is just a quarter of a second). In the fluency phase, you can increase the duration.

Phase 3. Fluency

Put the reinforcement on a gradually thinning variable ratio schedule, taking it back to continuous any time you introduce a new level of challenge or training parameter.

Once the behavior chain is approximately what you would like, transfer stimulus control from the luring motion, to a permanent cue. A vocal cue is most common because you can deliver it at a distance and it does not require the cat to be looking at you to be effective. Say "mat," and then lure toward the mat, at which point the cat will exhibit the behavior and you will reinforce the behavior. Repeat through several trials until you can drop the old temporary lure stimulus from the sequence. You can compliment this by making the lure motion less and less prominent with each trial. This also allows you to start working distance.

Next, begin proofing against the three D-parameters. If you begin with duration, remember to keep distance and distraction minimal, and to briefly reinstate a continuous reinforcement schedule. Instead of clicking immediately upon the cat getting to the mat, wait an extra second or two and then click and treat. Gradually work the duration up to about two minutes, and begin thinning an intermittent schedule of reinforcement again. Remember, the cat may engage in any response class form that is a component of the response class (remaining on the mat).

Once you have a reliable duration, begin working distance. Go back to continuous reinforcement and relax the duration parameter as you start from gradually increasing distance from the mat when you evoke the behavior. Ensure that you include the special feature of cueing the behavior when the mat is not directly within the cat's line of sight. This might mean that the cat is around a corner at first and eventually in another room entirely. Again, once you have the distance criterion reliable, put the behavior back on a thinning intermittent schedule.

Finally, you can proof the behavior against distraction. Start in the presence of minimally distracting stimuli, while relaxing the schedule of reinforcement, duration and distance parameters. Begin introducing incrementally more distracting stimuli into the environment, until the cat exhibits the behavior reliably, even under moderately to very distracting stimulation. Put the behavior back on a gradually thinning variable ratio schedule of reinforcement, and then begin working the features together.

Some cats will run smoothly through this process, while others will require remedial work at certain stages. Be prepared to track the progress of the training, and adjust as needed to ensure success. You should now be able to evoke the behaviors from a greater distance, under distracting circumstances, and the cat will reliably go to the mat and remain there until released. At this stage, replace the click with a release phrase.

Phase 4. Maintenance

Continue to introduce new distractions as they become apparent, thin the schedule of reinforcement, and keep up regular practice to ensure maintenance of the training. If at any point any component of the training seems to be deteriorating, refresh the training by building that parameter back up.

Loose Leash Walking

Phase 1. Preliminaries

Behavior Objective

S^{Ev}: Leash on.

Behavior: Walking adjustment behaviors such that the shoulders stay within 5 feet of the handler (Leash tightens at 6 feet indicating a non-criterion behavior).

Consequence: Treats and forward progress.

Criteria: Frequency: 100% through 10 trials; distance: maintain minimum 3 standard city blocks (approximately 1400 feet); distraction: various; time point: 20 sessions at ≈5min. each.

Walking on leash allows the cat to go outside and experience the various forms of stimulation unique to being outdoors. Ensure that the cat is up to date on vaccinations before venturing outside. A big part of loose-leash training is ensuring that the cat becomes adequately accustomed to the harness and leash and to being directed with regard to where they go and not go. The process must be gradual and cautious, because there are a great many novel and potentially fear-eliciting stimuli outside.

While some trainers tend to frame this behavior in a negative form, stating the behavior to be avoided, rather than the behavior to be achieved, the response class for walking on a loose leash can be stated positively as "walking close to the handler and responding appropriately to direction cues." These behaviors are a function of proximity and the required response class forms to achieve it. Trainers often frame loose leash walking in negative terms for a useful reason. A tightened leash is an indication that a mutually exclusive non-criterion behavior is occurring, and therefore the criterion behavior is not occurring. There are several ways to train loose-leash walking. I will describe one straightforward method below.

Identify the conditioned and unconditioned reinforcer you will use.

Phase 2. Acquisition

A graded approach works well for training loose-leash walking. Rather than providing the cat the opportunity to either pull or not, and reacting differentially to each with reinforcement and punishment, set both you and the cat up for success by initially setting the criteria low, then gradually and incrementally increasing the criteria, to keep the progress smooth and efficient, with a very high rate of success.

Stage 1. Start by leaving the harness and leash out for the cat to explore and become accustomed to. This does not usually take long. Consider using treats to reinforce confident exploration.

Stage 2. Have the cat target the harness and reinforce this through several trials. Then reinforce for tolerance if you put the harness up as though you are going to put it on but only part way and only briefly, not so far that the cat exhibits escape behavior. Repeat this through several trials and then gradually increase the approximations of putting it on the cat, until you can put it on them.

Stage 3. Put the harness on the cat without the leash attached and carry out some fun training or play for a brief period. This will distract the cat away from harness and allow them to become accustomed to wearing it. Wearing a harness bothers some cats and not others. Train through several brief sessions, pairing effective reinforcers such as treats and play with the harness. Never move to a new level of challenge until the cat responds with approach and contact enhancing behaviors (and certainly not escape behaviors). If the cat reacts with escape behaviors, this indicates that you have pushed too quickly and too far. You have not established a sufficiently strong history of reinforcement for tolerating the harness and you will need to introduce the cat to the harness much more gradually.

Stage 4. Next, attach the leash to the harness and allow the cat to drag it behind them. Prevent the leash from tangling or snaring on anything. These harness sessions should be approximately five minutes each, repeated daily or twice daily for about a week. Continue to reinforce acceptable tolerance related behaviors.

Stage 5. Carry out the same kinds of sessions discussed in stage four, except hold the leash. Allow the cat to generally go where they want. If they approach an area that is likely to entangle the leash, coax the cat vocally, and perhaps gently guide them away with slight pressure on the leash and then reinforce their change in direction. Gentle, slight tension on the leash will come to evoke the cat's looking to you for other directional cues.

Stage 6. In your first excursion outside, start by having the cat sit and wait while you open the door. The opening door might evoke retreat behaviors for some cats and if that is likely to be the case, you will have to take several graded steps to ensure a strong pairing between the opening door and reinforcers like treats or play. Once the cat tolerates an opening door, you may proceed.

Stage 7. Release the cat from the sit and allow them the opportunity to stay put, move away from the door back into the house or go outside on leash. If the cat is afraid of going out, it may take several sessions for them to become comfortable enough to explore outside. The trick here at first is not to force the cat into anything and do not allow anything aversive to occur to the cat. Continue to reinforce for maintaining a loose leash with every few steps. You can gradually increase the number of steps between reinforcement in a variable and seemingly random fashion. Also, reinforce for coming away from something that you call the cat away from and for accepting gentle guidance in the form of slight pressure on the leash.

Phase 3. Fluency

Continue to work toward more steps between reinforcers, gradually thinning a variable ratio schedule of added reinforcement as you proceed.

Next, begin proofing against the three D-parameters. You can use reinforcers found in the uncontrived environment as distractors.

Duration is the amount of time you walk before reinforcing. Work the duration gradually, and as usual, relax other D-parameters when you increase the duration.

Distance is not applicable to walking on loose leash since it is only an up-close behavior.

Phase 4. Maintenance

Continue to generalize the reinforcers utilizing the Premack principle, in order to help maintain control over the behavior with the opportunity to exhibit other more preferred behaviors. If at any point any component of the training seems to be deteriorating, refresh the training by building that parameter back up.

Cooperative Grooming Behaviors

Cooperative grooming behaviors can represent several behaviors that serve to allow for grooming and veterinary examinations. I will present the behaviors of presenting each paw, and remaining still and tolerating brushing below. The common element in each is to allow the hander to manipulate the body and have the cat remain still and relaxed throughout the process.

Paws

Phase 1. Preliminaries

Behavior Objective

S^{Ev}: Touching paw

Behavior: Operants required in allowing the paw to be manipulated, and remaining relaxed and tolerant throughout (tolerance-related behaviors).

Consequence: Treats and play.

Criteria: Latency: immediate; duration: remain relaxed until released, minimum 30sec.; frequency: 100% through 10 trials; distraction: various; time point: 14 sessions at ≈2min. each.

Allowing paw manipulation will make nail trimming or examination of feet for wounds much easier. This involves the various tolerance and balance-related behaviors. It is important to carry out the training with each foot.

Identify the conditioned and unconditioned reinforcer you will use. Licking wet food from a surface is easier for cats than chewing pieces of food, when they are in an awkward or atypical feeding position (Jacqueline Munera, November 25, 2014 personal communication). Your hands will be full and you may choose to use a tongue click rather than a clicker. In that case, condition it before hand.

Phase 2. Acquisition

Utilize the usual establishing operations.

The most effective and efficient approach to training paw presentation is with a graded approach. Start with one foot and once the behavior is exhibited fluently, carry out the training with each of the other feet in turn.

You may elect to have the cat stand for this procedure but this is not necessary; you can usually examine and clip nails when the cat is lying down as well. Touch the paw, immediately tongue click, and provide a treat for tolerance-related behaviors. That is, reinforce as long as the cat remains relaxed and does not engage in any escape behaviors. The most gentle and briefest of touches will rarely evoke escape behaviors, so this is a good place to start. If the cat pulls away or otherwise moves to escape the contact, you may need to start with a touch higher up the leg, perhaps even on the shoulder, to get relaxed tolerance-related behaviors. In this case, you can gradually and incrementally work your way down the leg to the paw, through several sessions. Repeat through several trials. Increase the force and duration of the touch very gradually and only increase that force or duration when the cat tolerates the level you are currently working at. This process may take some time, but if the cat continues to exhibit escape behaviors, medical problems might be causing pain and require veterinary intervention before you can proceed.

The next step is to begin increasing the magnitude and duration of the touch to approximate holding the paw. Work this phase gradually and incrementally, at a pace the cat will tolerate, reinforcing on a continuous reinforcement schedule until the cat allows you to hold their paw for several seconds and remain relaxed.

Once you have achieved this level of success, take the duration back to approximately five seconds, as much time as it takes to lift the paw off the ground. As soon as you have lifted the paw away from the floor and the cat exhibits the appropriate balance-shifting behaviors, while remaining relaxed and tolerant, tongue click, provide the treat, and gently replace the foot to reset for another trial. Repeat through several trials, perhaps through several sessions, gradually working the duration that you hold the paw up to 20 or 30 seconds. Ensure the sessions are short enough that the cat does not become too frustrated. Try to move at a pace that ensures the cat does not pull away (i.e., escape). Use highly effective reinforcers, as well as a high rate of reinforcement.

Introducing the nail trimmers comes next. Allow the cat to investigate the trimmers so their presence is not a surprise; provide treats as they investigate. Once the cat is accustomed to the trimmers do the same thing, but now with you holding the trimmers.

Once the cat is very accustomed to the trimmers, you can combine the paw holding with the trimmer holding. The sequence is: holding the trimmers in one hand, pick up the paw with the other hand, tongue click, provide a treat, and then gently replace the foot. Gradually bring the trimmers close to the picked up paw through several trials and once they are touching, work in manipulations to reveal the nails, all the while reinforcing appropriate balance and tolerance behaviors. Begin touching the trimmers to the nail and pressing the trimmers while it touches the nail, but is not yet positioned to cut the nail. Once the cat tolerates this, work in trimming a nail and reinforcing for tolerance. Work your way up to trimming more than one nail and eventually all of the nails on that paw. Snip just the smallest amount of the nail so that you may carry out more frequent training sessions and not reach the quick too soon. Always be very careful not to snip the quick. These instructions only required a few sentences but the process can take quite some time and several sessions, depending on how aversive nail trimming is for the cat.

Phase 3. Fluency

Repeat the process for each paw. You should find that with each successive paw, the process will go much quicker and smoother, but do not rush it.

Phase 4. Maintenance

Maintain this behavior on a continuous reinforcement schedule. If at any point any component of the training seems to be deteriorating, refresh the training by building that parameter back up.

Tolerating Brushing

Phase 1. Preliminaries

Behavior Objective

S^{Ev}: Presence and use of brush

Behavior: Tolerance and relaxation-related operants.

Consequence: Treats and play.

Criteria: Latency: immediate; duration: remain relaxed until released, minimum 3min.; frequency: 100% through 10 trials; distraction: various; time point: 10 sessions at ≈2min. each.

Identify the conditioned and unconditioned reinforcer you will use.

Phase 2. Acquisition

Utilize the usual establishing operations.

Training a cat to tolerate brushing, remaining relaxed throughout the process is much the same as training them to allow paw manipulation. Being similar to petting, many cats do remain relaxed through brushing, but some cats do not, particularly if there are mats to be worked out. Proceed in the same way that you did for nail trimming and paw handling. In this application of a graded approach, start by presenting the brush to the cat in the same way that you introduced the trimmers. Once the cat is comfortable with the brush, touch the cat with the brush only very briefly and gently. If it is likely that you will have a hard time getting the brush through the fur, start by using the back of the brush until the cat remains completely tolerant and relaxed with this process.

Phase 3. Fluency

As with trimming the nails, gradually and incrementally increase the intensity and duration of contacting the fur with the brush, reinforcing tolerance behaviors throughout. Work up to longer, deeper brush strokes and start with the least sensitive areas.

Phase 4. Maintenance

Maintain this behavior on a continuous reinforcement schedule. If at any point any component of the training seems to be deteriorating, refresh the training by building that parameter back up.

Continuing Education

Courses provided through The Companion Animal Sciences Institute at www.CASInstitute.com address all of the topics covered in this chapter.

The following is a useful resource in appreciating species-typical behavior patterns in cats:

- What is My Cat Saying? Feline Communication 101 PowerPoint Presentation on CD-ROM by Carol Byrnes and Jacqueline Munera

If you are looking for a good book to give to clients, consider these books:

- Getting Started Clicker Training for Cats by Karen Pryor

- Cat Training in 10 Minutes by Miriam Fields-Babineau

- How to Correct Behavior Problems in Your Adult Cat Staring from Scratch by Pam Johnson-Bennett

Jacqueline Munera's web site is at https://positivecattitudes.wordpress.com and any courses or products she generates would be useful, including the course that she and I co-instruct through CASI.

The Science and Technology of Animal Training

CHAPTER 9. TRAINING BIRDS

Behavioral Objectives

The objective of this chapter is to measurably expand the reader's repertoire of behaviors in relation to describing and relating the principles of behavior. Upon successfully integrating the concepts outlined in this chapter, the reader, where exposed to contingencies to do so, will accurately:

- Implement a training plan for common good manners-related behaviors, applying the systematic strategies previously discussed based on a systematic plan previously discussed.

- Select appropriate equipment to aid in training

The behaviors discussed below make everyday life easier among birds and people. The methods can be adapted to train tricks too.

Training Birds

Training birds is both very similar to, and very different than, training cats and dogs. Like cats and dogs, the principles of behavior remain the same, and the basic strategies and procedures apply well to training birds. However, most companion birds are prey animals, whereas dogs and cats have bodies structured by biological evolution to be predators. Consequently, birds tend to be more skittish, apprehensive and fearful than species such as cats and dogs (though this is far from universal). Birds share this general skittishness with horses, which are also prey animals. Because training birds is different in many important ways from training mammals, I will include some general information and care requirements that effective training necessitates for birds below.

Birds tend to be most active in the morning and evening, sleeping at night and resting at midday. The activity patterns of the bird are partly determined by ambient light. Birds should be provided a dark and quiet safe place to sleep at night and a nice bright (full-spectrum) lighted area to be during the day, where they can engage in various species-typical behaviors that will provide exercise and satiation for activity reinforcers ("enrichment" as some might put it). As with all animals, birds also require clean water and the best possible diet for their species. All of these practices will put the bird in the best possible position, physiologically, to participate in training. Birds are quite sensitive to environmental stressors such as drafts, chemicals, noise pollution etc. Observe carefully for potential stressors and eliminate or mitigate them to generate an effective training environment.

Whether one opts to keep flight feathers trimmed afterward or not, there are benefits to keeping the flight feathers trimmed throughout the timeframe in which you plan to engage in the most extensive training. The primary benefit is that the bird will be more likely to remain in the training area while you engage in training. Only an experienced professional should carry out flight feather trimming.

Here are some tips for training birds, partially from Johnson (2004):

- If you have multiple birds, start by training the "friendliest" and more food motivated bird first. Unlike with dogs and cats, birds tend to imitate the behaviors of others and you can condition them by allowing them to observe you training others, so keep the other bird in the cage nearby. If they are mating, sitting on eggs or raising young, avoid disturbing them all together.

- Start training in a minimally distracting environment where the bird is most comfortable. This might be on a table or chair near the cage or on top of the cage or a play structure.

- If the bird is cage-bound, you can deliver treats through the bars, or with a long handled spoon if biting is a problem.

- If the bird bites, you might find a location to engage in training where the bird does not bite, but if not, be patient and try to avoid the specific stimulation that evokes biting behaviors. Try to contribute as little reinforcement for biting as possible and not make a fuss about it, as drama and increased social distance can be powerful reinforcers for behavior by birds. Continue to additively reinforce behavior patterns that do not involve biting. A full behavior change program may be required to address this behavior.

- If the bird is fearful of you, take it slow and patiently. Use lots of added reinforcement and avoid using aversers. Gradually shape confidence behaviors and train the bird to "trust" you.

- Particularly at first, assign a single trainer to the bird. Once training is coming along well, others can participate. Ensure that everyone is on the same page in terms of the training plan and where you are in it to avoid frustration and inconsistencies.

- If the bird is cage-bound because of fearfulness, it is a good idea to start training some simple prop interaction behaviors with the bird in the cage to improve confidence. Use smooth gentle movements around the bird and you may need to deliver treats with a long spoon at first. You could start by shaping some kind of interaction with a toy, or to go to a specific location in the cage for instance. See (O'Heare, 2011) for details on rehabilitating such birds and encouraging creativity, persistence and resilience. You might work up to targeting and you may need to condition certain requisite behaviors for further training, such as evoking coming out of the cage confidently and stepping up on a perch. Once you have built a trusting relationship and conditioned confidence, you will be able to train a wider range of behaviors.

- Keep training fun! Be flexible with regards to the level of confidence the bird has with you and the level of stimulation that is productive but generally, praise as you go and maintain a fun and animated mood.

Equipment

You will require a clicker. You can purchase these in any pet supply store. It is best to introduce anything new to a bird, including toys and even the clicker, by placing it within view of the bird for a few days before bringing it closer to them. If the bird seems apprehensive with regard to the click sound, the newer button-style clickers tend to produce a softer click than the older hole-style clickers do. You can also keep the clicker covered with fabric (use a sock or keep it in your pocket) to soften the click. Alternatively, you can click a ballpoint pen or press on the metal lid of a glass bottle (e.g., a glass Snapple® beverage bottle or a glass baby food jar) to make a soft "clicking" sound.

You should have a pouch to keep the treats in as you train. You can often find treat pouches in pet supply stores although some of the best treat pouches are chalk bags used by rock climbers.

In most cases, guardians will know which treats will be the most effective reinforcers. Where this is in question, you can carry out a *preference assessment*. To carry out a preference assessment, provide a choice between several different treats and observe which one the bird consumes first. Repeat the assessment several times, in order to rule out "random" treat selection.

Small birds such as budgies (i.e., budgerigars) should receive treats that are approximately the size of a millet seed. Medium sized birds such as amazons or african greys should receive treats the size of half a sunflower seed. Large birds such as the largest macaws or cockatoos can receive treats about the size of a whole sunflower seed. Prepare the treats of appropriate size in advance, so that they are ready to deliver during your training sessions.

In many cases, a targeting stick can be useful. You can purchase targets made specifically for this kind of training with the benefits of being safe, durable and unique, so that it is not confused with other objects. You can buy targeting sticks with clickers built into them (e.g., the Clik Stik® clicker and target stick[73]). However, you can use any object as a target (e.g., a chopstick, drinking straw, car antenna, or pencil). Do not use a spoon as a target, if you might need to deliver treats with one at any point (such as if the bird may bite when handed treats or if they are cage bound). Also, do not use anything you might use as a perch when training "step up" later, such as a dowel or cage perch. For very small birds, you might use something as small as a toothpick. For medium sized birds, a pencil might work well. As with the clicker, place it within sight of the bird for a day or two before introducing it to the bird.

Occasionally, you may need a device to deliver treats without using your hands. A cocktail stir spoon can be useful as these spoons are longer than most spoons and tend to be slender. When I work with cage-bound aggressive birds, I usually install a tube with a hopper at the bottom that I can drop treats into and they pop out into the hopper for the bird to consume.

---

[73] http://store.clickertraining.com

Targeting (Grasping)

Phase 1. Preliminaries

Behavior Objective

S^{Ev}: Vocal "Touch."

Behavior: Lightly and briefly grasp the target object between the upper and lower beak ("grasping").

Consequence: Treats and social drama.

Criteria: Latency: 2sec.; frequency: 100% through 10 trials; distraction: various; time point: 3 sessions at ≈1min. each.

Targeting is an excellent behavior to begin training because it is relatively easy to train and you can use it in training many other behaviors. This can be a good behavior to start training for confidence as well. Use grasping the target as the behavior, rather than just a beak touch or nudge, because the grasp behavior will help later when training with a target stick for picking things up.

Phase 2. Acquisition

Increase the rate of responding by conducting training when the bird is not satiated with regard to either your social contact or food. Luckily, birds tend to enjoy lengthy training sessions, something you would not likely get away with for a cat for instance. Train at a time when the bird is not too tired, but also not hyperactive, in order to ensure effective concentration. These establishing operations prevent other contingencies from taking control over the behavior. Ensure that you begin training in a low distraction environment (i.e., there are minimal competing contingencies that could control the bird's behavior). You can allow for increased distraction later in the process—for now, set yourself and the bird up for success with a graded approach.

Begin by introducing the bird to the target stick in the same way that you should introduce them to just about anything new, by placing it close to their cage for a couple days so that they can get used to it. You might need to place it far away and get it a bit closer each day in an unceremonious manner.

Start by presenting the target oriented pointing at the bird's beak so that the bird will be less likely to grab the handle with a foot. Present it smoothly and slowly to avoid startling the bird and present it close to the bird's beak. Do not present it too low, in order to avoid having the bird investigate the target with a foot rather than solely with the beak or try to step up on it. If the bird moves his or her head away from the target or otherwise makes no immediate move toward the target, simply wait. Avoid pursuing the bird with the target. The bird will likely show some interest in the novel item and its presentation at some point, and they will grasp the target. Be ready! Click and treat immediately. In some cases, the bird might only nudge the target, so you may decide to shape the "grasp." In this case, start with looking at it or leaning toward it and shape the "grasp" through several approximations involving opening the beak to grasp at the object. Once you have achieved reinforcement of the grasping behavior, you should get the behavior more readily, and quickly establish a nice rate for the behavior, evoked by presenting the target stick itself. Do not worry if the bird grasps the target and does not let go right away at first. After the click and offering the treat, the bird should let go to accept the treat, and through successive trials, the bird will likely tend to begin letting go as soon as they have grasped the target because grasping the target sets the occasion for letting go to accept the treat. If they do not let go at first even to get the treat, you might need a tastier treat or you may simply need to give the bird a chance to examine the target before proceeding.

Phase 3. Fluency

Once the bird exhibits several grasping/targeting behaviors in a row, you can begin generalizing the behavior (i.e., establishing a wider range of circumstances in which the behavior can be evoked). You can present the target a little higher, so that the bird has to reach up to target it. You can also begin working on presenting the target far enough away that the bird has to take a step or two towards it in order to grasp the target stick. Work on reaching and approaching behaviors gradually. If you started with targeting inside of the cage, you can begin training the bird to exit the cage to grasp the target stick. If you started training outside of the cage, you can begin requiring the bird to go into the cage to target the stick.

Avoid subtractively punishing targeting behavior. For instance, avoid using targeting to get the bird to go into their cage at a time when this will subtract social or other reinforcers. Later, once you have trained the behavior well, you may be able to occasionally use it for these purposes, but it is best to avoid that as much as possible.

Once you are working outside of the cage easily, you can begin working on flat surfaces such as the floor or a tabletop.

The behavior is now fluent in its most simple form and you can begin thinning the schedule of reinforcement. Move from continuous reinforcement to a gradually thinning variable ratio schedule of added reinforcement.

Utilizing a graded approach, begin working form, latency and speed, one at a time, gradually and as needed, until each satisfies the behavior objective criteria. This might include decreasing the latency or training a quicker motion from beginning to reach or step to actually grasping the target (i.e., speed). Refining form involves shaping. If any components of the form are inadequate, identify exactly what movement needs to be changed and what the movement will look like when it meets the appropriate criterion. You may need a list of approximations if there is a significant difference between the current form and terminal form, but often, this kind of refinement simply requires slight shifts in form that you can achieve with one or two levels of approximations.

Once you have a reliable behavior trained with the appropriate form, latency and speed, exhibited under at least minimal levels of distraction, you can transfer stimulus control to a vocal cue, so that you will be able to use other objects in training where targeting will be helpful. To transfer stimulus control from the presentation of the target stick you have been using, simply present the vocal cue, such as "grasp," and then present the target stick. The bird will exhibit the behavior and you can reinforce. Repeat this through several trials, reinforcing each occurrence of the behavior when transferring stimulus control. It is important to remember not to backtrack on progress made with a form, latency or speed. When introducing a new feature, do not relax the requirements for the others, as is done when proofing against the three D-parameters.

Next, begin proofing the behavior against the three D-parameters. Duration involves the bird holding the "grasp" beyond the initial touch. Very gradually, require a longer duration of the "grasp" behavior, ensuring that you move slowly enough to maintain a high rate of reinforcement. This can be challenging, because the bird will usually let go quickly in order to access the treat. If you find that you simply cannot get enough of a hold to build on, you can train the bird to grasp and let go but to remain right at the target stick without moving off and gradually build up duration. Remember that where you feed treats matters. If you do not want the bird to grasp the target and then immediately move their head toward you looking for treats, ensure that you feed the treats right there where the bird stands. For this exercise, distance can involve the distance the bird needs to travel in order to grasp the target. Again, work this up gradually enough to ensure a high rate of reinforcement. Distraction, as always, refers to the presence of competing contingencies. This might mean the presence of other people, other birds, toys, and/or food in the environment. You can also think of working in different locations as distraction training. Work with new distractors gradually and at a pace to ensure a high rate of reinforcement. Work one parameter at a time and when you introduce a new one, relax the criteria for the others, to ensure success.

Next, begin discrimination training and bring the behavior under vocal control. Present the target stick without first presenting the vocal cue. The bird will likely grasp the target. Remove the target and do not reinforce occurrence of the behavior. After a few seconds, present the vocal cue and present the target stick. Reinforce occurrence of the behavior. Through the next several trials, vary presenting the vocal cue and not presenting the vocal cue in a seemingly random manner before presenting the target stick. You will reinforce occurrence of the behavior when you vocally cued it and extinguish occurrence of the behavior when you did not vocally cue the behavior. Eventually, the behavior will only occur when you present the vocal cue. Begin re-thinning the variable ratio schedule of reinforcement again but only once the vocal cue alone controls the behavior.

Phase 4. Maintenance

Once you have the final form, speed and latency criteria of the target behavior, it is under stimulus control, and reliably proofed through the three D-parameters, you can begin working toward maintenance. There is not always necessarily a clear line between fluency and maintenance phases. You will likely want to continue to develop proficiency in new locations or with new distractors etc., but once you are well along in the process, it is time to begin transitioning from the intensive training activities of the fluency development phase to less intensive maintenance of the fluency you have achieved.

Begin generalizing the reinforcers from just the treats, for instance, to praise or drama reinforcers.[74] Remember also that the clicker is just for acquisition and training toward fluency. Discontinue use of the clicker and instead use just the unconditioned reinforcers. If you incorporated duration into the "grasp" behavior, you can replace the click sound with a release stimulus. You might say "you're free" when the behavior is over and before you deliver the unconditioned reinforcer.[75] Also, begin using fewer trainer-mediated reinforcers. Use activity reinforcers via the Premack principle, in order to help you maintain control over the behavior. For example, if coming out of the cage is reinforcing, take the opportunity to require a target behavior before allowing the bird out of the cage. If going to the play/exercise area acts as a reinforcer, require a "grasp" while you open the door. The same goes for shower time and other reinforcing activities, etc. The goal in this phase is to work toward simply maintaining what you have achieved during training with minimal contrived activity outside of the evocative stimuli. If at any point any component of the training seems to be deteriorating, refresh the training by building that parameter back up.

Step Up

Phase 1. Preliminaries

Behavior Objective

S^{FA}: Perch item presented in front of legs just below crop.

S^{Ev}: Vocal "Step up."

Behavior: Step up onto the perch item with both feet.

Consequence: Treats, social drama and access to other areas.

Criteria: Latency: 2sec.; duration: until released, minimum 1min.; frequency: 100% through 10 trials; distraction: various; time point: 5 sessions at ≈1min. each.

The bird can step up onto the perch item one foot at a time starting with either foot. The goal is to have the bird perched on the item within two seconds and remain there.

The distraction criteria are vague in this behavior objective. This is partly because this is general, rather than specific to an individual, and for conciseness purposes. If you

---

[74] As used here, "drama reinforcers," refers to becoming very animated and excited. You should assess the bird's reaction to this. Some birds might be skittish with such stimulation and it would not function as a reinforcer at all. In other cases, it can be *the most* effective reinforcer available.

[75] You might have been thinking that the "you're free" will simply be another conditioned reinforcer and you would be right to the extent that it does undergo such conditioning. However, we are not using it as a conditioned reinforcer per se, but rather, antecedently, as a release stimulus. You will always use the release stimulus but you will not necessarily always follow that with an unconditioned reinforcer. The "release stimulus" is an antecedent stimulus that evokes a release from the behavior. Any other behavior the bird then exhibits may or may not then be reinforced, but just with the unconditioned reinforcer (the conditioned reinforcer has been eliminated).

allow a vague reference to distraction (or any other criterion) in your behavior objective, it is usually a good idea to include an addendum or footnote with some specific details. You will likely want the bird to step up when other people are engaged in various activities near by, when food or novel toys are available, in various locations, and when different types of perches are used. Identify and prepare a list of the specific requirements.

Most birds will have been conditioned to "step up" very early in their lives (and indeed very little conditioning would have been required for this species-typical behavior pattern). However, birds deprived of early training, or who were neglected, may need to be trained or retrained to "step up." In some cases, the bird is apprehensive or fearful of hands and your training will focus mainly on desensitizing the bird to hands, rather than solely focusing on the "step up" behavior.

Identify the first perch to you will use in training. You can use a finger or forearm, but you should consider also training the bird to step up onto a wooden perch (e.g., dowel) as well. In this example, a human appendage perch is used. However, if the bird is particularly fearful of hands, start with a wooden perch (e.g., a dowel). Once you have completed training with the wooden perch, you can begin working through the training with hands, but be sure to do so very carefully. Use a clicker and suitably sized treats.

Phase 2. Acquisition

Increase the rate of responding by conducting training when the bird is not satiated with regard to either your social presence or food. Train at a time when the bird is not too tired, but also not hyperactive, in order to ensure concentration. These establishing operations prevent other contingencies from taking control over the behavior. Ensure that you begin training in a low distraction environment, so that there are minimal competing contingencies that could control the bird's behavior. You can allow for increased distraction later in the process—for now, set yourself and the bird up for success with a graded approach.

In the case of a small bird such as a budgie, parrotlet, cockatiel or lovebird, you may use a finger as the perch. In this case, hold the index finger straight out and keep your thumb pressed into your palm so that it is not a distraction. In the case of a medium sized bird such as a quaker or even an amazon or african grey perhaps (depending on your strength), hold all of your fingers out and together like you would use for a "karate chop," but keep your thumb pressed tight to your palm. The target zone here for the step up will be around the knuckle of the top of the index finger. Lighter birds can go a bit further out onto the finger and heavier birds closer in toward the thumb and index finger knuckle. In the case of large birds such as cockatoos and the larger macaws, use the forearm, close to the wrist. Keep the hand in a closed fist to avoid distraction in this case. In all cases, the perch must remain solid and steady. You should also avoid quick jerky motions both as you approach the bird and while they are perched on you. Quick motions and unsteady perches will be aversive and disruptive to training. The training described here is for a small bird.

Slowly and steadily, approach the bird with the perch-positioned hand slowly to a position that is directly in front of and slightly above the bird's legs. The position of your hand should be similar to a step from the perch that they are currently on. If the bird is wary of the hand, try to maintain the hand very still and in position. Wait for the bird to become accustomed to the presence of the hand in front of them. If the bird shuffles away from your hand, you may need to take extra time to work on desensitizing the bird to your hand's presence before proceeding. Do not pursue the bird with your hand. Give it some time and begin shaping hand-approach behaviors.[76] This is where a shoot and hopper device comes in handy since you don't have to approach the bird to deliver the treat either with your hand or a spoon, either of which might be a problem in this situation. If you hold your hand up too high, the bird may use their beak to explore the hand before stepping up. Avoid this by holding the hand a little lower and closer to the bird's legs. If the hand seems unsteady or the bird is cautious, the bird might use their beak to steady him or herself before stepping up onto your hand. If the use of the beak is quick and painless to you and the bird steps up right away afterward, you might decide to allow it. If the bird is not too fearful but is only slightly hesitant for the step up, you can prompt the step up by *very gently* moving the perch to touch the bird above the legs, below the crop. Most birds will step up with this prompt, but if you can, avoid the initiated physical contact. Patience and shaping are preferable. Once the bird steps up onto the perch, remain steady, keep your perch hand in the same place, but click, and treat. You can use your other hand for clicking and treating or you can make a mouth clicking sound in place of the clicker and use the other hand just for handling treats. Very young birds might be a little unsteady. Be particularly careful not to let the bird fall. Next, gently move the bird so that the perch they were on originally will be the new step up perch. Ensure it is slightly forward of and above your finger perch. Click and treat when the bird steps back up onto that perch. A little drama reinforcer here is often effective as well; if there was any tension present, this kind of reinforcement helps relieve it.

Repeat through several more trials, in order to establish a strong history of reinforcement for this behavior. Avoid any type of aversive stimulation during this training and especially while the bird is perched on you. Stepping up should always be a reinforcing experience to maintain the contingency.

Phase 3. Fluency

You should now have the bird stepping up onto and off your finger reliably. If there is some part of the form, latency or speed of the behavior that needs improving, start working on that now. If the bird does not respond quickly to the presentation of your hand, you can set the latency criterion such that you will get most trials within the criterion time and some not. Reinforce the criterion-meeting latencies and not the slow ones. Do *not* pull your hand away as a subtracted punisher as this is simply too dangerous. Once most of the responses are within the interval, reset the criterion for a slightly shorter criterion interval and gradually improve the latency. If the bird begins to move as soon as you present the hand, but moves very slowly, you can use the same

---

[76] Approximations might be looking at hand, shifting weight toward hand, stepping/shuffling toward hand, raising a foot, touching perch with the foot, putting weight on perch, putting both feet on perch.

process for speed, incrementally adjusting the speed criterion until you have the behavior occurring quickly enough.

Begin gradually thinning the variable ratio schedule of reinforcement.

Once you have the form, speed and latency that meet the behavior objective, you can establish a vocal cue stimulus if you so wish. It can be useful to establish a vocal cue rather than allow the presence of your hand to continue evoking the behavior. You are likely to be using your hands around the bird very often (e.g., working with food, water, toys, etc.) and you do not want the bird to always try to step up on one of your hands, and this behavior could be dangerous. However, this training takes quite a bit of work because (a) a hand or finger presented as a perch is often strongly evocative of stepping up, a species-typical behavior that is difficult to extinguish and (b) we cannot remove the hand to prevent the behavior or subtractively punish it for safety reasons. In most cases, it seems easiest to use a very formal hand position for when you want the bird to step up so that discrimination occurs between this formal cue for stepping up and your hand merely being present for other purposes. The vocal "step up" can be used as a function-altering stimulus to promote discrimination.

However, establishing a verbal cue can be done and here is how to do that. To establish a vocal cue, as with target training, present the vocal cue, then present the finger perch, the bird steps up onto the perch, and you reinforce criterion responses on a continuous reinforcement schedule. Repeat this through several trials. After a number of vocally cued and reinforced responses, present the finger but without first presenting the vocal cue. The bird will step up. Do not reinforce this; rather have the bird step back onto the perch right away. You do not want to pull your hand away before the behavior occurs in order to prevent the behavior from occurring, as this risks a fall and injury. Instead, simply fail to follow "step-ups" that are not preceded by the vocal cue with contrived reinforcers and when you vocally cue the behavior. Be sure that you reinforce criterion responses. Carry this out through several trials, always reinforcing vocally cued "step-ups" and never reinforcing occurrences of the behavior that you did not vocally cue. Some reinforcement will come merely from the interaction, but the goal is to maintain the bird's safety and the reinforcing quality of the interaction. After several trials, maintain praise, but put the click and treats back on a gradually thinning variable ratio schedule again.

Once you have a reliable step under the control of a vocal cue or a formal step up hand position, you can begin proofing it against the three D-parameters. Start with duration. Gradually increase the time that you require the bird to remain perched on your finger. Vary the intervals in a seemingly random manner, raising the criterion at a pace that ensures success as well as progress. Distance usually refers to the distance between the bird and the trainer when you cue a behavior. However, in this case, consider the distance from the original perch to address this parameter. Gradually, increase the distance that you take the bird from the original perch and include various routes to help the behavior of staying perched on the finger generalize to different environments. For distraction, start using the cue when the bird is perched in different locations and with increasing social, food and toy distractors present.

Phase 4. Maintenance

Once you have the final form, speed and latency criteria of the target behavior, it is under stimulus control, and reliably proofed through the three D-parameters, you can begin working toward maintenance. There is not always necessarily a clear line between fluency and maintenance phases. You will likely want to continue to develop proficiency in new locations or with new distractors etc., but once you are well along in training, it is time to begin transitioning from the intensive training activities of the fluency development phase to less intensive maintenance of the fluency you have achieved.

As with target training, begin generalizing the reinforcers. The variable ratio schedule should be quite thin. Because you have no new parameters to train, you will discontinue using the clicker. Instead, for reinforcement targeted trials, use praise and drama reinforcers, as well as activity reinforcers (e.g., going to the play station or for a shower or to socialize, assuming these activities are reinforcers for the bird). The goal in this phase is to work toward simply maintaining what you have achieved in training with minimal contrived activity outside of the cues. If at any point any component of the training seems to be deteriorating, refresh the training by building that parameter back up.

Exiting Cage

Phase 1. Preliminaries

Behavior Objective

S^{Ev}_1: Vocal "Let's go."

Behavior₁: Approach cage door.

Consequence: Treats and social drama but eventually opportunity to exhibit other link.

Criteria: Latency: 2sec.; speed: within 3 seconds; frequency: 100% through 10 trials; distraction: various.

S^{Ev}_2: Arrives at open cage door.

Behavior₂: Steps up on presented finger.

Consequence: Treats, social drama and access to other areas.

Criteria: Latency: 2sec.; duration: maintain until released, minimum 1min.; frequency: 100% through 10 trials; distraction: various; time point: 4 sessions at ≈1min. each.

I will only describe the steps for this project briefly since the training process and steps described in the previous two projects cover much of what you will need here. This project is the first behavior chain I will describe specifically for birds. You will evoke two distinct behaviors in this behavior chain. Below, I will describe the chaining of these two behaviors.

As before, the distraction is stated vaguely and you can be more specific in your list depending on the requirements you have for the project.

Identify the conditioned and unconditioned reinforcers you will use, and gather and prepare your equipment.

Phase 2. Acquisition

As in all new training projects, conduct the training when the bird is most receptive to it. Choose a time of day when the reinforcers will be effective and the bird is not too tired or excessively energetic.

This will be a particularly easy behavior chain to train since you have already trained the bird to target and to step up. Prompt coming to the cage door with the target stick, remove the target stick as the bird arrives, and cue stepping up on your finger (or hand/wrist/forearm, depending on the size of the bird) by presenting the finger in a perching position. Repeat through several trials until the bird exhibits the behaviors smoothly in sequence. Maintain a continuous reinforcement schedule.

Phase 3. Fluency

You should now have the bird coming to the cage door and stepping up onto your finger, each evoked separately. The first thing to do in the fluency stage is to chain the two behaviors together. If you use your formal hand position as the cue to step up, then this really becomes a case of sequencing. If you use a vocal cue for step up, then this is a chain. In either case, you can proceed similarly. Start by introducing a vocal cue for the behavior chain. Say "let's go" and then present the target stick, remove the target stick as the bird reaches the cage door, present the perch finger, and reinforce coming to and stepping up on your finger. Repeat through several trials so that the new vocal cue takes on stimulus control over the sequence of behaviors. Next, fade the target stick. Make the target stick less and less obvious in each trial until the target is not necessary for the bird to exhibit approach behavior. Begin presenting the perch finger immediately following the vocal cue "let's go." The bird should approach and immediately step up onto the finger. Reinforce this and repeat through several trials.

Put the behavior on a gradually thinning variable ratio schedule of added reinforcement.

Once you have the behavior chain under the control of a single vocal cue, refine the form, latency and speed as needed, so that the behavior chain meets the criteria. Work as described for the previous behaviors, being sure to work one at a time and gradually shifting the criterion until the behaviors are adequate.

Once the behavior meets criteria, is under appropriate stimulus control under minimal distraction, begin proofing against the D-parameters. The primary parameter for this behavior is distraction. Begin calling the bird away from low-level distractors (e.g., playing with an old toy or near the end of eating or grooming). Work up to the bird playing with new toys, beginning to eat, and engaging in social interactions.

It is likely that you will go to the bird for various reasons, including cleaning, taking, or presenting food or new toys etc. Work discrimination training into these activities. Reinforce coming to the cage door to come out only when you have vocalized "let's go" and soon, your mere approach will not evoke the behavior chain (or at least not as reliably as the vocal cue does).

Phase 4. Maintenance

Once you have the final form, speed and latency criteria for the behavior chain, it is under stimulus control, and reliably proofed against distraction, begin working toward maintenance. Continue to develop proficiency in new locations or with new distractions.

As with target training, begin generalizing the reinforcers. The variable ratio schedule should be quite thin. Because you have no new parameters to train, you will discontinue using the clicker. Instead, for reinforcement targeted trials, use praise and drama reinforcers, as well as activity reinforcers (e.g., going to the play station or for a shower or to socialize, assuming these are reinforcers for the bird). If at any point any component of the training seems to be deteriorating, refresh the training by building that parameter back up.

Holding a Wing Up

Phase 1. Preliminaries

Behavior Objective

S^{Ev}: Vocal "Left wing."

Behavior: Raises left wing up above head and spread out.

Consequence: Treats and social drama.

Criteria: Latency: 2sec.; speed: within 2 seconds; duration: maintain until released, minimum 1min.; frequency: 100% through 10 trials; distraction: various; time point: 6 sessions at ≈1min. each.

Raising a wing can be useful for examinations, flight feather trimming and for showers. I will present the steps for raising the left wing here, but you should also separately train "right wing" as well.

As before, distraction criteria are vague and you can be more specific in your list depending on the requirements you have for the project.

Identify the conditioned and unconditioned reinforcers you will use, gather, and prepare your equipment.

Phase 2. Acquisition

Instate the usual antecedent control conditions.

It is difficult to prompt wing-raising. The easiest approach to training this behavior is usually to shape it when the nontrainer-contrived environment evokes the behavior. This requires some patience. Start by carefully observing the bird to identify the stimuli that evoke wing-raising. You may be able to determine this, but in some cases, the behavior evoking stimuli occur inside of the bird. If you can determine conditions when the behavior will be more likely, this will make your job a lot easier because you can either choose to train when these conditions are present or you can arrange for the conditions to be present.

Begin by preparing a list of approximations. It may look like this (with the units of measurement adjusted for the size of bird):

- Separating wing from body by 1/16th of an inch

- Separating wing from body by 1/8th of an inch

- Separating wing from body by 1/2 of an inch and raised by 1/16th of an inch

- Separating wing from body by 1 inch and raised by 1/2 of an inch

- Separating wing from body by 2 inches and raised by 1 inch

- Extending wing out and up 3 inches

- Extending wing out and up 4 inches

- Fully extending wing out and up above head

As indicated in the above approximations, you will reinforce separating the wing from the body and then reinforce successive approximations of raising the wing out and up above the head, the terminal behavior.

Begin by being prepared with the clicker and treats. Be patient. When the bird exhibits the first approximation behavior, click and treat. Timing is very important. Try to be boring and exhibit minimal animation or verbal behavior, except for the reinforcement event, in order to provide sufficient contrast. The first few sessions can be several minutes long, as the initial stages of shaping such a behavior without contrived prompts is minimally intrusive and once you begin to achieve some momentum in the reinforcement, it is best to continue the training session for awhile. It may take several minutes before you get the first approximation behavior. Another strategy is to start with a finer grained initial approximation. You might break the first behavior down into a few behavior approximations. In other words, the initial behavior is only the slightest of movements of the left wing. This can then generate momentum. Once things get rolling, you should find the rate of the behavior increases quickly.

Phase 3. Fluency

Once you have the initial approximation behavior occurring reliably, move to the next approximation behavior, extinguishing the approximation before it. This might be a good time to end that session.

In the next session, remember to review the initial training to get the momentum built back up. Continue to work your way through the list of approximations, using the size of approximation to keep the pace of conditioning moving smoothly. The initial shaping project will help you achieve the form of the behavior.

Once you have the form, you can work on speed and latency as necessary, to meet the criteria you have set for it.

Once you have achieved the form of the behavior exhibited at the appropriate speed and with a suitable latency, you can bring the behavior under stimulus control. To do so, present the vocal cue immediately before the behavior is exhibited and reinforce each instance, until the vocal cue alone reliably evokes the behavior without the presence of prompts, including the training context that might involve the presence of clickers and treat pouches etc.

Once you have achieved the form, speed and latency, and the behavior is under stimulus control, you can begin gradually thinning the variable ratio schedule of reinforcement and begin proofing against the D-parameters. The duration involves the amount of time the bird holds the wing in position. The distance will involve how far away you are from the bird when you evoke the behavior. You will likely not require a great distance for this behavior. Distraction might include social activity (including novel people such as veterinarians) and the presence of spray bottles or novel environments such as an exam room. Remember to bring the schedule of reinforcement back to continuous one when introducing a new D-parameter and re-thinning it as you train.

Phase 4. Maintenance

Once you have the final form, speed and latency criteria of the terminal behavior, it is under stimulus control, and reliably proofed against the appropriate D-parameters, begin working toward maintenance. Continue to develop proficiency in new locations or with new distractions, including in veterinary exam rooms.

As with target training, begin generalizing the reinforcers. The variable ratio schedule should be thin. Carry out the other maintenance related procedures as described for previous behaviors. If at any point any component of the training seems to be deteriorating, refresh the training by building that parameter back up.

Leave It

Phase 1. Preliminaries

Behavior Objective

S^{Ev}: Vocal "Leave it."

Behavior: Looks away from whatever was being attended/oriented to when the evocative stimulus is presented and ceases approaching.

Consequence: Treats, social drama and access to objects.

Criteria: Latency: 1sec.; speed: within 1 seconds; duration: maintain until released, minimum 10sec.; frequency: 100% through 10 trials; distraction: various; time point: 8 sessions at ≈3min. each.

"Leave it" can be a very handy behavior to have under stimulus control for those times when the bird gets into something that is not safe for them. It is easy to distract the bird with a treat or toy when they get into something troublesome and bribe them to disengage from the dangerous/problematic situation by showing them a treat but this inadvertently causes two problems. First, bribing[77] can cause the behavior to become dependent on the reinforcer being present. Second, it can condition a behavior chain, reinforcing the problem behavior. In other words, by "redirecting" the bird, as some would put it, the chain of attending to the problem stimulus, which then makes available the opportunity to disengage from it, which then contacts the reinforcer is conditioned. You can then expect an increase in attending and contacting the problem stimulus as it leads to eventual reinforcement. To avoid these problems, shape the disengagement.

Identify the conditioned and unconditioned reinforcers you will use, gather, and prepare your equipment.

Phase 2. Acquisition

Utilize antecedent control procedures to establish the conditions most conducive to successful training.

Start with a favored treat or toy, whatever will definitely get and maintain the bird's attention. Have a clicker and treats ready. Hold the item out to your side with an extended arm. The bird will attend to the item. This will set the occasion for shaping leave it. Have a list of approximations prepared that starts with glancing away from the item even for a fraction of a second, proceeding gradually to a few seconds. It is not

---

[77] Recall from a previous use of the word bribe that it is broadly defined as paying someone to do something wrong. However, in the animal training field, it almost universally refers to showing the unconditioned reinforcer before the behavior in order to motivate the subject to exhibit the behavior, which does not need to be a "bad" behavior. This practice creates problems as the text describes and I use the term bribe here in that context.

always easy to see exactly where some species of bird are looking so watch the eyes and head position carefully.

Phase 3. Fluency

The initial shaping project helps you achieve the form of the behavior. Once you have the form, you can work on speed and latency as necessary to meet the criteria you have set for it.

Once you have achieved the form of the behavior, exhibited at the appropriate speed and with a suitable latency, bring the behavior under stimulus control. To establish the evocative stimulus, present the vocal cue immediately before the behavior is exhibited, which is usually right when they see your hand going out to the side, and reinforce each instance until the vocal cue alone reliably evokes the behavior without the presence of prompts or other stimuli. Other stimuli include the training context that might involve the presence of clickers and treat pouches etc.

Once you have achieved the form, speed and latency, and the behavior is under stimulus control, begin gradually thinning the variable ratio schedule of reinforcement and begin proofing against the D-parameters. The duration involves the amount of time the bird maintains attention away from the item without looking back at it or approaching it. Gradually, build the duration up to 10 seconds. Distraction might include the effectiveness of the item as a reinforcer (i.e., how interesting and attractive the item is to the bird), as well as what is going on in the environment around the bird at the time you present the vocal cue. Remember to bring the schedule of reinforcement back to continuous when introducing a new D-parameter and re-thinning it as you train.

Begin applying the "leave it" cue to instances when the bird is attending to something that you are not holding.

Phase 4. Maintenance

Once you have the final form, speed and latency criteria of the terminal behavior, it is under stimulus control, and reliably proofed against the D-parameters, begin working toward maintenance. Continue to develop proficiency in new locations or with new distractions. Work the features of maintenance as described for previous behaviors.

Special Note on Avoiding and Eliminating Problem Vocal Behaviors

Many species of bird can be conditioned to speak. Although we are not focusing on resolving problematic behaviors, it is instructive to comment on problem vocal behaviors exhibited by birds. Birds frequently become conditioned to say words that can be offensive to certain audiences. The problem is usually two fold. First, when people swear, they commonly do so with animation and special emphasis, they swear emphatically and these are exactly the kinds of words that birds are most likely to repeat. Second, the reaction people often have to a swearing bird is usually a highly effective reinforcer. Whether you find it funny and laugh or you get upset and try to "scold" the bird, these are all likely drama reinforcers that will readily maintain the behavior. The best advice is of course not to say anything around the bird that you do not want them repeating. However, if the bird does take to repeating a problematic word or phrase, it is best to use differential reinforcement to eliminate it.

First, avoid saying the offending words from that point on. Make sure everyone in contact with the bird knows about the existence of the problem words and reacts the same way, every single time the bird says them. Generally, ensure that you recognize and reinforce any acceptable vocal behaviors to ensure a generally high rate of reinforcement throughout the day. If you were currently not interacting with the bird at the time, do not respond *at all* to the word. Provide as little reinforcement as possible for it. If you were interacting with the bird in some way at the time, immediately (like a light switch), turn away and discontinue interacting with the bird. One caveat here—this assumes that your interaction was reinforcing. Assuming it was, immediately ceasing interaction with the bird will subtractively punish the vocalization. If your interaction with the bird was aversive for whatever reason, do not let swearing "chaise you off." The point is to extinguish and/or subtractively punish any undesired vocalizations and that requires accurate identification of the actual reinforcer maintaining the behavior. Remember, ignoring is *not* the same as extinguishing! On the flip side, find fun words and phrases to say in an animated manner and encourage the bird to repeat them; reinforce each instance with drama and treats. The strategy outlined here is consistent with a book written by behaviorologist Glen Latham called "The Power of Positive Parenting," which I believe many parrot guardians would find useful.

Continuing Education

Courses provided through The Companion Animal Sciences Institute at www.CASInstitute.com address all of the topics covered in this chapter.

Courses provided through Behavior Works at www.behaviorworks.org address topics found in this chapter.

If you are looking for a good book to give to clients, consider these books:

- Getting Started Clicker Training for Birds by Melinda Johnson

- Good Bird! A Guide to Solving Behavioral Problems in Companion Parrots! by Barbara Heidenreich

Believe it or not, if you are looking for some reading material that is a little different from the usual, but potentially very useful in working with birds (and any other species), consider the book below:

- The Power of Positive Parenting by Glen Latham

CHAPTER 10. TRAINING HORSES

Behavioral Objectives

The objective of this chapter is to measurably expand the reader's repertoire of behaviors in relation to describing and relating the principles of behavior. Upon successfully integrating the concepts outlined in this chapter, the reader, where exposed to contingencies to do so, will accurately:

- Implement a training plan for common ground manners-related behaviors, applying the systematic strategies previously discussed based on a systematic plan previously discussed.

The behaviors discussed below are common behaviors trained for good manners of companion horses, referred to broadly as "ground manners." These behaviors include a sampling of behaviors that will make daily interaction with companion horses a more enjoyable and safe experience.[78]

Training Horses

Training horses is, as you would expect, similar to and quite different from training cats, dogs and birds. Whereas you might, for instance, be able to get away with physically manipulating or manhandling a dog to get them to do what you want (although this approach is definitely *not* recommended), you will not get away with this with a horse. In fact, a big part of training horses is geared to control the horse's movements (notably, the hooves and head), as well as to have the horse readily take direction, respond to various stimuli without fear, and to act safely around people and other animals (e.g., prevent them from pushing you over or knocking you unconscious). Considering the size, strength and speed of horses, trainers must consider safety at all times. Training horses is similar to training other species in that all of the same laws and principles of behavior apply, and in general, we use the same strategies and procedures. We refer to much of the training described here as "ground manners," and it is intended to allow for safe handling and to guide the horse's movements. These are foundational training projects, because they will be applicable to a wide range of activities with the horse and will provide the basics that will allow for training more advanced behaviors. These behaviors are also foundational because the strategy and techniques demonstrated with these behaviors can easily be adapted to executing numerous other training projects.

As with parrots, but not cats or dogs, horses are prey animals. As such, horses are easily spooked and escape behaviors are easily evoked. When training horses, it is

---

[78] I would like to thank Amanda Clase, PH.D., DIP.CBST, Jelena Kallay, DIP.ABT, KPA CTP, Kathleen Kemp, B.SC., CERT.PDTST, for reviewing a draft of this chapter and providing feedback that improved the original draft chapter.

imperative to control sudden loud noises and movements that could startle the horse. The escape behaviors they evoke may generalize to stimuli present during training, including the trainer him or herself.

There are different approaches to achieving training of complex goals. You could train a large number of discrete behaviors and then apply them in sequences to get the job done or you could train through a complex goal, in the process, breaking it down into steps and in the end, achieving the training for several discrete behaviors. I will adopt the latter approach here—each grand behavior objective is broken down into minor behavior objectives or steps as per a graded approach. The training of these minor behavior objectives is applicable to a wider range of situations.

Equipment

In order to train a horse, some basic equipment is required. First, you need a closed pen area that is secure enough to keep the horse safe. As with all training, you need a clicker and treat pouch. You will also require at least one target stick; either a telescoping stick that you can adjust or use two sticks of different lengths. You can purchase an excellent target stick from the On Target™ Training's website[79], along with other very useful tools. You will also need a lunge (or "Longe") line. This is a long rope (nylon or cotton) used to keep the horse on a line, but allow enough room for the horse to move freely. A lunging cavesson is also helpful. A cavesson is a type of head halter that accommodates attachment of the lunge line in various locations. It is important that the lunging cavesson fits correctly, otherwise it can slip on the horse's face and cause irritation, discomfort or pain. You will also need a halter and bridle for one of the training projects covered below. If you use a metal bit with the bridle, you should have a rubber bit available, as well.

Most horse guardians will be able to inform you as to what their horse's preferred treats (i.e., most effective reinforcers) are for training. You can also perform a *preference assessment* by presenting two favored treats and taking note of which the horse takes first. You can compare multiple pairs in this way and come up with a rank order list of the most effective reinforcers. Carrot and apple pieces work well, as horses usually like the taste of these items and they are not harmful to a horse's digestive system. However, some horses are not used to taking food in the hard form of carrots and apples. In these cases, grate the carrot or apple at first and then gradually increase the size of the treats until the horse will accept larger pieces. If weight control is a concern, you can use about a teaspoon of the horse's regular grain as a treat, training before the horse eats to prevent satiation and subtract the total amount provided from the meal allowance.

---

[79] http://www.on-target-training.com/products-tools.php

Behaviors

Before discussing specific training projects, we will explore training to prevent or eliminate "mugging" behaviors below. Although the training to minimize mugging behaviors is more of a behavior change program to change a behavioral excess, our emphasis here is on mild cases or prevention more than the resolution of significantly problematic and longstanding or entrenched problems (the topic for another book). The line between training and behavior change programming is not always clear and precise since in some ways it is rather arbitrary.

Mugging

"Mugging" refers to response class forms associated with crowding, shoving/pushing or rooting at people in search of food. Horses are very big and strong, and treats tend to evoke behaviors that function to enhance contact with them. The problem is that during training, you have a bag of treats strapped to your waist, and unless you train otherwise, the horse can easily become conditioned to mug you for the treats. It is a simple matter of training. Worse, many people carry treats in a pocket and horses become conditioned to mug any person who has pockets, which can be quite dangerous. It can result in ripping of clothes and/or injuries. The behavior that functions to contact the treats will become more likely, whether you consider the behavior polite or impolite.

Mugging behaviors include crowding, shoving, and pocket-rooting, so these are all to be targeted for extinction. This problematic contingency is quite simple: pockets, treats or even just people evoke crowding, pushing and pocket-rooting behaviors.

$$\text{People} \rightarrow \text{Mugging behaviors * } \rightarrow \text{Treats}^{(+R)}$$

* Crowding or pushing people and/or pocket rooting.

To minimize this behavioral excess and prevent it from becoming significantly problematic, we instate a differential reinforcement procedure. The response class in this case is any response class form that involves taking treats that have not been offered. The response class forms targeted for extinction will be any of these "mugging" behaviors and the response class forms targeted for reinforcement will be relaxed distance maintaining behaviors. The problematic response class forms (now targeted for extinction) are replaced with mutually exclusive response class forms. The horse cannot be engaged in mugging behaviors if he or she is relaxed and calm, and maintain at least a one–foot distance between their head and the trainer. The replacement contingency is as follows:

$$\text{People} \rightarrow \text{Relaxed 1–foot distance maintenance} \rightarrow \text{Treats \& social contact}^{(+R)}$$

Including relaxed behaviors generally promotes calmness and lack of agitation. Indications of relaxation in horses include lowered head, relaxed muscles in face, ears and neck, a hind foot resting on toe, and a relaxed tail position. The horse is conditioned to contact treats (and continued social contact) with this relaxed distance-maintaining behavior rather than mugging behavior.

The strategy is quite simple. Start with the horse in the stall with a rope across the door, or in a small pen, with the trainer on the other side of the door/pen. Avoid tying the horse down, as the horse might become frustrated due to the restriction of their head movement. This frustration can result in behaviors such as head swishing or biting, in order to get to the treats in possession of the trainer. Approach the horse, but stay approximately three feet away. It is best to have the treat pouch attached behind your back, at waist level, in order to reduce the likelihood of evoking mugging behavior. Start when the horse is somewhat satiated with regard to food, to make mugging behaviors less likely. To deliver the feed, stretch your hand out completely, opening the palm at this location, allowing access to the treats only once your hand is in that location. Make sure to stand as far away as you can and to outstretch your hand away from you when you feed the horse. This prevents the encouragement of crowding behaviors and conditions the horse such that the source of treats is always away from the trainer. Reinforce relaxed, distance-maintaining behaviors and extinguish mugging behaviors. As long as the horse remains calm/relaxed and their head remains outside of the distance criterion, click and treat, and generally continue to provide treats every few seconds for as long as the horse maintains the criterion behavior. If the horse crowds (i.e., enters the distance criterion), say "Oops" and step back and face slightly away (such that you can still see the horse in your peripheral vision) for five seconds before resetting the environment and trying again. Alternatively, you can leave the area, stay out of the horse's sight for three to five seconds, and then return to try again. The contrast between the steady stream of treats and the loss of the opportunity to get treats will promote quick conditioning.

The first task is to generate the behavior. Remember, you are looking for a combination of relaxed/calm posture as well as maintaining distance with the head when they could otherwise enter the criterion "personal space." As mentioned, how you deliver treats is very important to this training. When the horse takes feed from your open palm, it should be with their head positioned outside of the distance criterion. Stand about three feet from the horse. The horse will likely orient toward you, especially if you have food. Adjust the distance between yourself and the horse to ensure safety and the possibility of breaching the one–foot "personal space" but far enough away to make it less likely. You can work your way closer to the horse when it is less likely to evoke orienting behaviors. You may need to either shape the behavior or use a target stick to prompt the behavior at first and fade the target-stick as soon as practicable. As described above, continue to reinforce for criterion behaviors repeatedly until the horse exhibits no mugging or agitation related behaviors.

Once you have achieved a high rate of responding and a very low frequency of mugging behaviors, you may work on the three D-parameters. In this case, duration relates to the time you can remain within potential reach without the horse exhibiting mugging behaviors. Distraction would involve the presence of various effective reinforcer treat items as well as other common distractors. Practice this training from various orientations including standing in front of the horse, facing the horse, facing slightly away from the horse, standing on the other side of the horse, and on the other side of the horse while faced slightly away from the horse (i.e., be sure to practice working on both sides of the horse).[80] Include trials with the treats on you and viewable to the horse. Reserve this for when you have a good solid history of reinforcement for non-mugging behaviors though. Distance might involve reducing the distance between yourself and the horse. Although you might get close to the horse (say, a foot and half), continue to feed the horse away from your body. Work one D-parameter at a time gradually, relaxing the criteria for others when you introduce another one and combine them in a gradual manner once you have them trained separately.

Maintain these rules of food presentation and the differential reinforcement procedure of extinguishing mugging and reinforcing relaxation-related distance-maintaining behaviors in *all* interactions with the horse, in order to ensure that the behavior generalizes appropriately and remains strongly conditioned. You can also continue to carry out this training while engaged in other training projects.

Now that you will be ensuring everyone's safety by conditioning non-mugging-related behaviors, you can move on to other training projects. Pay particular attention to the first training project, targeting, as more detail is included in this description than the following projects in order to prevent tedious redundancy.

Targeting

Phase 1. Preliminaries

Behavior Objective

S^{Ev}: Presented target item.

Behavior: Nudge the target item with nose.

Consequence: Treats and social contact.

Criteria: Latency: 2sec.; frequency: 100% through 10 trials; distraction: various; time point: 1 session at ≈20min.

Target training is one of the most important and fundamental behaviors you can train. Touching a target can be used to guide the horse's head—where the head goes, the rest of the horse goes. Target training is useful for training horses to go to a particular place, remain at a particular place (e.g., for grooming, to stay at the door and not to rush through until released), and even to focus on maintaining a behavior if startled rather than engage in various escape behaviors that might be disruptive at best or dangerous at worst. It also conditions taking direction in general. These outcomes are useful because you can apply them to many other training projects. For instance, we do not explore training a horse to enter a trailer here, but having target training in your

---

[80] Any reference to facing away from a horse in this book should be taken to mean facing away from the horse at approximately a 30 degree angle, where you are not looking directly at the horse, but such that you can still see the horse in your peripheral vision. It can be dangerous to turn your back on a horse that is within reach of you or could be within a few seconds.

repertoire will allow you to easily train trailer loading. Target training is also another good opportunity to simultaneously apply mugging training.

There are a few different types of targets that you will find useful. A commonly used stationary target is a six–inch tall cone (similar to orange traffic cones, but very small). This target is particularly useful because you do not need to hold it. A handheld target stick 10 to 14 inches long is useful, although longer handheld target sticks are also available and useful, as well. A telescoping handheld target is very useful because you can adjust its length to meet your needs for a particular project. Target sticks available through On Target™ Training's website[81] are very useful. However, any target stick that is safe for the horse and you is suitable for these training projects.

Identify the most effective reinforcers for the horse. It may be pieces of apple or carrots, or even just a small amount of grain. Carry out the training when the horse has not recently accessed the reinforcer, to ensure he or she is not satiated and that it indeed will be an effective reinforcer. Gather and prepare your equipment.

Phase 2. Acquisition

Increase rate of responding by conducting training when the horse is not satiated with regard to either your social contact[82] or food. Train at a time when the horse is most receptive to the interaction. This tends to be more important with horses than with many other species. Training when the horse is "not in the mood" can be disastrous. These motivative operations prevent other, disruptive, contingencies from exerting control over the behavior. Ensure that you begin training in a low distraction environment, so that there are few competing contingencies that could control the horse's behavior. You can allow for increased distraction later in the process—for now, set yourself and the horse up for success with a graded approach.

It can be useful to begin with the horse in their stall, a small yard, or a round pen to limit distraction and prevent mugging or other disruptive behaviors. Start with the cone. As horses tend to spook easily, it is important to allow the horse to become accustomed to the sight of the cone. Place it just outside of the horses paddock for a few days before introducing it into the horse's stall or yard. Begin by presenting the target at a distance that requires a reach for the horse to touch it but not so far that it is out of reach or requires them to take a step—this should just be a few to several inches. Click and provide a treat and praise, moving the target out of the horse's sight while the horse takes the treat. Take care to present the treat clearly in an open palm close to the horse's mouth, away from you. If the horse becomes obsessed with the treat pouch at this point, say "Oops," and walk away. Then, wait three to five seconds out of the horse's sight, before returning to try again, as per the mugging training. Present the target again and reinforce nudging it again right away. Repeat through several trials, until the horse will readily and quickly nudge the target in this minimally distracting environment and with the target presented approximately in the same location.

---

[81] www.on-target-training.com/products-tools.php
[82] Social contact does not necessarily refer to touch. It can refer to simply being close by so the horse can see and/or hear you. Horses tend to find simply being around others comforting.

Phase 3. Fluency

Once the behavior is smooth and reliable in these minimally demanding conditions, begin moving the target around, to require the horse to move in different directions to nudge the target and even take a step to contact it.

Put the behavior on a gradually thinning variable ratio schedule of added reinforcement. You will take the schedule back to continuous reinforcement when introducing any new features or changes in criteria.

If you need to change the form, chances are that it will be a small adjustment. In this case, have a clear picture of exactly what form adjustment you need and shape that change by differentially reinforcing approximations toward that form. If the latency is too long, gradually reset a shorter latency criterion, reinforcing criterion responses and extinguishing non-criterion responses until you achieve the appropriate latency. If the speed of the behavior is too slow, do the same thing, gradually resetting the speed criterion until you achieve a suitable speed.

For the next part of the training, it is a good idea to have an assistant to hold the lunge line, particularly if you are in a large open area. Take the horse from the stall and try the same targeting exercises with your assistant holding the lunge line to encourage the horse to remain near by (which should not be difficult, as you are the source of such great reinforcers), and to be able to prevent mugging if necessary. Remember to extinguish or subtractively punish mugging behaviors. You can step out of range, face slightly away while keeping the horse in your peripheral vision, or even leave the yard for three seconds before returning and providing the opportunity to carry on with training without pushy behaviors. Remember, any time you use subtracted punishment such as this, it assumes that your presence and the activity you are engaged in with the horse is reinforcing. If the training is aversive for the horse, then your subtracting yourself from the environment will subtractively *reinforce* that problem behavior. Always keep a close eye on the rate of the problem behavior if you use subtracted reinforcement to ensure you are punishing and not reinforcing the problem behavior. Also, make sure the training is reinforcing. Minimize distraction at first. Carry out the training as before through several trials until it is smooth and reliable in this new environment. If you were not outside in these trials move outside in the next training session and carry out the exercises again. Remember, if you are getting more than the occasional extinction or subtracted punishment trial, you are failing to set the horse and yourself up for success. Go back to an easier level of expectation and build up more gradually with an eye to preventing non-criterion behavior.

Utilizing a graded approach, begin working form, latency and speed, one at a time, gradually and as needed, until each satisfies the behavior objective requirements. Assess the features you need to change based on current proficiencies and the target criteria. This might include decreasing the latency or training a quicker motion from beginning to reach or step to actually nudging the target (i.e., speed). Work through one feature at a time. Refining form involves shaping. If any components of the form are inadequate, identify exactly what movements need to change and what they will look like when they will meet the criterion. You may need a list of approximations if there is a significant difference between the current motion and terminal motion, but often, this kind of refinement simply requires slight shifts in form that can be obtained with one or two levels of approximations.

Once you have a reliable behavior trained with the appropriate form, latency and speed exhibited under minimal levels of distraction, transfer stimulus control to a vocal cue, if you elect to do so. At this point in training, presentation of the target stick evokes the behavior and this is suitable for most circumstances. However, if you wish to bring the behavior under the control of a vocal cue, simply present the vocal cue, such as "touch" and then present the target. The horse will exhibit the behavior and you can reinforce this. Repeat this through several trials, reinforcing each occurrence of the behavior.

Once the behavior is reliable with an appropriate form, latency and speed, begin proofing against the three D-parameters. Duration is not generally applicable in this case, unless you are also working on training the horse to remain at the target for an extended interval. In this case, you can gradually increase duration. If you choose to work on duration, do not forget to use a release phrase, such as "you're free," when the duration is completed. Next, begin working distance by gradually requiring the horse to walk farther, in order to nudge the cone. You should have it up to a few steps at this point, but vary the distance in a seemingly random manner, gradually working your way up to 10 to 15 feet. Watch the horse for indications that you are moving too quickly. If the latency begins to increase, you will know that you are moving too quickly through distance. Once you have distance worked up, begin working on distraction. Have other people present and engaged in gradually increasing levels of distracting conditions. Move at a pace that ensures success.

If you chose to bring the behavior under the stimulus control of a vocal cue, begin discrimination training. Reinstate continuous reinforcement for vocally cued responses and instate extinction for responses not preceded by the vocal cue. Present the target without first presenting the evocative stimulus. The horse will likely nudge it. Say "Oops," remove the target and do not reinforce occurrence of the behavior. After three to five seconds, present the vocal cue and immediately present the target. Reinforce occurrence of the behavior. Through the next several trials, alternate in a seemingly random manner between vocally cueing the behavior and not vocally cueing the behavior immediately before presenting the target stick. Reinforce occurrence of the behavior when vocally cued, and extinguish occurrence of the behavior when not vocally cued. Eventually, the behavior will only occur when vocally cued. Begin re-thinning the schedule of reinforcement again.

Repeat the training, this time with a handheld target stick. You should find that the training goes more quickly, but do not expect targeting to transfer to a different object automatically just because you present the vocal cue. If you present the evocative stimulus holding a handheld target without further training, the horse will likely exhibit looking-around-for-the-cone behaviors. So, begin by presenting it close, just as you did for the cone, and reinforcing the nudge. After a few trials like this, you should find that the process goes much more quickly and smoothly and you can bring the behavior under control of the vocal cue.

Phase 4. Maintenance

Once you have the final form, speed and latency criteria of the target behavior, it is under stimulus control, and reliably proofed through the three D-parameters, begin working toward maintenance. There is not always necessarily a clear line between fluency and maintenance phases. You will likely want to continue to develop proficiency in new locations or with new distractors etc., but once you are well along in training, it is time to begin transitioning from the intensive training activities of the fluency development phase to less intensive maintenance of the fluency you have achieved.

Begin generalizing the reinforcers from just treats, for instance, to praise and petting reinforcers.[83] Remember also that the clicker is just for acquisition and training toward fluency. Discontinue use of the clicker and instead use just the unconditioned reinforcers, or use the release phrase and unconditioned reinforcers if you extend duration. Also, begin using fewer trainer-mediated reinforcers. Use activity reinforcers via the Premack principle, in order to help you maintain control over the behavior. For example, if coming out of the stall is reinforcing, take the opportunity to require a target nudge before allowing the horse to leave their stall. The goal in this phase is to work toward simply maintaining what you have achieved during training, with minimal contrived activity outside of the evocative stimulus. If, at any point, any component of the training seems to be deteriorating, refresh the training by building that parameter back up.

In training targeting, you have actually trained several things. You have conditioned taking direction, attending to trainer-mediated cues, and you have trained the horse not to be too pushy, as well. Every time you train something, you are actually training a more cooperative relationship between you and the horse. This will pay off when it comes time to train other behaviors.

General Handling

Behavior Objective

S^{Ev}: Brushing mane

Behavior: Remain still and relaxed *

Consequence: Treats and social contact.

Criteria: Duration: 5min.; frequency: 100% through 10 trials; distraction: various; time point: 1 session at ≈20min.

* As before, indications of relaxation in horses include lowered head, relaxed muscles in face, ears and neck, a hind foot resting on toe, and a relaxed tail position.

Handling is necessary for most interactions with companion horses from attaching and removing equipment to grooming or examining various parts of the horse, and even

---

[83] Remember that praise and petting are at least partly conditioned reinforcers and that we maintain those through repeated pairing with other established reinforcers.

using touch as a reinforcer. Mane brushing is described here as an example, but you can apply the same procedure and principles for handling the feet, tail or other body parts. The general strategy is most important.

In this training project, we introduce another major behavior change strategy. If you are replacing agitation and stress-related behaviors with relaxation-related behaviors, problematic emotional behaviors are also involved. The primary target behavior is operant, but the procedure utilized, if carried out properly, will simultaneously achieve both operant conditioning of the operant target behavior and respondent conditioning of the respondent emotional behaviors that play a motivational role in the contingency. We frame the procedure in operant terms, but the beneficial respondent conditioning will occur as a byproduct, because the procedure also comports with respondent conditioning principles. Through repeatedly pairing the sight of the brush with added reinforcers, and not with aversers, in our operant focused procedure, any stress or anxiety related emotional behaviors will become extinguished (eliminated) and/or counter-conditioned (replaced). In other words, repeatedly pair the brush with treats and praise, rather than anything "unpleasant," and the brush will come to elicit relaxation or even emotional behaviors experienced as "pleasant" instead. However, for clarity, we externalize the contingencies and frame the procedure in terms of operant conditioning and the primary target behavior is operant. The respondent conditioning of emotional behaviors occurs simultaneously because, while operant behaviors are being reinforced, brushes are presented at approximately the same time as treats (referred to procedurally as "pairing"), resulting in respondent conditioning of respondent emotional behaviors (i.e., the chemical and neural processes involved in emotions). You can only achieve this with procedures that utilize added reinforcers solely, and ideally with a high rate of reinforcement. For a fuller discussion of resolving problem behaviors that involve emotional responses, see O'Heare (2015).

Identify the unconditioned reinforcers you will use and gather the other equipment needed (e.g., a clicker and a soft brush or mane comb). Carry out the training when the horse has not recently accessed the reinforcer, in order to ensure they are not satiated and that it indeed will be an effective reinforcer. On the other hand, if a hungry horse tends to obsess over the treats, this can be counterproductive and you may need to train when the horse is more satiated.

As is usually the case, a graded approach is most effective. Break the training down into smaller, more manageable, steps. In this case, sight of the brush can be one step, and touching the horse with the brush can be another step. You may find it necessary to break the project into even smaller steps, if the horse is particularly fearful or apprehensive.

Step 1. Response to Brush

Phase 1. Preliminaries

Behavior Objective

S^{Ev}: Sight of brush

Behavior: Remain still and relaxed.

Consequence: Treats and social contact.

Criteria: Duration: 3min.; frequency: 100% through 10 trials; distraction: various; time point: 3 session at ≈20min.

Phase 2. Acquisition

It can be useful to begin with the horse in an open pen, perhaps holding a lunge line while an assistant manages the presentations of the brush or vice versa. The horse will have a freer range of movements and be less likely to panic and injure him, herself, or you.

Begin by having the brush within sight of the horse's stall for a few days, so they become accustomed to having it in their environment, but without anyone touching it (as with the target cone training). Determine precisely what the appropriate initial exposure level should be. The starting exposure must not evoke escape behaviors or stress-related behaviors. Indications of stress in horses include, but are not limited to a raised head, muscle tension in neck, face and back, ears that are held stiffly or pinned back, holding breath, whale eye (where you see a flash of the whites of the horse's eye), moving away, the tail being held tight to hindquarters, and sharp tail swishing. One of the primary dimensions for manipulating the intensity of exposure is distance. If you need to present the brush at a distance greater than you can manage while being near the horse for reinforcing behavior, you may need to have an assistant manage either the exposures or the reinforcement. While you observe the horse, very carefully approach it with the brush in your hand (but try not to act "weird" about it or the horse will be "weirded out" by this). If you can walk up to the horse without evoking any distress-related behaviors, place the brush down near by and start there, keeping your safety in mind. If, during the approach, the horse becomes tense, back off and take note of that distance. In this case, you will need to start further away from the horse than that point. If this was outside of reaching distance, employ the help of an assistant. You may need to begin even before you handle the brush, in which case reaching for the brush will be an early level of exposure. Determine the form or intensity of exposure that will not immediately evoke agitation-related behaviors.

Begin by presenting the brush at the first level of intensity and immediately click and provide the treat, provided the horse remained still, quiet and relaxed rather than being agitated and trying to escape. If escape behaviors occur, it indicates that your level of exposure is too intense and you need to find a less intense exposure level. At first, do not expect long exposures—click and treat immediately. Repeat through several trials in succession, until the horse exhibits interest in the treats when you present the brush (but not agitation related-behaviors or excessive interest-related behaviors, such as mugging; the target behaviors here are being still, quiet and relaxed). When the horse is completely relaxed and even demonstrates interest when the brush is presented at that intensity of exposure then you are ready to increase the intensity of exposure a small amount and build further on that foundation. You should treat each level in the same way.

If you are familiar with behavior change programming to resolve problem behaviors, you might have been thinking that this procedure resembles what we call **_systematic desensitization_**. You would be correct in that it does indeed comport with that procedure. However, the behavior of concern is operant (whereas systematic desensitization is said to address respondent emotional arousal) and the procedure utilized is differential reinforcement. Pairing a stress-eliciting stimulus with a "pleasure"-eliciting stimulus in a gradual and incremental manner until the previously problematic stimulus comes to elicit the pleasure related emotional arousal looks exactly like a graded differential reinforcement procedure. However, it is possible to more readily observe and quantify the operant behavior of concern than emotional arousal, so we externalize the contingency and frame the procedure in operant terms. As discussed above, this procedure generates beneficial respondent conditioning as a byproduct of the operant procedure. Behaviorologists do not ignore emotional behaviors as is sometimes suggested. The emotional arousal and the problematic operant behavior share a cause and in addressing the operant behavior, we can address the emotional arousal as well so long as we utilize a graded added reinforcement-emphasized approach.

Phase 3. Fluency

Often, the fluency phase begins with putting the behavior on a gradually thinning variable ratio schedule of added reinforcement. In this case, we use a high rate of reinforcement in order to achieve the respondent conditioning objectives. We will work in various new criteria almost continuously. Thus, the schedule of reinforcement should remain continuous, for now.

Once you have increased the rate of still, quiet and relaxed behaviors in response to sight of the brush at a given distance for approximately one second, begin gradually decreasing the distance between the horse and the brush in this same incremental manner, very gradually increasing the intensity of exposure and only when the previous level evokes solely relaxed behaviors. Once the horse is responding appropriately each time that you present the brush at the distance you have been working, incrementally decrease the distance and train through several trials at that new distance. Do not reduce the distance again until the horse has worked through several trials at the current distance and only exhibits the target behaviors. Ensure a strong history of conditioning at each criterion level before changing the level. Work in this manner until you can present the brush within reaching distance of the horse and the horse remains still, quiet and relaxed.

Next, begin extending the duration of exposure. Duration is usually easier to work on than distance. Relax the distance a little, in order to ensure success and begin gradually, incrementally increasing the duration that you present the brush to the horse. Repeat through several trials until the horse exhibits still, quiet and relaxed behaviors each time that you present the brush to them. Set the horse up for success. Remember that you want to avoid extinction or subtracted punishment trials and that you want a high rate of reinforcement. This is the only way to also ensure respondent conditioning of the emotional behaviors. Manipulate the training by manipulating the criterion changes (i.e., jumps). Use smaller, rather than larger jumps in difficulty and ensure a solid history of conditioning at one level before moving to the next level of difficult, just as was done with the distance parameter. Once you have achieved a series of approximately 10 still, quiet and relaxed responses to the brush for a specified duration level, try again, with a slightly increased duration. At this point, begin varying the trial durations by increasing the duration very gradually, but in a seemingly random manner. For instance, first require half of a second, then one and a half seconds, then half second, then two seconds, then one second, then two seconds, then half of a second, then one second, and then two and a half seconds, etc. Gradually continue to work at a pace dictated by the success of that pace to ensure a high rate of reinforcement and minimal extinction trials, to reach a duration of the behavior of three minutes. This may require a few training sessions.

If the horse becomes agitated during any presentation of the brush, do not reinforce the behavior. If it seems likely that the horse will calm down within a few seconds, wait and reinforce once the horse is still, quiet and relaxed. Otherwise, remove the brush. This indeed will subtractively reinforce agitation related-behaviors but there is little choice, as continued exposure will only generate more agitation and stress. This situation would definitely be a setback in training. Evaluate the reasons why the trial deteriorated or was unsuccessful. How did you increase the intensity of exposure in that trial? Make a strong effort to effectively avoid the situation from occurring again.

Work to the point where you have a contingency wherein the brush evokes calm interest-related behaviors but also still, quiet and relaxed behaviors, and the horse can maintain these relaxed behaviors for three minutes while the brush is present. Take this opportunity to gradually work back to the distance parameter you achieved previously (remember, it was slightly reduced to start training the duration parameter). Now that you have the duration to criterion, you can work on generalizing the response. Start by resetting the duration criterion back to a half of a second. Present the brush in a slightly different way. If you took it from a shelf, then perhaps remove it from a bag. If you presented, maintained and then removed the brush at the same distance, work on presenting the brush at a greater distance and approaching the horse with it (i.e., approaching the horse with the brush in your hand is the new dimension).

Work one parameter at a time and relax the duration while you do so. Once you have that criterion established, you can begin gradually increasing duration again. Then, once you have these parameters combined, you can work to increase another parameter in the same way. Continue through this process of presenting the brush in new ways until you can present it in any way likely to occur, with a reasonable duration of still, quiet and relaxed behaviors by the horse.

Because one of our goals was to counter-condition (replace) problematic emotional arousal, maintain a rich schedule of reinforcement all the way through the fluency stage.

Phase 4. Maintenance

Once the behavior is fluent, put the behavior on a gradually thinning variable ratio schedule of added reinforcement. Start to vary the reinforcer types as well, including the use of praise and petting (assuming that these are reinforcers for the horse). Maintain gentle praise on a rich schedule, but thin the treat schedule.

Step 2. Touching the Horse with Brush

Phase 1. Preliminaries

Behavior Objective

S^{Ev}: Touch and brush with brush.

Behavior: Remain still, quiet and relaxed.

Consequence: Treats and social contact.

Criteria: Duration: 3min.; frequency: 100% through 10 trials; distraction—various; time point: 3 session at ≈20min.

Identify the most effective reinforcers for the horse and train when the horse is not satiated with regard to your presence or the treats. Gather and prepare your equipment.

Phase 2. Acquisition

This is a big jump in difficulty, so go slowly and ensure you establish a solid foundation of conditioning before proceeding to make it more challenging. Allow the horse to examine the brush each time you approach with it; allowing the horse to sniff the brush will help ensure they become comfortable with its presence before the contact occurs. You may need to work approaching the mane with the brush as a sub-step. If you do so, gradually work toward touching the mane with the brush in a manner consistent with the approach described above. You may also need to begin the touch the mane with the back of the brush, so that there is no accidentally pulling of tangles and it feels more like a petting than brushing. You can turn the brush around and gradually increase the intensity of brushing after you have established a solid history of reinforcement for remaining still, quiet and relaxed for short, smooth strokes. Another trick is to use massage before touching the horse with the brush. Massage is quite relaxing for most horses who are accustomed to being touched. Massaging the horse right before touching with the brush can relax the horse. We also pair the massage with the brush and improve the horse's general emotional reaction to the brush.

Phase 3. Fluency

Once you can make contact with the mane with the brush and the horse remains still, quiet and relaxed, work in more lengthy strokes and increase the number of strokes. Remember that each of these is a new parameter and when you introduce a new one or increase one, you should relax other parameters (e.g., duration). Gradually, work until you are able to present the brush, and approach the horse with it and then brush the horse's mane in a way that meets your criterion for a few minutes, and the horse remains still, quiet and relaxed for the entire time.

Phase 4. Maintenance

Begin thinning the schedule of reinforcement and begin relying less on treats and more on various other kinds of reinforcers, especially praise and petting. Remember also that the clicker is just for the acquisition of behaviors and not for maintenance so discontinue use of the clicker and use just the unconditioned reinforcers.

Bridling / Haltering and Hackamores

Before this step, and ideally before any steps in this training project are carried out, the horse must be accustomed to, and tolerant of, being touched in all of the ways that will be required to put the bridle on. If you can touch all parts of the horse's head, including in their mouth and the horse has no injuries or other problems causing a problematic reaction to the touch, you can proceed. If there are any problems with any of these kinds of contact, you will need to ensure the horse is healthy and then work through handling exercises before proceeding with the bridling process.

A graded approach involves working through the training using a properly fitted bridle with no bit (e.g., a hackamore) or reins. Start by working your way through the training, first without the bit and reins attached and then go back and work through with a properly fitted bitless bridle and no reins. Then work through the training with a rubber bitted bridle and no reins. Then work through the training with a bridle with a metal bit (assuming you will be using one) and no reins. Finally, work through the training with reins attached to a bridle with a metal bit. This involves a fair bit of training, but it will generate the best results and you should find that the latter stages tend to go more smoothly and quickly than the early stages. It is important to ensure that the halter and bridle fit properly. If they do not, they may slip and cause discomfort. This may make it more difficult to get the bridle on and off, which will be disruptive to the training (Karrasch, 2012, pp. 83-84). If the halter and bridle are not fitted right away, you can work adjusting the fit at the appropriate stage in the training once you are putting them on the horse and working on distraction and duration at that stage.

As with handling, this training project is broken down into smaller steps that are more manageable as part of a graded approach. The primary goal is to put the bridle on the horse while the horse remains still, presenting his or her head low enough and taking the bit when presented. This will be a much easier process if it is broken down into smaller projects. The final behavior of "bridling" is an example of sequencing, as it involves a series of discrete behaviors that the horse exhibits when each is cued—the behaviors will occur in the same order each time but your participation at each link makes it sequencing rather than chaining.

Instate all of the usual antecedent control measures described under previous behaviors above. It can be useful to begin with the horse in their stall to limit distraction and prevent mugging or other disruptive behaviors. However, always be extremely careful when working with a horse in their stall to ensure they do not panic, as this can be quite dangerous for the horse and you.

The description of each project below follows the same general approach as handling.

Step 1. Response to Bridle

Phase 1. Preliminaries

Behavior Objective

S^{Ev}: Sight of bridle.

Behavior: Remain still.

Consequence: Treats and social contact.

Criteria: Duration: 1min.; Frequency: 100% through 10 trials; distraction: various; time point: 3 session at ≈15min.

Instate appropriate antecedent control measures, and gather and prepare your equipment.

Phase 2. Acquisition

Start with the first form of stimulation associated with the bridling procedure. As with the general handling training described above, you may be replacing agitation and stress-related behaviors with relaxation-related behaviors. Therefore, you will be reinforcing calm behaviors in place of agitation-related behaviors. At the same time, you will be changing the horse's emotional reaction to the bridle as a byproduct of the operant conditioning techniques, assuming that you maintain a high rate of added reinforcement. If you are not replacing agitation behaviors with relaxation related behaviors, the process will go much more quickly and smoothly. Start by finding an intensity of exposure to the bridle that does not evoke agitation-related behaviors. This might mean that you present the bridle at a distance and that distance gradually decreased. It may also mean not touching the bridle at first and gradually working that parameter. You may need an assistant, if the bridle needs to be far away; someone will arrange the presentations and the other will reinforce the behaviors.

Present the bridle and immediately click and treat for remaining still. Repeat through several trials until the horse is very relaxed in the presence of the bridle. Gradually increase the duration of exposure in a variable manner, until the horse remains still for one minute after you present the bridle.

Phase 3. Fluency

Next, gradually decrease the distance from the horse that you present the bridle. At this stage, no one approaches the horse with the bridle. The goal is to generate and reinforce maintaining stillness at incrementally closer presentations of the bridle. Approaching with the bridle is the next step. As always, do not increase the difficulty level until the horse is completely relaxed through the current level. Of course, the bridle does need to get closer to the horse in order to carry out this training, but do so in a minimally obvious manner, picking it up from behind you and presenting it in front of you without taking steps toward the horse. The next level of difficulty will involve walking toward the horse with it.

Once you can present the bridle within reaching distance of the horse and the horse remains still for 1 minute, move to step 2. You can instate the maintenance phase for the final step.

Step 2. Approach with Bridle

Phase 1. Preliminaries

Behavior Objective

S^{Ev}: Approach with bridle.

Behavior: Remain still.

Consequence: Treats and social contact.

Criteria: Duration: 1min.; Frequency: 100% through 10 trials; distraction: various; time point: 3 session at ≈15min.

Phase 2. Acquisition

As in the handling project, relax the duration criterion and begin gradually introducing someone approaching the horse with the bridle in his or her hand. Start by presenting the bridle at a distance that will be common for the brindle to be carried after training is complete and take a step or two (depending on how many you can get away with and the horse remain relaxed and still) toward the horse with the bridle in hand. Click and treat if the horse remained still and relaxed. Repeat through several trials.

Phase 3. Fluency

Next, repeat again through several trials but approach a little closer in a gradual and incremental manner until you can get all the way to the horse with the bridle in hand. Once you are approaching from the expected distance and the horse maintains the target behavior through the approach, and for one minute after you have arrived at the horse, you are ready to move on to the next project.

Step 3. Presenting Bridle to be Put On

Phase 1. Preliminaries

Behavior Objective

S^{Ev}: Presenting bridle.

Behavior: Lower head to facilitate putting the bridle on.

Consequence: Treats and social contact.

Criteria: Frequency: 100% through 10 trials; distraction: various; time point: 3 session at ≈15min.

Phase 2. Acquisition

Unlike in the handling project, putting on a bridle involves a sequence of behaviors. The horse is to generally remain still, but in this step, when the handler reaches out to the horse with the bridle to put it on, the reaching evokes head lowering that facilitates having the bridle put on. This can be achieved by training the horse to gently target their nose into the opened bridle and then remaining still. Since you have already conditioned the horse to target generally, this is usually the best approach here, as well. The target will be the noseband inside the bridle opening, or if there is no noseband, simply target the opening of the bridle as you hold it up in the opened position. By training the horse to help facilitate bridling, they are more engaged in the process, which makes the process easier.

You will need an assistant for this project, as someone will need to manage the bridle and someone will handle the reinforcement. First, ensure that your foundation is solid and that you can present the bridle, opened up at the height you want to ultimately put it on the horse, at a foot or two away from the horse. Once this evokes remaining still, rather than agitation and escape behaviors, you are ready to begin training the horse to target the inside of the bridle.

You have a few options at this point. You could hold the bridle up and have an assistant hold the target stick right at the bridle such that it requires the horse to insert their nose to touch the target and reinforce then, fading the target stick afterward. Alternatively, you could start by training the horse to target the opening of the bridle. The former approach will usually make the training initially faster, but it requires fading the target. The latter method may take a bit more time and work to get the training started, but you will not have to fade the target stick. Assuming you use just the bridle opening for targeting, proceed in a manner similar to that used to condition nudging the target stick. Hold the opened bridle close to the horse's head, but positioned somewhat lower than head-level, so that it is closer to the location where you ultimately want to put it on the horse. Be patient and wait. The horse will move their head around, and at some point, they will nudge the inside of the bridle, assuming they are not anxious about the bridle's presence. When they nudge the inside of the bridle, click and treat. Repeat through several trials until the horse is readily targeting the opening of the bridle. You might also practice presenting it at different angles, in order to ensure that you conditioned the targeting well in various presentation circumstances. If the horse does not nudge the inside of the bridle, shape it.

Phase 3. Fluency

If you need to work on form, speed or latency, this is the appropriate time to do so. In any of these cases, work gradually by starting with a reasonable criterion and once it is smooth, require the behavior to be closer to your required ultimate criterion. You may need the horse to insert their nose into the bridle a bit more, to move more quickly when they are moving the nose, or to begin moving toward the bridle sooner.

Gradually and incrementally, increase the duration that the horse must remain touching the inside of the bridle up to one minute. To establish a vocal cue, say "bridle" immediately before presenting the bridle. Reinforce the behavior and repeat through several trials to solidify the transfer of stimulus control from merely presenting the bridle to the vocal cue. Test the stimulus control by vocally cueing the behavior and waiting a couple seconds. If the horse exhibits the behavior, reinforce and repeat. If they did not, present the target and then either repeat through several more trials before trying again, or gradually fade the target stick. Fade the target stick by making it less and less obvious with each trial until the horse lowers their head and remains still for one minute without the target stick present at all.

Step 4. Taking the Bit

Phase 1. Preliminaries

Behavior Objective

S^{Ev}: Presenting the bit.

Behavior: Lower head to bit and take it in mouth, remaining still otherwise.

Consequence: Treats and social contact.

Criteria: Frequency: 100% through 10 trials; duration: 1min.; distraction: various; time point: 4 session at ≈15min.

This step assumes you use a bridle that has a bit component. If you use a hackamore, a halter, or bridle with the bit removed, you can skip this section and move directly to step 5, which deals with attachment of the throatlatch buckle of the bridle, etc.

Phase 2. Acquisition

The least aversive way to train taking the bit is through shaping, so that the horse reactively takes the bit in their mouth, rather than merely passively accepting manipulation of their mouth. Ideally, start with a rubber bit. Since this exercise primarily involves working with the bit, you can fold the rest of the bridle back away from the horse's head, so that it is not in your way. Start by presenting the bit a few inches away from the horse's mouth. Once you have the horse targeting the bit, you will be able to lower it to the height where you want the horse to take it in his or her mouth. The terminal behavior will be for the horse to open their mouth and facilitate its placement on their bars (i.e., the interdental space between the front teeth and the molars), and then closing their mouth on the bit. As usual, when you see "and then," you know there is two distinct behaviors. Indeed, opening the mouth and then closing the mouth and two distinct behaviors. This will therefore be a sequence of two distinct behaviors. Approach with the bridle will evoke mouth opening and placing the bit on the bars will evoke mouth closing. You can usually train both behaviors together quite easily since they will always occur in this specific order.

The approximations for the first behavior might look like this:
1. Target the bit with nose
2. Any movement of lips to separate
3. Any movement of lips to mouth bit
4. Mouthing the bit
5. Mouthing further into the mouth

You may then place the bit on bars and shape or reinforce closing the mouth with the bit in place.

You may need to break some of these approximations down into finer approximations. At each approximation, click before removing the bit and then immediately provide the treat. When you get to the point where you place the bit on the horse's bars, continue to remove the bit before feeding, even when you are increasing the duration. Because you have not tied the bridle, the bit can slip out and injure the horse's teeth. Once you get to the point where you have the bridle attached (step 5 below), you can begin feeding the treats while the horse is wearing the bridle. Work your way through shaping the terminal behavior exhibited under minimal criterion conditions.

Once you have shaped the behaviors and the horse is not biting down on the bit with their teeth, switch to a metal bit, if you will be using one. Throughout the training, *be careful* not to hit the horse's teeth when you are bridling or unbridling the horse.

Phase 3. Fluency

Once the terminal behaviors are smooth and reliable, you can refine the form, latency and speed, as needed for each. If you want to bring the behaviors under vocal stimulus control, this is the ideal time to do this, and as usual, present the vocal cue, followed by presentation of the bit and then reinforce the behaviors through several trials. You may want to simply allow bit presentation to evoke the initial bit-taking behavior. You can then put the sequence (bit-taking and mouth closing behaviors) on a gradually thinning variable ratio schedule of added reinforcement and begin proofing against duration and distraction. Remember to go back to continuous reinforcement any time you introduce a new parameter to the training and re-thin the schedule as you work through it.

Step 5. Attaching the Bridle

Phase 1. Preliminaries

Behavior Objective

S^{Ev}: Attaching the bridle.

Behavior: Maintain lowered head, remaining still.

Consequence: Treats and social contact.

Criteria: Duration: 3min.; Frequency: 100% through 10 trials; distraction: various; time point: 3 session at ≈15min.

Phase 2. Acquisition

The next step involves taking the bridle (hackamore or halter) over the horse's ears and attaching the mechanisms (be they on the back of the neck or at the throat, such as the buckle at the throatlatch for bridles that have them). Guestimate the fit for the bridle and err on the side of it being loose for now. You will work proper fitting in later. When you begin working with the bridle, you should work without reins attached. You can then work through the behavior with the reins attached. Always use a graded approach and break a project down into as many steps as necessary. You should have the bridle almost on and the duration worked up, so it is a short step to fasten it. If you can achieve this sequence in one motion while the horse maintains the target behaviors, reinforce, and repeat through several trials until it is smooth and reliable. If the horse is head shy and becomes agitated, and/or exhibits any escape behaviors (e.g., pulling away), you will need to break this process down into more steps at the very least. If this is the case, and you skipped working through with a halter and then a bitless bridle etc., you should go back and work through these previous behaviors before working through with the bridle and bit. Use as many steps as necessary, inserting finer scaled steps when you encounter any resistance during training.

271

As soon as the bridle is on, click and provide the treat. If you use a bit, use treats that are not too big, such as large carrot pieces, to prevent the treat from becoming stuck in the bit and causing problems with swallowing or presenting a choking hazard. Some horses will be sloppier with eating treats with a bit in place than others will. Allow the horse to finish eating and then remove the bridle.

Taking the bridle off while the horse remains still with a lowered head is a separate project. Putting the bridle on provides the opportunity to remove it so it is a good idea to work on unbridling at the same time as you are working on putting the bridling on. When removing the bit, ensure that you do not knock the teeth with it!

Next, work through the training process with reins attached to the bridle. Timing is vital here.

You should now have the horse wearing the bridle with a bit and reins, whether you had to work through these different components together or one at a time. Although it could easily be another step, you should also work in placing the reins over the horse's head. Many horses spook when this is done so very gradually work up to placing the reins over the horse's head, continuously reinforcing for calm relaxed behaviors and maintaining the head in the lowered position.

Phase 3. Fluency

Once you have the bridle on, you can refine the form and latency as needed and build up the duration to at least three minutes. Make sure to pair wearing the bridle with any suitable reinforcing stimulation (e.g., leaving the stall or praise and petting) for the horse.

You should now work on increasing duration and distraction. Start by making adjustments to the fit of the bridle if it was not properly fitted before starting, which is likely was not if the horse is new to bridling. Take each trial as an opportunity to refine the fit until the bridle fits the horse perfectly. This has the dual effect of allowing you to fit the bridle and to get the horse used to these kinds of adjustments.

Phase 4. Maintenance

Begin gradually thinning the variable ratio schedule of added reinforcement. Also, vary the reinforcer types, including the use of praise and petting (assuming these are reinforcers for the horse). Maintain gentle praise on a rich schedule, but thin the treats. If at any point the behaviors begin to degrade, refresh the training.

Head-Lowering

Phase 1. Preliminaries

Behavior Objective

S^{Ev}: Vocal "lower."

Behavior: Lower head and then remain still.

Consequence: Treats and social contact.

Criteria: Frequency: 100% through 10 trials; distraction: various; time point: 3 session at ≈15min.

Training the horse to lower their head can be useful for grooming and several other activities, and it is a relatively simple behavior to train.

Phase 2. Acquisition

You can use the target stick to generate the lowered-head position and then fade the target stick, or you can shape the head-lowering behavior, taking advantage of the common variable head movements that horses make, and build on that. If you use the target, you simply present the target at the appropriate spot to evoke lowering the head and repeat through several trials until the behavior is smooth and reliable. Then, gradually increase the duration, then fade the prompt, and finally transfer stimulus control to a vocal cue. If you shape head lowering, you will not need to fade the target prompt. In that case, you simply begin shaping an incrementally lower head position and the training environment and behavioral momentum controls the behavior. You can then attach a temporary vocal cue and increase the duration. Then, once the horse exhibits the behavior in its final form, you establish the permanent vocal cue. Targeting is usually quicker and easier.

Phase 3. Fluency

Once you get to this level, increase the duration that the horse maintains their head in the lowered position. This will help ensure that the horse remains still, quiet and relaxed throughout the process and it sets you up for the next step.

Phase 4. Maintenance

Begin gradually thinning the variable ratio schedule of added reinforcement. Also, vary the reinforcer types, including the use of praise and petting (assuming these are effective reinforcers). Maintain gentle praise on a rich schedule but thin the treats. If at any point the behaviors begin to degrade, refresh the training.

Loose-Lead Walking

There are many ways to train a horse to walk well on lead and to take direction. The process described here is similar to the way that we would train a dog to walk on a loose leash, with the exception of emphasizing finer scale movements (e.g., forward, back, halt and turn). Once conditioned, these behaviors will be useful in a wide variety of situations for the rest of the horse's life.

Start with the horse on halter with a six-foot lead attached. If the horse is young and uncomfortable with the halter, you will need to address halter wearing before beginning this training project. Start with the horse beside you facing the same direction as you are. Brace the lead in your hand against your waistband. It is helpful to keep a thumb in the waistband of your pants to help brace your hand; if the horse lunges away, the thumb will release, preventing injury to your hand.

Many horses will react violently to pressure and restriction on the halter. This is especially the case with young horses that have not been sufficiently trained to yield to pressure and horses that have previously been trained with coercive methods. If you are facing either of these situations, you will need to counter-condition this reaction before proceeding with this exercise. Treat this similarly to another kind of handling exercise as described above. Very gradually, differentially reinforce tolerance and relaxation related behaviors with a graded approach and a high rate of reinforcement. Only once the horse will tolerate pressure on the halter, should you proceed with loose lead walking.

Forward

Phase 1. Preliminaries

Behavior Objective

S^{Ev}: Vocal "Forward."

Behavior: Walking forward at a pace to remain beside and slightly behind handler.

Consequence: Treats and social contact.

Criteria: Latency: 1sec.; frequency: 100% through 10 trials; distraction: various; time point: 1 session at ≈20min.

Instate all of the appropriate antecedent control conditions to ensure success, and gather and prepare your equipment. Initially, begin work in a small yard, so that the horse cannot get far away from you.

Phase 2. Acquisition

Begin by taking a couple steps forward. The horse should take a step or two forward to keep up with you. If they do not, you may begin the training with your handheld target stick to generate the forward movement. You might also simply try some verbal prompting (i.e., encouragement). Once the horse takes a step when you start walking forward, you will click and provide a treat and praise. One step is great at the beginning. You will build the number of steps up later and it will be much easier once you have the first step conditioned. Make sure the click occurs *while the horse is moving* even though it is just one step at first, rather than clicking for stopping (timing is important). When you click and present the treat (remember, to present the food away from your body), the horse will stop to eat the treat, which provides you another opportunity to engage in another trial. The horse may simply tend to keep walking and so it is important to follow the click with stopping yourself and providing the treat in a way that the horse must stop to take the treat. Try to stay about a foot ahead of the horse. This will allow the horse to easily see you. In addition, you present your directional cues toward them, which would be impossible if the horse is ahead of you.

Phase 3. Fluency

Once the horse moves forward to keep up with you, and your moving forward while holding their lead evokes forward motion, you can transfer stimulus control to a vocal cue. You can say "forward" or "let's go," or whatever you prefer. Present the vocal cue immediately before beginning to walk forward. After several trials, you should be able to present the vocal cue and the horse will begin to move forward right away. If you used a target stick to generate the forward walking behaviors, you can transfer stimulus control, from that prompt to the vocal cue in the same way.

Once this is smooth and reliable, begin adding a few more steps. Gradually and incrementally, increase the number of steps you take before reinforcing walking close to you in the four or eight o'clock position. Vary the number of steps around a gradually increasing average number of steps. Be careful not to increase the number of steps too quickly.

If, at any point, the horse veers off in another direction or barges forward, prompt them back verbally or with the handheld target stick. If this seems ineffective, say "Oops," plant yourself solidly and remain still, waiting for a three-second loose lead before walking again. Remember that if the horse will react violently to pressure on the halter, address that issue *before* continuing with this training. Once you are back on track, ensure that the horse walks nicely for at least a few steps before reinforcing again, in order to minimize the chance of conditioning a problem behavior chain. In this case, also examine why you failed to achieve the target behavior. What went wrong? Is the reinforcer adequately reinforcing? Is the horse satiated with regard to the treats? Barging or veering occurs when distraction is too strong and/or reinforcers are ineffective. Make whatever adjustments are required to get back on track. Work toward achieving four or five steps at this point in the training.

Refine form, latency and speed as needed. Put the behavior on a gradually thinning variable ratio schedule of added reinforcement and practice in various locations with various forms of distraction. You should also practice walking at various speeds to ensure that the horse will keep up and maintain the same speed as you.

Phase 4. Maintenance

Continue to thin the schedule of reinforcement and practice under various conditions. Begin using various kinds of reinforcers including praise alone, or providing access to the opportunity to engage in fun activities.

Run through the training again, leading from the other side of the horse. This should progress more quickly than the first training process.

Halt

Phase 1. Preliminaries

Behavior Objective

S^{Ev}: Vocal "Halt."

Behavior: Stops walking and remains in position.

Consequence: Treats and social contact.

Criteria: Latency: 1sec.; duration: 3min.; frequency: 100% through 10 trials; distraction: various; time point: 1 session at ≈20min.

Instate all of the appropriate antecedent control conditions to ensure success, and gather and prepare your equipment.

Phase 2. Acquisition

Once the horse will moving forward for several steps, you can begin training the horse to stop when you do. Thus far, you have clicked and provided treats when you reach the criterion number of steps and the horse is still walking close to you rather than going elsewhere. When you click and provide the treat, you would have stopped moving but you need a cue that will evoke the stop (clickers are not cues). Training is usually straightforward with a handheld target stick. Present the target within view in front of the horse as you walk (a component of moving forward training), but then stop walking and hold the target stationary in space. In other words, the horse will walk into the target if they do not stop. The target should be positioned forward of the horse, at about their chest level, so that their head will be faced down to touch the target if they walk too far forward, but it will not be too far out of the horse's reach if they stop well short of it. Repeat through several trials until that is smooth and reliable. Alternatively, you can slow down yourself and if this is adequate to prompt the horse to slow down as well, you can shape slowing down and eventually stopping.

Phase 3. Fluency

Once you are able to get the horse to stop walking when you stop walking, you can fade the target prompt and transfer stimulus control to a vocal cue. To do so, say "whoa," "halt," or whatever you wish to use as the vocal cue, right before you stop moving forward. Reinforce each successful trial. After several successful trials, the vocal cue will begin to take on control over the behavior.

If the horse does not stop when you present the vocal cue for halting, say "Oops" and brace yourself. Administer extinction and subtracted punishment before beginning again.

Refine form, latency and speed as needed.

Put the behavior on a gradually thinning variable ratio schedule of added reinforcement and practice in various locations with various forms of distraction.

Begin proofing the behavior against duration and distraction. Duration is particularly important, as it is useful in other training projects. Gradually increase the duration. Replace the click with a release stimulus, such as "you're free." Build up the duration to three minutes. Distraction could involve your moving around the horse as well as away and back toward the horse, while the horse continues to be relaxed and remains in position.

Phase 4. Maintenance

Continue to thin the schedule of reinforcement and practice under various conditions. Begin using various kinds of reinforcers including praise alone, or providing access to the opportunity to engage in fun activities.

Turning

Phase 1. Preliminaries

Behavior Objective

S^{Ev}: Vocal "Left" or "Right."

Behavior: Turn to walk in a way that maintains position.

Consequence: Treats and social contact.

Criteria: Latency: 1sec.; frequency: 100% through 10 trials; distraction: various; time point: 2 sessions at ≈15min.

Instate all of the appropriate antecedent control conditions to ensure success, and gather and prepare your equipment.

Phase 2. Acquisition

Turning with you is usually easy to generate, but if needed, a target stick can help. Start with training one direction and work on the other direction once you have trained the first direction. This will make transfer of stimulus control easier. As you are walking forward, with the previous training in place, you might now be able to make small turns and have the horse also make the turns with you. Start with five degrees and gradually work your way to sharper turns. If the horse does not adequately change directions to walk with you, use a target stick. Present the target stick as you walk in front of the horse and incrementally move it to one side, so that the horse must change directions in order to touch the target. Once the horse changes direction, click and treat. Repeat through several trials.

Phase 3. Fluency

Once this is smooth and reliable, you can begin increasing the angle of your turn. Do this gradually, varying the angle around an average number of degrees, at a pace that assures a high rate of reinforcement, and low rate of extinction or subtracted punishment. To transfer control from the target to a vocal cue, present the vocal cue "left" or "right," depending on which direction you started working first, immediately before targeting the behavior and reinforce through several trials. Often, you can simply allow your changing direction to control the horse's turning behavior.

Refine form, latency and speed, as needed.

Put the behavior on a gradually thinning variable ratio schedule of added reinforcement and practice in various locations with various forms of distraction.

Phase 4. Maintenance

Continue to thin the schedule of reinforcement and practice under various conditions. Begin using various kinds of reinforcers including praise alone, or providing access to the opportunity to engage in fun activities.

Once you have trained the horse to turn in one direction, train the horse to turn the other direction. Once you have completed this, engage in inter-stimulus discrimination training, changing directions in a seemingly random manner to solidify the evocative stimulus-behavior functional relation.

Backing Up

Phase 1. Preliminaries

Behavior Objective

S^{Ev}: Vocal "Back."

Behavior: Walks backward.

Consequence: Treats and social contact.

Criteria: Latency: 1sec.; frequency—100% through 10 trials; distraction: various; time point: 1 session at ≈20min.

Instate all of the appropriate antecedent control conditions to ensure success, and gather and prepare your equipment.

Phase 2. Acquisition

You can also train backing with the target stick although it is a little tricky since the horse's body is in the way of the direction you want them to move. However, it can be done. Present and place the target stick just in front of the horse's lower chest below their head and evoke touching it. The horse will likely backup in order to touch the target more comfortably. The moment the horse takes that step back, click and provide the treat. Repeat through several trials until the behavior is smooth and reliable. A little verbal encouragement can help prompt the behavior further, if needed. Taking steps back can sometimes be challenging. If you are having a hard time generating the behavior, reinforce even the slightest shift backward and incrementally shape steps backward.

Phase 3. Fluency

Once this is smooth and reliable, you can begin increasing the number of steps back required for reinforcement. Do this gradually, at a pace that assures a high rate of reinforcement and low rate of extinction or subtracted punishment. To condition a vocal cue and fade the target prompt, present the vocal cue "back" immediately before targeting the behavior and reinforce through several trials. You can gradually decrease the prominence of the target stick in the process.

Refine form, latency and speed as needed.

Put the behavior on a gradually thinning variable ratio schedule of added reinforcement and practice in various locations with various forms of distraction.

Phase 4. Maintenance

Continue to thin the schedule of reinforcement and practice under various conditions. Begin using various kinds of reinforcers including praise alone, or providing access to the opportunity to engage in fun activities.

Remaining Aligned with Mounting Block

One of the major foundational applications of "stay" is getting a horse to stand next to a mounting block. You will be introducing another position change behavior to the horse's repertoire in this project as well—bringing the hips around. Training the horse to remain positioned at the mounting block will simply be comprised of a "halt" or "stay" along with a "pivot" in the hips, in case it is required to align the horse with the mounting block. For this training, the mounting block will come to contribute to evoking remaining aligned with and still at the block.

Phase 1. Preliminaries

Behavior Objective

S^{Ev}: Vocal "Stay" (when aligned with mounting block).

Behavior: Remain aligned parallel with the mounting block and still.

Consequence: Treats and social contact.

Criteria: Latency: 1sec.; Duration: 3min.; frequency: 100% through 10 trials; distraction: various; time point: 1 session at ≈20min.

Instate all of the appropriate antecedent control conditions to ensure success, and gather and prepare your equipment.

Phase 2. Acquisition

Begin by walking the horse to the mounting block. When you get to the correct placement, use the previously conditioned "halt" behavior. There are several applications for remaining still and "stay" can be useful for these rather than saying "halt." Begin by establishing the vocal cue "stay." Say "stay," followed by "halt" and then reinforce the behavior of remaining still (including not lifting feet). Repeat through several trials as you begin training until the vocal cue "stay" evokes remaining still reliably and smoothly.

Occasionally, the horse will swing their rear end away from the mounting block. Of course, this is a non-criterion behavior. This behavior should evoke you instating extinction or subtracted punishment. Usually, horses tend to shift away as an escape behavior or if spooked by something. Therefore, any time this behavior occurs, you should identify the stimulation causing the problem and work on desensitizing the horse to it. By desensitizing, I mean training the horse to exhibit and maintain the target operant behavior without indications of stress or escape related behaviors.

You should also consider training the horse to pivot back in towards the mounting block. You can shape this with a series of approximations geared toward getting the rear end to pivot in and bringing the terminal behavior under stimulus control.

Phase 3. Fluency

If you need to improve latency, do that before moving on to more complex training. As you begin applying "stay" to remaining aligned with the mounting block, use a graded approach to gradually and incrementally work in more distracting activities that occur around the horse at the time. Begin by building the duration up to three minutes. Then relax the duration and begin introducing your moving away from and back toward the horse. Work in walking around the horse, as well as out of and back into sight. Work in manipulation of equipment, including the reins. Also, work in your stepping up onto and off of the mounting block. Work on all of these activities separately and then some combined (as appropriate), at a minimal duration and then work duration back up as well. You are training the horse to remain still and aligned with the mounting block, even though you may be carrying out various other actions.

Put the behavior on a gradually thinning variable ratio schedule of added reinforcement and practice in various locations with various forms of distraction.

Phase 4. Maintenance

Continue to thin the schedule of reinforcement and practice under various conditions. Begin using various kinds of reinforcers including praise alone, or providing access to the opportunity to engage in fun activities.

Foot Handling

You need two distinct behaviors for handling feet. In one instance, you will require that the horse bend their leg and present the bottom of the foot so that you may handle it in that position (e.g., so that you can clean the area). In another instance, when grooming for example, you want to handle the leg and feet but you need the horse to keep their foot planted on the ground. You do not want the horse to be confused about which behavior to exhibit, so you will need a specific cue for each behavior and solid inter-stimulus discrimination. Begin training with as assistant holding the horse. Horses spook easily, so having an assistant watching the horse's reactions to you handling their feet is imperative for safety reasons.

Phase 1. Preliminaries

Behavior Objective

S^{Ev}: Handling feet.

Behavior: Remain still and relaxed.

Consequence: Treats and social contact.

Criteria: Latency: 1sec.; duration: 3min.; frequency: 100% through 10 trials; distraction: various; time point: 1 session at ≈20min.

Instate all of the appropriate antecedent control conditions to ensure success, and gather and prepare your equipment.

The horse must be comfortable with a person handling their feet in order to train them to plant their foot or to pick it up for you. If they are fidgeting and fussing every time you reach for a foot, shuffling away or kicking at you, then you are not ready to begin training these foot behaviors. Indeed, in this case, you will need to do some careful desensitization[84] work prior to doing these exercises.

Your first project will be for the horse to allow handling of the feet and being still, without exhibiting agitation or escape behaviors. To achieve this, use a graded approach. Break the goal down into several minor goals. Your final goal will be to handle all parts of the feet, while the horse remains still and relaxed throughout. For the target behavior, remaining still is the main criterion. However, if the horse is still, also look for behaviors associated with either relaxation or agitation/fear. Your goal is to work at a pace that assures the horse will remain still, *as well as relaxed*. Reinforce when the still horse is exhibiting relaxation-related behaviors. Start with one foot (one of the front feet) and work the others in turn. This is how you will desensitize the horse to the touch stimulation.

Phase 2. Acquisition

Begin any handling of feet with rubbing the shoulder (or hip) and gradually rubbing down the leg toward the foot. This will help prevent the horse from startling when you handle the foot. It will also indicate to you at what point, if there is a point, at which the horse becomes uncomfortable with handling. This can then be your starting point for extra careful desensitization work. Touch the shoulder or hip gently and briefly. Repeat through several trials, reinforcing for remaining still and relaxed.

Phase 3. Fluency

Begin very gradually increasing the duration of your touch and/or the magnitude of pressure of the touch at one time, continuously reinforcing for remaining still and relaxed. Be careful not to pull the back of the knee of the front legs or the back of the foot or ankle of the back feet, as this may prompt lifting the foot. You do not want the horse to lift a foot right now.

Next, relax the duration and magnitude of the touch. Start by rubbing the shoulder or hip, as always, and touch the horse just a little lower down the leg. Take small steps; do not rush this, as foot handling rests on this foundation. Again, gradually increase the

---

[84] Recall that by desensitizing, I mean training the subject to exhibit and maintain the target operant behavior without indications of stress or escape related behaviors. One achieves this by taking a very careful graded approach to exposing the subject to the stimulus and additively reinforcing relaxation-related behaviors occurring during that exposure as well as the target operant behaviors.

duration of the touch and the magnitude of the touch at this level. Go through this process of touching lower just briefly and lightly at one level, gradually increasing the duration and magnitude of the touch before relaxing these parameters and working a little lower until you can handle all parts of the foot for several seconds with significant pressure.

Phase 4. Maintenance

Put the reinforcement on a gradually thinning variable ratio schedule of added reinforcement and refresh the handling as required to maintain the target behavior.

Now you may begin the process, working with another foot. You will find that the process will go quicker than with the first foot, but do not rush the process. Once you have worked through all feet, maintain regular handling of the feet and reinforce on a variable ratio schedule, in order to maintain the conditioning. You are now able to train the horse more effectively to keep still for handling etc., or to present the foot for handling.

Keeping Foot Still

Phase 1. Preliminaries

Behavior Objective

S^{Ev}: Vocal "Stay" while handling feet.

Behavior: Remain still (including foot remaining in place) and relaxed.

Consequence: Treats and social contact.

Criteria: Latency: 1sec.; duration: 3min.; frequency: 100% through 10 trials; distraction: various; time point: 1 session at ≈20min.

Instate all of the appropriate antecedent control conditions to ensure success, and gather and prepare your equipment.

Phase 2. Acquisition

The most obvious place to start is with training the horse to keep the foot still since that is what you already trained in the handling project above. Since you have already trained the behavior and you are merely bringing the behavior under stimulus control, you can move directly to the fluency phase.

Phase 3. Fluency

Stimulus control and discrimination training are important in this training project because both behaviors share a number of evocative stimulus features. In both instances, you will be handling the feet. This general context can promote the occurrence of either remaining still or the mutually exclusive behavior of lifting the foot, and so "confusion" is possible. Confusion is really a matter of ineffective stimulus control. Proper stimulus control will eliminate this confusion and allow for reliable evocation of each behavior, even though they both share some contextual similarities. The cue you choose for each behavior should be distinct and its evocative strength consistently maintained, in order to prevent confusion. In this case, you may use the vocal cue "stay." Recall that we already trained this behavior in the context of remaining aligned with the mounting box. Stay meant to remain still and relaxed (including keeping the feet planted) and so it works well for this application of the same behavior. Since foot handling already evokes the behavior, it is a simple matter to transfer stimulus control to this vocal cue. Simply say "stay" immediately before handling the foot each time you move through several trials. Be sure to reinforce still and relaxed behaviors, including maintaining the handled foot on the ground. A distinct vocal cue for lifting the foot and holding it there will ensure reliability.

The behavior should already be on a gradually thinning variable ratio schedule of added reinforcement from the handling project.

Phase 4. Maintenance

As always, generalize the reinforcers and maintain the behavior.

Presenting Foot

Phase 1. Preliminaries

Behavior Objective

S^{Ev}: Vocal "Lift."

Behavior: Raise handled foot and keeping it lifted.

Consequence: Treats and social contact.

Criteria: Latency: 1sec.; duration: 3min.; frequency: 100% through 10 trials; distraction: various; time point: 1 session at ≈20min.

Instate all of the appropriate antecedent control conditions to ensure success, and gather and prepare your equipment.

You can now handle the horse's feet and the horse will keep the foot planted on the ground without exhibiting escape behaviors. You can use the verbal "stay" to help ensure reliability. You have been careful not to pull on the front knees or the back of rear feet.

Phase 2. Acquisition

Start with a front leg and work each leg in turn. Begin, as always, by rubbing the shoulder or hip and gently working your hand down to the feet, to prevent startle responses. For the front leg, stand to the side, and when you get to the back of the knee, hold one hand behind the knee and the other near the front of the foot to help support it. You do not say "Stay" in this case. Gently pull the knee forward. This ought to evoke a shift in weight for balance and allow you to lift the foot. As soon as you get the foot off the ground, an assistant can reinforce while you hold the foot. Then carefully place the foot back on the ground. Do not just release the foot. Repeat through several trials, ensuring the horse is comfortable with the process and participating in lifting the foot.

Phase 3. Fluency

Next, gradually increase duration. Once you can hold the foot off the ground with the knee bent for several seconds, relax the duration criterion and gradually work in more handling of the foot. The more handling you engage in, the greater the duration will be, so you will be working these back up together. Lifting the back leg is a little different in terms of how the leg bends but the key, in either case, is to expose the bottom of the foot.

Once you have achieved the objective of the horse lifting the foot and tolerating various kinds of handling for several seconds, establish a vocal cue for the behavior. Immediately before gently pulling on the knee to generate the behavior, say "lift," and reinforce leg lifting. Repeat through several trials. You might find that the horse begins to lift the foot when cued before you begin prompting with contact behind the knee. That is good! Reinforce this. If not, fade the prompt by gradually decreasing the pressure (prompt) of the contact through repeated trials.

Put the behavior on a gradually thinning variable ratio schedule of added reinforcement.

Thoroughly fading the prompt and engaging in discrimination training will help prevent confusion. Once lifting the foot is under stimulus control of the vocal cue, begin handling the feet as before but ensure that you use a vocal cue for "stay" or "lift" each time you do so. Reinforce for criterion responses, and extinguish and subtractively punish for non-criterion responses in order to allow clear discrimination between both evocative stimuli.

Phase 4. Maintenance

As always, generalize the reinforcers and maintain the behavior.

Continuing Education

The following book could be useful to professionals as well as guardians:

- You Can Train a Horse to do Anything by Shawna and Vinton Karrasch

CHAPTER 11. TRAINING HUMANS

Behavioral Objectives

The objective of this chapter is to measurably expand the reader's repertoire of behaviors in relation to describing and relating the principles of behavior. Upon successfully integrating the concepts outlined in this chapter, the reader, where exposed to contingencies to do so, will effectively:

- Define social behavior and verbal behavior and differentiate between them.
- Define verbal community and the relation between verbalizers and mediators.
- Work your way through training clients' new behaviors in a systematic manner.

Verbal Behavior

The vast majority of work that professional animal trainers exhibit involves training humans to train their companion animals. Only a very small percentage of efforts go directly to training companion animals. Of course, professional trainers train their own companion animals, and under certain circumstances, train clients' companion animals, but, by and large, they merely demonstrate procedures as they train the human client to carry out training the client's animal.

Training humans is both similar to and quite different from training other species. The same laws and principles of behavior apply to humans and, the same strategies and procedures apply. On the other hand, humans engage in extensive use of some very complex verbal behavior that we call language when engaged in social interactions. *Social behavior* is behavior, the consequences of which are mediated by other organisms. *Verbal behavior* is behavior, the consequences of which are mediated by an organism when the other organism is behaving in ways that have been conditioned and maintained in the same verbal environment (Ledoux, 2014, p. 443; Fraley, 2008, p. 949; Moore, 2008, p. 162). The added component is that the socially mediated consequence has been previously conditioned within a verbal community. Much of the verbal behavior exhibited by humans is characterized as "language." *Language* refers to a relatively stable pattern of verbal behavior that is conditioned and maintained by the contingencies that are in place within that verbal community (Fraley, 2008, p. 951).

A *verbal community* simply refers to a group of individuals whose mutual mediating reinforcements condition the verbal and mediating behavior of the individuals in the group, because of the benefits that accrue to the group from generating and maintaining these verbal behaviors. Importantly, (a) verbal behavior involves a lot more than many people suspect (Ledoux, 2014, chap. 20), (b) individuals belong to shared verbal communities, and (c) our training of clients is largely carried out as verbal behavior. Of course vocal behavior is verbal, but so too is sign language, as well as various private behaviors commonly referred as "thinking" (Fraley, 2008, p. 949). Notice that I defined verbal behavior above as reinforced "by *an* organism" as opposed to "by *another* organism."[85] Private verbal behavior is still verbal behavior even though there is only one organism involved; the same organism mediates the verbal behavior and the reinforcement based on conditioning that has taken place within a verbal community. Even a dog scratching at a door and the guardian opening the door to let the dog out is an instance of verbal behavior (Ledoux, 2014, p. 443). The dog and the person constitute at least part of a verbal community in which certain behaviors have come to be conditioned by others within that community. The dog scratches because it will has historically resulted in a human opening the door, allowing the dog to go outside. That is verbal behavior. In the study of verbal behavior, we refer to the speaker as the **verbalizer** and the listener as the **mediator** (because they mediate the reinforcement)[86] to allow for a broader range of verbal behaviors than just spoken or vocalized verbal behavior.

From a psychology perspective, verbal behavior is explained by postulating underlying mental processes within the mind and treating it as the transfer of information from mind to mind. Verbal behavior is said to be the expression of the agent as in the statement "he is expressing himself."

Behaviorologists study verbal behaviors functionally, as the operant behaviors they are. Just like other operant behaviors, verbal behaviors are conditioned and maintained in accordance with the laws and principles of behavior. Antecedent stimuli evoke them, reinforcement strengthens them, punishment suppresses them, and extinction eliminates them.

The main point of this section (the "take home message") is that verbal behavior is operant behavior, and just like any other operant behavior, it can be prompted, evoked, reinforced, punished and extinguished, and it can be used to prompt, evoke, reinforce, punish or extinguish the verbal behavior of others. Verbal behavior is not some mysterious/magical "transfer of information" from "mind to mind" as psychology's communication theory would suggest. Although it is extremely complicated, it nevertheless is perfectly natural, and operates in an orderly and lawful manner just like all other behaviors. By appreciating this fact and conceptualizing verbal behavior behaviorologically, you can more readily appreciate how to change the verbal behavior of others.

---

[85] Many sources define verbal behavior as behavior reinforced "by another organism" and they proceed to stipulate that the subject can act as both the verbalizer as well as the other organism mediating the reinforcement. This seems confusing and almost nonsensical to me. That is why I bypass such confusion by defining verbal behavior as behavior for which the consequences are mediated by "*an* organism" rather than "by *another* organism" However, take note of these differing approaches to defining verbal behavior and allowing for private verbal behaviors.

[86] I prefer the word "consequator" to "mediator" because (a) consequator more accurately describes the "listener's" role in the contingency as providing the consequence that maintains the verbal behavior, and (b) the term mediator used to refer to the listener can be confusing since while they mediate the consequence, the verbalizer mediates the verbal behavior. However, my preference is not standard terminology within behaviorology and I defer to the verbalizer/mediator distinction.

Consequating Client Verbal Behavior

At different stages of the professional relationship between you and a client, you will have different behavior objectives to achieve. You will not likely write them out, however, you will need the client to honestly and freely provide data regarding the subject's behavior and their interactions with the animal. Later, you may need to convince a client to carry out specific tasks. At some point, you may need to address concerns or persuade the client to accept a particular point of view. On a moment by moment basis, you may need to extinguish counterproductive client verbal behavior and instead reinforce more productive verbal behavior. Clients may exhibit very little verbal behavior or perhaps too much. Some clients argue every point, while others seek attention and sympathy. It is easy to forget, when working with clients, that the professional's reactions to client behavior are consequences for the client's behavior, but appreciating it as such can help keep the discussion on track and help you obtain accurate data, and avoid or escape certain problematic interactions. Set the occasion for and reinforce client sharing with undivided attention. Maintain soft eye contact, gently lean in and interject appropriate prompts such as "uh huh," "okay," "I see" or "go on." Avoid checking your watch, looking away, checking your nails, or the like. If the client veers off topic, avoid punitive reactions that may cause problem side effects. Perhaps interject "Okay, tell me more about such-and-such," as a way to get the client back on track, evoking more on-topic discussion. Acknowledge the client's appropriate feelings and concerns. If the client engages in counterproductive behaviors such as excessive complaining, criticism of previous trainers, excessive sympathy seeking, argumentative verbal behavior, excessive criticism and so on, provide as little extra reinforcement as possible for these behaviors, prompt the discussion back to more productive topics and reinforcement these productive verbal behaviors.

General Strategy for Training Humans

Just as with dogs, cats, parrots, and horses, the best way to train anyone to do anything is to break the objectives down into manageable sub-objectives and then arrange the environment in such a way that the subject is very likely to succeed. Then additively reinforce success, repeat through an adequate number of trials, and gradually work toward the ultimate goal. This graded approach can involve various strategies and techniques. In even the simplest cases, various client behaviors need to be conditioned and brought under appropriate controls. One of the main training opportunities with clients involves training the client to exhibit certain specific training behaviors with their companion animal. The following steps provide a general strategy for training clients.

Establish Objective

Start by identifying the objective. For instance, this might simply involve explaining to the client that the goal will be for the client to train their dog to sit (criteria specified) by the next time you meet with them (time specified). I will use training the client to train their dog to sit as an example. This is a long-term gold and will require training the client to exhibit several specific behaviors in order to achieve this long-term goal. Be specific and precise with your goals, rather than vague and open ended. Some objectives might involve much more elaborate training and multiple tasks for the client to engage in. Examples might include complex behavior chains or working through a behavior change program to resolve a problem behavior. In these cases, set realistically achievable sub-goals. For example, in a lengthy behavior chain, you might decide to train the client to achieve the training of the first behavior as their first goal. Then you can follow up with training other behaviors one at a time and then finally, on chaining them together. In resolving a problem behavior, you might set two tasks for the client to achieve in the coming week: (a) avoid allowing any of the evocative stimuli listed on the behavior change document you provide the client, so as to avoid evoking the problem behavior, and (b) train the subject to exhibit a specific behavior on cue, completely outside of the problem evocative context, which will be used later as a replacement behavior. Once you achieve that objective, you can move on to the next objective.

Of course, at the same time that you are training the client, you and the client are training the companion animal. Establish a baseline for the subject's behavior and begin also training the client on how to quantitatively track the subject's behavior. Perhaps you will draw a rough graph and train the client how to identify the trials and mark the dots on the graph (using this same method described in this chapter). This is important for training the subject, but it is actually also immensely helpful in training the client as well. The line on the graph will show, "in salient black and white," how well the training is progressing, which is much better than simply remembering "what it was like" before training. You will of course, arrange the circumstances so that progress is assured with the subject, setting the subject and client up for success and achieving results right from the start of their training program. The clear evidence of progress will be a very powerful reinforcer for the client's training behaviors.

Describe, Explain and Demonstrate

Describe the target behavior, demonstrate (model) the behavior, and explain the purpose for the behavior. In the example of training a dog to sit, you might begin by describing how to lure the behavior, and then when the criterion behavior is exhibited, clicking and immediately follow up with providing a treat. You will explain the function of the clicker and the treat as a reinforcers and how this process will help increase the rate of the behavior. It is a good idea to repeat descriptions and explanations, finding a new way of describing the process, to help clients comprehend what you are telling them. Follow up your description and explanation with a demonstration of the behavior. You may identify any potential pitfalls and the solutions as you demonstrate. For example, you might point out that if the client holds the treat too high above the dog's head, the dog might jump up instead of sit, or that some dogs tend to shuffle back rather than sit. Ask the client if they have any questions and to describe back to you exactly what is to be done. Clients will frequently nod in agreement as you describe and explain processes to them, even if they do not comprehend the description. They may also say they understand if you simply ask them if they understand the instructions, even if they do not. They may indeed believe that they do understand when in fact they do not. Clients will often avoid asking questions or suggesting they are not following your instructions. Ensure that you prompt and reinforce question-asking and honest concern-sharing as you progress in your work with your clients. When they describe the behaviors to you, reinforce as needed.

Avoid excessive jargon in your explanations of behavior and conditioning processes. You can usually explain the basics of the ABCs of behavior without using excess jargon. Put the concepts in simple terms, and if you do introduce any technical terms, simply explain what they mean and then reiterate them a couple more times during the discussion.

Remember that as you train clients to train their companion animals, you are also training yourself and the client is training you! The clients' verbal behaviors and their successes will reinforce your effective training/coaching behaviors. Not all conditioning will be as direct as this however. Successes and failures might fail to meet the contiguity requirements for effective conditioning. When there is a problem in training, identify why and verbally tie the ineffective behaviors (mistakes) to the failure in order to condition more effective training. This will create a covert verbal contingency that will reduce the future likelihood of the problem training behaviors. Tie this also to solutions, more effective training behaviors that, when successful, you can tie to the success and hence reinforce solution-finding behaviors. That was a little abstract so allow me to present an example to make it clearer. I may train a client to carry out the luring motion to generate a sit from the subject. If I fail to demonstrate how high to hold the treat over the dog's head and this failure on my part results in frustration (and related behaviors) on the client's part, this may not change my behavior without verbal supplementation. That is because my behavior is separated in time from the consequence, making the contingency ineffective on its own (lack of contiguity). I may verbally tie these events together by thinking to myself "my failure to demonstrate this height feature of training resulted in the client's frustration." This creates a contiguous contingency. I may even follow-up with a "verbal rule" such as "when training clients to lure a sit, explain that height of the lure is important." Then in the future, training a client to lure a sit will evoke my thinking of the rule, which will evoke my more effective demonstration of the procedure. This "rule-governed behavior" demonstrates how our environment effectively conditions our behavior as well. Make sure to also recognize successes and tie those to effective training behaviors in the same manner.

Remember that aversive methods cause just as much harm to the client as they do to their companion animals. Indeed, aversive private verbal behavior can cause problematic side effects as well. Emphasize added reinforcement throughout the training program. This goes for the conditioning of your *own* behaviors as well!

Assess Proficiency

Have the client carry out the behavior as you carefully observe them. Once they have done so, reinforce with appropriate praise. If it went really well, you can move forward. It is important to emphasize added reinforcement here. If a client does something "wrong," there is no need to draw excessive attention to that with punishment. Instead, pick out what can be additively reinforced and then rearrange the circumstances to promote criterion responses over the incorrect responses. You might say "Okay great. Your timing was right on! How about we do it again, and this time, hold the treat a little closer to the dog's head? That'll make it less likely for him to jump up." Rather than tell the client what to do, it is usually more effective to make a recommendation instead, as in the example above. As you carry out the observation, you will be assessing the client's proficiency with the various components of the skill you are training them to carry out. Repeat until the client can adequately demonstrate the requisite behaviors. If there are too many failures, you need to back up a bit and establish smaller scale objectives to ensure a high rate of success and reinforcement. Does this sound familiar? This is exactly how we handle training companion animals. Once you have achieved success, you may then move on to other behaviors. For instance, perhaps you will also describe, explain, demonstrate and assess for luring a down and stand. Try not to add too many different skills on at one time, but you obviously want to provide the client with enough to work on at home toward obtaining a useful objective the next time you meet with them (i.e., their homework). You might also explain and demonstrate the training techniques that they will exhibit once they achieve the initial goal. For instance, you might explain that once the dog will sit reliably and smoothly after several trials, that they might start fading the treats-in-hand prompt and work toward fading the hand motion to a hand signal movement. You might also explain how to start increasing duration. Demonstrate, and assess as described above in each instance. It is also a good idea to give the client a handout for each of the training tasks you trained them to execute. They can refer back to this as a "prompt," reminding them of the behaviors to mediate. Ideally, the client should be able to contact you, in case they have any questions about the training.

Follow-up

The next time you meet with the client, there will be a number of tasks to carry out. Start by simply asking how things went and seeing if there were any problems. This will give you a bit of a "heads-up" for when you get started. Have a look at the chart or graph and point out the progress as a reinforcer. If there were failures, frame this for the client as an opportunity to fine tune the training.

Start training with a review of what was previously covered. Next, have the client demonstrate the current state of the training. Make sure to reinforce their successes or improvements in their skill proficiency. Be excited and pleased with the progress. As before, go back to describe, explain, demonstrate and assess for any deficient components, always being careful to frame these as opportunities to progress rather than "failures." Once you have achieved the objective or sub-objective for this segment of the training, you can then move on to the next objective. Continue to establish solid foundations for requisite behaviors and progress from there with new, more advanced, objectives until the ultimate objective that the professional relationship was established for is met.

Training Groups of People

Training group classes follows the same general outline above, but can pose a number of challenges not present when working with one individual. The virtue of training at the subject's pace is always important, but when you have a group of subjects that you are simultaneously training, you might be training at a pace appropriate to some class members, but that is either too fast or too slow for others. Those working ahead of you will be bored and those working behind you will be frustrated and discouraged. I will discuss some guidelines that can help you reach as many people as possible below.

Between four and eight students seems to be a good range for group classes if you are training alone. With less experience, aim for four to six students and if you have lots of experience, aim for six to eight. For puppy classes, aim a bit lower and for basic good-manners classes, you can usually aim a little higher. For specialty classes such as for reactive dogs, aim lower, and for sports with experienced guardians, you can usually aim a little higher. If you will have an assistant, you can involve another student or two and if you have two assistants, you might be able to include as many as three or four more students. Having an assistant allows you to continue on training. If someone needs closer supervision and advice, the assistant can work individually with them and get them caught up. Even with assistants to help you in class, make sure that you address each person individually and by name at least once during the class, ideally to include added reinforcement as well as some personalized instruction.

Try to keep the mood light, relaxed, and fun. Clients are likely nervous that they might "look bad" or that they won't be able to keep up, or their dog might do something embarrassing like urinate on the floor or jump up on someone, or just fail to pay any attention to them when they try to follow the directions. Use humor[87] and an easygoing demeanor to help them feel more comfortable. Anxious clients, like anxious dogs, are distracted. Make sure to start with particularly quick and easy tasks, so that each client achieves some quick success to reinforce their training behaviors. You can increase the pace of the class as appropriate, to keep everyone participating and progressing. Keep a close eye on everyone and if it looks like someone is having a hard time, an assistant can discretely approach and quietly help him or her through it. If more than a couple people are having trouble, back up a little and begin again to give everyone a chance to catch up and everyone else a chance to review. Make sure to reinforce particularly

---

[87] Be careful with humor! Do not make anyone feel like he or she is the butt of a joke. Only use humor in a way that ensures the client feels at ease (e.g., the incident is no "big deal" at all, normal and common, in fact, and that you can all laugh together about it). When in doubt, avoid humor and simply assure the client that whatever it is, it is not a problem.

excellent breakthrough moments. Even just a quick smile and acknowledgement can really keep clients engaged and encouraged.

Wilde (2003, pp. 39–40) points out that there will often be clients that can disrupt or otherwise set the pace of training back. She warns of "Rambling-Rose" who asks a question and then launches into an extended off-topic story, "Look-at-me-Leah" who "acts out" for attention, "Know-it-all-Ned," and "Argumentative-Al," both of whom who will eat up time engaging you in disruptive discussions, if permitted. Keep the class on track and provide as little reinforcement as possible to disruptive behavior, instead reinforcing cooperative behavior wherever/whenever it occurs.

If you use games, or candy or ribbon reinforcers etc., to highlight good work, be careful. Competition involves winners and losers, and it is no fun being the loser. Likewise, if you provide a grandiose reinforcer to someone, then this may highlight that there may not be one for someone else. If you use games, try to make sure there are no losers and reinforce everyone's effective training behavior. Be sure to deliver reinforcers to all of the class members.

Ensure that you adequately describe, explain and demonstrate skills for your clients and then assess their execution of the skills, remediating where necessary. Provide homework and follow-up in the next class with a review, plus any required remediation, and then progress to the next set of skills.

Continuing Education

Courses provided through The Companion Animal Sciences Institute at www.CASInstitute.com address all of the topics covered in this chapter.

Wilde (2003) provides good advice for those considering a career in animal training.

Ledoux (2014) provides an excellent introduction to verbal behavior.

APPENDIX 1. TRAINER EXERCISES AND SKILLS DEVELOPMENT

Behavioral Objectives

The objective of this appendix is to measurably expand the reader's repertoire of behaviors in relation to describing and relating the principles of behavior. Upon successfully integrating the concepts outlined in this appendix, the reader, where exposed to contingencies to do so, will accurately:

- Analyze their training behaviors for effectiveness and efficiency, providing data by which trainers may reinforce effective behaviors, extinguish ineffective behaviors and shape their training chops.

This appendix is composed of a series of exercises that trainers may find helpful in expanding their repertoire of effective training behaviors. These exercises build upon one another—early exercises provide the repertoire expansion necessary for the later exercises. Thus, in order to maximize the benefit from these exercises, sequentially work through them from first to last.

Video-record all of your exercise sessions. A big part of the conditioning experience will come from observing and scrutinizing the videos, rather than just carrying out the training exercises. It is a good idea to take note of any deficiencies that you can improve upon and repeat the exercises, to ensure that improvement has occurred. Recognize these improvements in order to reinforce your own effective behaviors. Continue to carry out the exercises, observe the video, and repeat this sequence until you are demonstratively proficient with the exercise. Reinforce your own improvements and successes, literally. This will improve your conditioning.

Exercise #1. The Bouncing Ball Exercise

This exercise will help you improve these behaviors or features of behaviors:

- Timing

- Concentration

- Clicker accuracy/dexterity

- Quantitative tracking behaviors (basic)

Begin with the basic Bouncing Ball Exercise, and move to the more advanced Tossed Ball Exercise following demonstrated proficiency here. You will need:

- an assistant;
- a ball;
- a clicker; and
- video recording equipment.

Video-record all of your training sessions.

Have your assistant bounce a ball with a variable interval (with a range of around one to ten seconds) between bounces. Ideally, they should toss the ball up a foot or two, allowing it to fall and bounce (as opposed to simply throwing the ball to the ground). Your assistant should bounce the ball ten times, take a 30 second break, bounce the ball ten more times, take another 30 second break, and then bounce the ball ten more times. Your objective will be to click the clicker at the exact moment the ball bounces on the ground for each repetition. Carry out this exercise with your favored hand, and repeat the exercise with your other hand. Confirm your accuracy by observing the video. The click should occur precisely at the moment the ball contacts the ground. Use a table to help you record criterion behaviors (perfect timing) and non-criterion behaviors (off timing). For any off timing behaviors, is there a trend in terms of whether you click after or before the bounce? If so, use this to help you improve your timing for the next round.

Repeat the exercise until you have achieved 29 out of 30 precisely timed clicks with each hand. Once you have completed this exercise, literally pat yourself on the back, and follow up by treating yourself in some way!

Exercise #2. The Tossed Ball Exercise

This exercise will help you improve these behaviors or features of behaviors:

- Timing
- Concentration
- Clicker accuracy/dexterity
- Quantitative tracking behaviors (basic)

You will need:

- an assistant;
- a ball;
- a clicker; and
- video recording equipment.

For the Tossed Ball Exercise, you will carry out the sequences exactly the same as in the Bouncing Ball Exercise except that the ball will be tossed between one and four feet into the air instead of bounced on the ground. This exercise is more advanced because the bouncing is approximately a fixed distance, whereas the tossed ball will have a much more variable distance, as well as a lack of auditory feedback as the ball contacts the surface when the click is supposed to occur. Click the clicker at the precise instant the ball reaches the apex of its arc.

Once you have completed this exercise with a success rate of 29 out of 30 with each hand, pat yourself on the back, and follow up by treating yourself in some way!

Exercise #3. The Ball and Treat Exercise

This exercise will help you improve these behaviors or features of behaviors:

- Timing

- Concentration

- Clicker accuracy/dexterity

- Quantitative tracking behaviors (basic)

- Accurate treat delivery to stationary target

This exercise builds on the ball exercises and introduces another behavior set. This training requires that a number of behaviors be exhibited, either consecutively or sequentially, in a short period of time. Thus, this exercise will help increase the number and level of difficulty of behaviors you can exhibit effectively.

You will need:

- an assistant;

- ball;

- clicker;

- standard drinking glass;

- treat pouch;

- treats; and

- video recording equipment.

Video-record all of your training sessions. Carry out this exercise exactly as in the basic Bouncing Ball Exercise previously described, except this time, you will be incorporating treat delivery as well. Instruct your assistant to allow you a maximum of one full second ("one alligator") to deliver your treat before they begin counting silently toward their inter-trial interval. Carry out the exercises right beside a table at about waist height. Have the drinking glass placed on its side on the table propped with erasers, blocks, or something similar, so that the glass does not roll away. Your objective is to click precisely at the instant the ball bounces on the floor, and within one second of clicking, retrieve a treat from the pouch and place it inside the glass, without knocking the glass out of position.

Use the video recording to observe and quantify your accuracy for both click timing and treat delivery. Record the number of times you clicked precisely when the ball bounced, and track the number of treats administered into the glass within one second of your click. Missing the glass and/or delivering the treat outside of the one-second interval are non-criterion behaviors.

Repeat the exercise until you have achieved 29 out of 30 precisely timed clicks, and 29 out of 30 accurate treat deliveries. Once you have completed this exercise, pat yourself on the back, and follow up by treating yourself in some way!

Exercise #4. Planning List of Behavior Approximations for Shaping

This exercise will help you improve these behaviors or features of behaviors:

- Judging appropriate behavior approximation size based on how difficult the different components will be

- Breaking approximations into smaller sub-approximations

- Planning for acceptable operant approximations

Planning a shaping project requires a repertoire of various behaviors, and one of them is planning your list of behavior approximations. Although one must remain flexible in training to quickly accommodate unexpected events, prevent frustration, reduced rate of responding, or to take advantage of sudden leaps in progress, one can only really be truly flexible if one prepares a plan that allows for such contingencies. In this exercise, you will practice planning a list of behavior approximations for a training project, including allowing for acceptable deviations.

You will need:

- a pen or pencil; and

- paper

Choose a behavior to shape. This should be a behavior that is more or less unlikely to occur frequently in its final form, and that you cannot readily prompt in its final form. An example of a relatively simple behavior is turning in a full circle.

First, prepare a list of behavior approximations for the behavior. The approximations should be large enough so that the subject is not rapidly skipping multiple steps at a time, but small enough that progress remains smooth and efficient. Ensure that each approximation describes specific body part movements—that they are operational. Second, for each approximation, break it down into three small sub-approximations, in case you need to quickly utilize them to prevent frustration. Third, look at each approximation. For the particular behavior identified, is there an acceptable alternative response class forms that still progresses toward the terminal behavior? In other words, prepare yourself now for any step along the way where you might not immediately get the specific approximation you listed, but you might get another response class form that is still a suitable approximation toward the terminal behavior. This may not be possible for all terminal behaviors and is usually more applicable to terminal behaviors that are more complex. If you need to continue a new branch of approximations from there to your terminal behavior, do so. Otherwise, note where it connects back to your initial list of approximations.

Once you have completed your list, you should be well prepared for deviations and challenges, such as the subject progressing more quickly than anticipated, or becoming frustrated due to too large a step, or bored by too small a step. Making these judgments in the training situation requires another set of skills, but the repertoire of behaviors practiced in this exercise will help ensure you are prepared, setting both you and the subject up for success. This exercise helps establish a stronger training foundation and demonstrates the benefits of planning.

Consider doing this exercise with two other simple behaviors and then perhaps one more sophisticated behavior to challenge you.

Once you have completed this exercise, pat yourself on the back, and follow up by treating yourself in some way!

Exercise #5. Free-Shaping a Friend

This exercise will help you improve these behaviors or features of behaviors:

- Timing

- Concentration

- Clicker accuracy/dexterity

- Relying less on trainer-mediated prompting

- Planning behavior approximations for a shaping procedure

- Identifying target behavior suitable for trainer proficiency

- Maintaining high rate of reinforcement to avoid reduced rate of responding and frustration (on-the-fly judgment-related behaviors)

- Treat handling

This exercise is an excellent way to practice shaping without subjecting a companion animal to the frustration and confusion often associated with novice trainers carrying out an advanced training procedure. It is largely based on an excellent game (the "Training Game") described in *Don't Shoot the Dog* by Karen Pryor, with the addition of a concerted shaping plan before the game, and an exercise analysis stage at the end (debriefing). As always, video-record your training exercises.

You will need:

- a clicker;

- at least one assistant for the exercise (although it is more fun with two or more);

- reinforcers (e.g., quarters that can be given to the subject in order to simulate treat delivery);

- piece of paper; and

- pen or pencil.

Planning Stage (You as Trainer)

There are two projects in this exercise, one in which you are the trainer and the other in which you are the subject. In the first project, your assistant is the subject and you are the trainer. The subject leaves the room for the planning stage. Identify an operationalized behavior to train the subject to exhibit. Write down your target behavior and prepare a plan of approximations just as you would if you were planning a training project (as described in Exercise #4).

Training Stage (You as Trainer)

Once you have a plan, have the subject come back into the room and you begin the training. No speaking or contrived prompting, including subtle head nods and noises or other facial expressions are allowed. You can justify occasional non-vocal prompts in this exercise in order to work through a particularly frustrating series of trials, but try not to use them unless you absolutely need to in order to be successful. Rather, focus on preventing the need for them—working through these challenges with adjustments to approximation size and pace.

Click for successive approximations of the target behavior in accordance with your training plan, but be prepared to adjust your tactics where appropriate. Maintain a high rate of reinforcement to ensure smooth training and minimal frustration or confusion for the subject. In place of the unconditioned reinforcer, give the subject quarters or some other physical reinforcer for this exercise to simulate treat delivery. When the subject finally exhibits the terminal behavior, the game is over.

Exercise Analysis (You as Trainer)

Once you achieve the terminal behavior, observe the video several times, each time looking for specific things.

First, observe for clicker accuracy. Did you click at precisely the correct time? How many clicks out of the total number would you judge to be precisely accurate?

Second, observe again for pace management. Did you set the approximations at suitably sized steps to maintain smooth progress and prevent frustration? Where you did not, had you planned appropriately ahead of time for these occurrences? In the training session, did you remain flexible, adapting quickly and effectively to adjust the approximation size and reinforcement rate to manage the frustration? Did any superstitious behaviors occur, and if so, did you work through them or use extinction to prevent them from becoming too disruptive?

Third, observe again, now looking for prompts. Did you rely solely on reinforcing from among variations exhibited without contrived trainer-mediated prompts, or did you prompt? Although this exercise features avoidance of prompts in order to focus on other skills, you can and should use prompts where they prevent frustration. Did the prompts you used function to move past a particularly troubling series of extinction trials or were they unjustified? Were they successful in moving you past a particularly frustrating part? Ask yourself whether you could have avoided any of the prompts and still staved off frustration? How could you have done so?

Next, ask your assistant for their observations on the experience as a subject. Ask them if they were confused or frustrated at any point, and if so, why they believe they became confused and/or frustrated. Ask them open-ended questions about their general evaluation of your skills as a trainer, including your clicking and "treat" delivery skills. Ask them what improvements they believe might help. Ask any other observers these questions as well—they may have useful observations regarding both the subject and the trainer.

Finally, identify at least three less than proficient skills in your training, and for each, identify specific ways you can improve upon them. This might include what you could have done differently, but it also could include how you will change your training practices to adjust and refine these particular skills. For any skills that you believe definitely need more practice, identify specific ways that you can improve your performance.

Planning Stage (You as Subject)

Carry out the exercise again, this time with your assistant carrying out the training and you acting as the subject. Have your assistant plan a shaping project for you to engage in as the subject. Coach them how to go about doing this. You may want to video-record your coaching so that it can be included in your analysis of the exercise.

Training Stage (You as Subject)

Your assistant may or may not be a proficient trainer. If they are not, take this opportunity to coach them on the basics of choosing a suitable behavior and suitable behavior approximations, as well as timing and the rules regarding reinforcement while avoiding prompts. This can be a repertoire-expanding conditioning experience. If your assistant is already a proficient trainer, ensure they know what is required from the exercise and proceed. Be sure to video-record this session.

Exercise Analysis (You as Subject)

Once you have exhibited the terminal behavior, you have the opportunity to evaluate many of the same skills, but this time for someone other than yourself. This change in perspective can be an exceptional conditioning experience.

First, take note of your feelings. Feelings are your experience or awareness related behaviors in response to emotional arousal (which itself is the physiological processes going on inside you). Do you feel exhilarated and excited or bored? Do you feel frustrated? Attempt to identify likely causes for these feelings. What exactly about the experience elicited these feelings? Translate these into accurate descriptions of training proficiencies and deficiencies. Once you have a list of training deficiencies, for each one, identify what actions should have been exhibited instead. In addition, separately identify how a trainer should have handled the situation as soon as it became clear that the subject was becoming frustrated or confused.

Next, observe the video for the same type of things you observed in your own training video. Look for timing and management of pace, etc., and make notes about each of the factors.

Once you have completed this exercise, pat yourself on the back, and follow up by treating yourself in some way! You may wish to also treat the people who helped you with the exercise with something that is reinforcing to them.

Exercise #6. Targeting

This exercise will help you improve these behaviors or features of behaviors:

- Timing
- Concentration
- Clicker accuracy/dexterity
- Location of treat delivery
- Handling multiple items at once

For this exercise, you will need:

- a targeting stick (or a suitable substitute, such as a pencil);
- treats;
- treat pouch;
- clicker; and

- a reinforcer-deprived companion animal that has been clicker trained but has not been target trained.

This simple exercise is a good place to start expanding your skills in training companion animals, because, while it is simple, it also involves several key skills in training. One new skill this exercise presents is choosing a location to deliver the treat, which will affect how the behavior comes to be exhibited—this will be discussed below. Another new skill that this exercise introduces is simultaneously handling multiple pieces of equipment. In this case, you will be handling the clicker, the target stick and treats at the same time, requiring you to use at least one hand for more than one piece of equipment. Hold the targeting stick and clicker in the same hand, using your other hand to retrieve and deliver treats.

Planning Stage

As with any training project, prepare a plan ahead of time. In this case, you will likely be able to achieve the final form of the behavior, so you do not need to shape this behavior. You might shape a little bit to tighten up the spot touched or the latency, but you can leave this aside for now and treat this as a simple differential reinforcement procedure. Write down your target behavior in operational terms. You can plan to measure the frequency of the behavior, since it is most informative to measure the number of times the subject exhibits the behavior in comparison with the number of opportunities they are provided to exhibit it (i.e., you can determine the percentage of times the behavior is exhibited). You can record this when you watch the video of your sessions. Gather and place all of your equipment and have it ready to go.

Training Stage

Present the targeting stick to the subject within a few inches of his or her nose/beak. The subject will likely sniff it. If they do not contact the tip of the targeting stick within a few seconds, pull it away, turn away for a few seconds to reset and then present it again. If this is not working, you should increase the salience of the stimulus by rubbing treats on the tip of the stick (although this is rarely necessary). The instant the subject touches the tip of the stick, click, and retrieve and deliver the treat to the subject's mouth at their head height. The location of treat delivery is important. You could just click and drop a treat to the ground, but this will affect the behavior you are training. If you deliver treats on the ground after the targeting behavior, you will likely notice that the subject responds to presentation of the stick with touching it and then quickly lowering their head down toward the ground. This might not be a problem in some situations, but this exercise provides the opportunity for you to practice delivering treats at a specific spot in order to promote a nice clean targeting behavior without subsequent by-product behaviors.

Repeat through a few more trials. You should see the behavior becoming smoother and more reliable as you continue through the process. Begin presenting the stick about five or six inches away from the subject's nose/beak, and once that is reliable, present it far enough away that it requires that the subject take a few steps to touch the target stick. Once you have this trained reliably, begin to randomly present the stick on either side of the body. You should very quickly have this behavior exhibited at a frequency of 100%. Once you have achieved ten consecutive behaviors, in which the target stick is touched within a couple seconds of presentation regardless of the location that the target stick is presented, you can consider the project a success. Be sure that you video-record all of your training sessions.

Exercise Analysis

Now observe the video. Take note of your timing. Was it precise each time? Were you able to deliver the treat within one second every time? Did you accidentally click any times when the subject did not exhibit the behavior? Were there any extinction or subtracted punishment trials. If so, why? Did you increase the difficulty level too quickly? How did you work through this on the next trial? Finally, find something that you could improve upon. Write that down, as well as exactly what you can do to improve that aspect of your training.

Once you have completed this exercise, pat yourself on the back, and follow up by treating yourself! Perhaps purchasing one of the target sticks with a built-in clicker (e.g., a Clik Stik® clicker and target stick) would be a nice reinforcer.

Exercise #7. Simple Shaping Project

This exercise will help you improve these behaviors or features of behaviors:

- Timing

- Concentration

- Clicker accuracy/dexterity

- Relying less on trainer-mediated prompting

- Planning behavior approximations for a shaping procedure

- Identifying target behavior suitable for trainer proficiency

- Maintaining suitable energy/enthusiasm level

- Maintaining high rate of reinforcement to avoid reduced rate of responding and frustration (on-the-fly judgment-related behaviors)

- Setting criteria for next behavior approximation to avoid excess extinction but still reliably responding at the next approximation

You will need:

- a clicker;

- treats;

- treat pouch;

- paper;

- pen;

- a reinforcer-deprived companion animal that has been clicker trained but has not been target trained; and

- video equipment.

Building on your work in previous exercises to shape a human behavior, you are now going to transfer these skills to working with nonhuman companion animals. In this exercise, you will shape a simple behavior with your companion animals. This will build on the exercises you have already completed and introduce the challenge of shaping. Choose a behavior that suits your situation; for this exercise, it will be something related to a cardboard box. If the box is large and low enough, in relation to the subject, you might shape stepping into the box with all of their feet. If the box is small enough, you might train the subject to pick the box up in their mouth/beak or to push the box with their nose/beak for a short distance. You may choose another behavior, but ensure whatever you choose is a single discrete behavior and not a sequence of behaviors.

Planning Stage

Begin by defining your target terminal behavior operationally, as well as the behavior approximations you will use to achieve it. Remember to plan for the necessity to break a step down into smaller steps.

Training Stage

Begin training, utilizing the skills you have been practicing, including precise timing of the click, treat delivery within one second, and accurate placement of the treat to avoid post-reinforcement behaviors that might disrupt training. Maintain a high rate of reinforcement to avoid a reduced rate of responding and frustration. Be sure to video-record all of your training sessions.

Once you have achieved stable responding with a behavior approximation, move to the next approximation. Take care at this point; you will be evaluating this skill later. If you wait too long after achieving reliable responding at one approximation step before setting the next approximation, the behavior will have become well conditioned or entrenched, and this will require more extensive extinction when you reset the criteria for the next step. If on the other hand, you reset the criteria for the next step too quickly before the behavior is stable, the subject may fail to exhibit the next approximation at all. You need the step to be small enough that a little variability in responding will lead to the next approximation. It is a constant balancing act of criterion size and at what point you change the criterion that represents a major difference between the "chops" of a professional and those of a novice. Emphasize this on-the-fly judgment in this exercise by paying close attention to the rate of responding and reinforcement, and any disruptions to smooth continuous conditioning. Also, be sure to use an appropriate level of animation or energy/enthusiasm to keep the process fun and smooth.

Continue training until the subject exhibits the criterion terminal behavior five times in a row when the box is presented. You should keep sessions generally short and stop while the training is still fun for all concerned. You will likely be able to complete this training in a single session, but if you need to take two sessions, do so. In that case, remember to end a session on a positive note, and when you start a new session, review earlier steps before proceeding further into the list of approximations.

Exercise Analysis

Observe the video several times, looking for specific features each time. Take note of all the things that you previously practiced, such as the precise timing of your click, whether you retrieved and delivered the treat within one second, whether you delivered the treat precisely, whether you missed any opportunities to click, and how many times you clicked for a non-criterion response. It should become habit for you to evaluate all of these factors every time you evaluate your training skills. Note how well you planned your approximations. Were all of the approximation steps a suitable size, or did you need to adjust them, and if you did, how might you plan more effectively next time?

For this exercise, you also need to evaluate your energy level—that is, how you use encouragement, tone of voice, and how animated you move in order to maintain the subject's rate of responding at an appropriate level. Were you animated enough or perhaps too much? Take special care to look for your proficiency with the last two behavioral objectives identified for this exercise.

Observe the video at least once with an eye to assess whether you maintained a high rate of reinforcement by correctly adjusting the approximation size. Did the subject become frustrated or bored at any point? If yes, why? What can you do to avoid this in the future? If it occurs, what can you do quickly to get back on track?

Observe the video at least one time with eye to assess your timing, regarding instating the extinction criterion for the present approximation step, and instating the reinforcement criterion for the next step. Were the first few responses at the new step confusing for the subject? Did the subject continue to exhibit the previous approximation behavior more than a few times, or did behavioral variability occur, but the subject failed to exhibit the next approximation behavior quickly? These are disruptions to the smoothness and efficiency of training. Ask yourself how you could have planned more effectively to avoid this situation or how, when you changed from one approximation to the next, that might have influenced this disruption. Specifically, what could you do to avoid, and to respond to, these disruptions. Write your questions and answers down on paper. If this exercise did not go perfectly smoothly, try again, this time with a different behavior.

Once you have completed this exercise, pat yourself on the back, and follow up by treating yourself in some way!

Exercise #8. Task Analysis for Chaining

This exercise will help you improve these behaviors or features of behaviors:

- Identifying discrete behaviors in a behavioral situation involving a series of behaviors exhibited one after the other

You will need:

- paper; and

- a pen.

Even many simple behavior episodes that you seek to train involve more than one discrete behavior. To carry out an effective chaining project, you need to be able to effectively plan the project. To plan a sequence of behaviors in a chaining project, you conduct a task analysis. In most cases, you will observe animals exhibiting the behavior you plan to train, and from that, identify each of the discrete behaviors that comprise the overall behavior. In other cases, you may be able to simply visualize the occurrence of the behavior.

Part of the skill in carrying out task analysis and planning a chaining project involves determining what the discrete behaviors are. Any behavior can be broken down into multiple finer motor actions. Thus, part of the judgment regarding what is a discrete behavior is context related and depends on the appropriate scale for the task. For example, taking a sip of water from a glass that you are already grasping might seem to be a single behavior but it might better be analyzed to be two behaviors: raising the glass and then taking a sip. You might even break it down into several discrete behaviors involving finer motor actions involving the lifting of the glass, opening the mouth, generating a firm seal with the glass with the mouth, moving the tongue to allow water into the mouth, closing the mouth, removing the glass away from your face, and finally swallowing the water. However, overanalyzing can become counterproductive at a certain point, depending upon the context of your purpose. Determining appropriate discrete behavior demarcation points, for training the specific chain you plan to train, requires balancing the scale of specificity.

In this exercise, you analyze a complex behavioral event, breaking it down into appropriately sized discrete behaviors. First, identify a complex behavioral episode on which you can carry out a task analysis. Start by observing a member of the same species (or genus) exhibit the behavior chain, or visualizing it. What are the discrete behaviors in this chain? When shall you end the chain? Decide on all of the specific operational details for this exercise. Carry out your analysis until you are satisfied that it contains sufficient but not excessive individual links.

Once you have completed this exercise, pat yourself on the back, and follow up by treating yourself in some way!

Exercise #9. Simple Chaining Exercise.

This exercise will help you improve these behaviors or features of behaviors:

- Discriminating between suitability of different chaining procedures
- Determining whether shaping or simple differential reinforcement is most suitable to a task
- Transferring stimulus control to attach links in the chain

You will need:

- a clicker;
- treats;
- treat pouch;
- task analysis from exercise #8;
- a reinforcer-deprived subject that is clicker trained; and
- video equipment.

In this exercise, you will build on the task analysis you developed in exercise #8 and actually carry out the training required to achieve this chain. The exercises are becoming more challenging and advanced, requiring many more proficient environment-behavior relations combined in one project.

Planning Stage

You already have a task analysis completed, so this will be the basis for your chain. One of your next goals will be to decide which particular chaining strategy will be most suitable to your project; will you utilize forward or backward chaining? You may want to review the section on chaining in the chapter of advanced training procedures. Consider your project, and write down your choice of strategy and the reasons why this strategy is more appropriate to your project than the others.

Now that you have your basic strategy decided upon, you will need to consider how you will train each behavior. Start with a formal behavior objective for each behavior (including form, speed and latency features as well as distance, duration and distraction features). For each discrete behavior, write out how you plan to train it. What procedure will you use? This should only take a single sentence per behavior, since you are more or less deciding whether to shape it or just differentially reinforce it, and whether to prompt it or not, and if so, how.

Next, plan specifically how you will fade prompts where appropriate and transfer stimulus control of each behavior to completion of the behavior before it. Write this down as well. Now you should have a basic plan of action.

Training Stage

Your next task is to carry out the training. Handle it systematically, training each behavior or segment as required and appropriate to your plan, including its various features. Treat each discrete behavior as a separate training project until it is time to chain them. Be sure to video-record all of your training sessions.

Once the subject exhibits the entire sequence of behaviors, without any interjected prompts or cues, five times in a row, perfectly to criteria, you are finished with this phase of the exercise.

Exercise Analysis

Review your video several times. As before, review the video for the usual set of accuracies described in previous training exercises. After this, consider any difficulties or disruptions that may have occurred. What were they and what caused them? What changes to your plan or your reaction to the disruptions would generate more effective and efficient training? Write these questions and their answers down. The skills you are evaluating are broader than in the initial simpler exercises and so our questions are generally broader as well. Your goal will be to evaluate your planning and execution of the training project objectively, and identify your strengths and weaknesses in planning and executing that plan. Recognize particular excellences and the behaviors you could improve. The key to making them a conditioning experience that will result in an expansion of your repertoire of effective and efficient training behaviors is to identify deficiencies, explicitly explain what caused them to be deficiencies, explicitly identify what you can do to make them more efficient, and finally integrate these changes into your training and evaluate whether it was indeed effective. This will be your task in the exercise analysis aspect of this exercise.

Once you have completed this exercise, pat yourself on the back, and follow up by treating yourself in some way! This was a big and advanced project and you deserve some pride reinforcement.

Exercise #10. Discrimination and Generalization Training

This exercise will help you improve these behaviors or features of behaviors:

- Planning for an appropriate level of discrimination and stimulus generalization

- Carrying out discrimination and stimulus generalization training

You will need:

- a clicker;

- treats;

- treat pouch;

- a reinforcer-deprived nonhuman companion animal that is clicker trained;

- and video equipment.

Most of the previous exercises have focused on the planning and acquisition stages of training. In this, the final exercise, the focus is on some behaviors that are more associated with the fluency and maintenance stages of training. Recall that in discrimination training, trainers oppose the general tendency toward stimulus generalization by narrowing the range of stimuli that will evoke the target behavior. Decide on the specific stimulus to control the behavior and rule out similar stimuli. How narrow or tight you make this depends on your objectives for the behavior. For example, you may want the word "sit," and not similar stimuli such as "sip," to evoke sitting behavior, but you likely want "sit" verbalized by other people as well as you, to evoke sitting. This is an instance of allowing a certain amount of stimulus generalization. If generalization is too narrow or too broad, the evocative stimulus becomes less useful.

Training Stage

Start by choosing a behavior that has not yet undergone any discrimination and generalization training. You might have a behavior you have been working on that is just not at that stage yet, or you may need to train a behavior from the beginning. In that case, choose a simple behavior such as spinning in a circle, rolling over, or high five. The simplicity of the behavior does not matter for this exercise, so choose something that can you can quickly and easily train to this level. Once the behavior is reliable, and under stimulus control, you can begin discrimination training, ruling out similar stimuli. Reinforce for the target stimulus and not for other similar stimuli. Take it to a level that is appropriate for the behavior in question. This includes trials allowing for appropriate stimulus generalization (such as having other people to present the evocative stimulus, or you cueing it while facing away from the subject or lying down, rather than standing). The key in this exercise is to rule *in* all applicable stimulus conditions and rule *out* all inapplicable stimulus conditions. Once you can reliably evoke the behavior with appropriate stimuli, and several similar stimulus conditions will not, you are finished with the training component of the exercise and you can move to the exercise analysis stage.

Exercise Analysis

Observe the video of the training several times. As usual, observe for the common mechanical and judgment skills evaluated in previous exercises. Then, observe how effectively and efficiently you conducted your discrimination and stimulus generalization training. Did it run smoothly, or did anything cause frustration or otherwise disrupt training? What exactly were they? What exactly could you have done to prevent these disruptions? What can you do next time in response to these disruptions that would minimize them and get back on track? Write these things down. Consider this exercise as a whole, and identify any behaviors that might improve your efficiency in performing this kind of training.

Once you have completed this exercise, pat yourself on the back, and follow up by treating yourself in some way!

Continuing Education

Courses provided through The Companion Animal Sciences Institute at www.CASInstitute.com address all of the topics covered in this chapter.

BIBLIOGRAPHY

Alexander, S. (2003). *The quicker clicker kit!*

American heritage dictionary of the English language. (Ed.) (n.d.). *Mind.*

American Psychological Association. (2013). *APA Handbook of behavior analysis.* Retrieved February 27, 2015, from http://www.apa.org/pubs/books/4311509.aspx

American Psychological Association. (2014). *How does the APA define "psychology"?* Retrieved February 19, 2014, from http://www.apa.org/support/about/apa/psychology.aspx#answer

American Psychological Association. (n.d.). *Division 25 Behavior Analysis.* Retrieved February 27, 2015, from http://www.apadivisions.org/division-25/index.aspx?__utma=12968039.1944350039.1425037411.1425037411.1425037411.1&__utmb=12968039.1.10.1425037411&__utmc=12968039&__utmx=-&__utmz=12968039.1425037411.1.1.utmcsr=bing|utmccn=(organic)|utmcmd=organic|utmctr=apa%20division%2025&__utmv=-&__utmk=124795658

Association of Animal Behavior Professionals (2008). *Professional practice guidelines* Retrieved November 11, 2008, from www.associationofanimalbehaviorprofessionals.com/guidelines.html

Azrin, N. H. (1956). Some effects of two intermittent schedules of immediate and non-immediate punishment. *Journal of Psychology, 42*, 3–21.

Behavior Analyst Certification Board (2010). *Guidelines for responsible conduct for behavior analysts.* Retrieved from http://www.bacb.com/index.php?page=85

Bailey, J. S., & Burch, M. R. (2005). *Ethics for behavior analysts: A practical guide to the Behavior Analyst Certification Board guidelines for responsible conduct.* Mahwah: Lawrence Erlbaum Associates.

Balaban, M. T., Rhodes, D. L., & Neuringer, A. (1990). Orienting and defense responses to punishment: Effects on learning. *Biological Psychology, 30*, 203–217.

Bond, C. (2007). *The effect of jackpots on response.* Unpublished research project for Diploma of Canine Behavioral Sciences. *Companion Animal Sciences Institute.*

Catania, A. C. (1998). *Learning* (4th ed.). Upper Saddle River: Prentice Hall.

Catania, A. C. (2013). *Learning* (5th ed.). Cornwall on Hudson, NY: Sloan Publishing.

Chance, P. (2009). *Learning and behavior* (6th ed.). Belmont: Thomson Wadsworth.

Clark, T. (2008). *Worldview naturalism in a nutshell.* Retrieved October 14, 2014, 2014, from http://centerfornaturalism.blogspot.ca/2008/11/worldview-naturalism-in-nutshell.html

Cooper, J. O., Heron, T. E., & Heward, W. L. (2007). *Applied behavior analysis* (2nd ed.). Upper Saddle River: Merril Prentice Hall.

Darwin, C. (1859). *On the origin of species by means of natural selection or the preservation of favoured races in the struggle for life.* London: John Murray.

Delta Society (2001). *Professional standards for dog trainers: Effective, humane principles.* Naches AVE SW: Delta Society.

Donaldson, J. (2005). *The culture clash* (2nd. ed.). Oakville, Ontario: James and Kenneth. [This is an excellent popular book geared toward dog guardians.]

Estes, W. K. (1944). An experimental study of punishment. *Psychological Monographs, 57*(3), 1–40.

Ferreira, J. B. (2013). Stream of energy: Using elementary principles of behaviorology to describe Progressive Neural Emotional Therapy (PNET). *Journal of Behaviorology, 16*(2), 11-17.

Fraley (2008). *General behaviorology: The natural science of human behavior.* Canton: ABCs. [This 1600 page, three-course book presents an in-depth treatment of all of the major areas of interest in behaviorology. Perhaps the most important book ever written.]

Fraley, L. F. (2014). Behaviorological science and the complexity of unfathomable variation. *Journal of Behaviorology, 17*(1), 13–18.

Fraley, L. F. (2015). What is reality to an organic unit of matter? Some physics of behavior with implications for sentience and sociability. *Journal of Behaviorology, 18*(1), in press.

Fraley, L. E., & Ledoux, S. F. (2002). Origins, status, and mission of behaviorology. In S. F. Ledoux (Ed.), *Origins and components of behaviorology* (Second ed., pp. 33–169). Canton: ABCs.

Hiby, E. F., Rooney, N. J., & Bradshaw, J. W. S. (2004). Dog training methods: their use, effectiveness and interactions with behaviour and welfare. *Animal Welfare, 13*, 63–69.

Karrasch, S. (2012). *You can train your horse to do anything!* San Bernardino: Self-Published.

Kurland, A. (2007). *Clicker training for your horse.* Waltham: Sunshine Books, Inc.

Laraway, S., Snycerski, S., Michael, J., & Poling, A. (2003). Motivating operations and terms to describe them: Some further refinements. *Journal of Applied Behavior Analysis, 36*(3), 407–414.

Latham, G. I. (1994). *The power of positive parenting.* North Logan: P & T Inc.

Ledoux, S. F. (2002a). An introduction to the origins, status, and mission of behaviorology: An established science with developed applications and a new name. In S. F. Ledoux (Ed.), *Origins and components of behaviorology* (Second ed., pp. 3–24). Canton: ABCs.

Ledoux, S. F. (2002b). Increasing tact control and student comprehension through such new postcedent terms as added and subtracted reinforcers and punishers. In S. F. Ledoux (Ed.), *Origins and components of behaviorology* (Second ed., pp. 199–204). Canton: ABCs.

Ledoux, S. F. (2012a). Behaviorism at 100 unabridged. *Behaviorology Today, 15* (1), 3–22.

Ledoux, S. F. (2012b). Behaviorism at 100. *American Scientist, 100* (1), 60–65.

Ledoux, S. F. (2014). *Running out of time: Introducing behaviorology to help solve global problems.* Ottawa, Ontario. [This is an excellent introduction to behaviorology. Dr. Ledoux is a very eloquent writer and has a knack for making very challenging concepts easier to comprehend. I personally believe this book should earn Dr. Ledoux a Nodel Prize. It makes a very compelling argument for the widespread adoption of a natural science of behavior and the dismissal of mysticism.]

Ledoux, S. F. (2015). *Origins and components of behaviorology: A review of the history of this comprehensive natural science plus samples of its disciplinary components (Third edition)*. Ottawa, Ontario: BehaveTech Publishing.

Ledoux, S. F., & O'Heare, J. (2015). Elements of the history of the behaviorology discipline. In S. F. Ledoux (Ed.), *Origins and components of behaviorology* (Second ed., pp. 199–204). Ottawa: BehaveTech Publishing.

Lerman, D. C., & Vorndran, C. M. (2002). On the status of knowledge for using punishment: implications for treating behavior disorders. *Journal of Applied Behavior Analysis, 35*, 431–464.

London, K. B., & McConnell, P. B. (2001). *Feeling outnumbered? How to manage and enjoy your multi-dog household*. Black Earth: Dog's best Friend, Ltd.

Michael, J. (1993). Establishing operations. *The Behavior Analyst*, 16, 191–206.

Miltenberger, R. G. (2008). *Behavior modification: Principles and procedures* (4th ed.). Belmont: Thomson Wadsworth.

Moore, J. (2008). *Conceptual foundations of radical behaviorism*. Cornwall-on-Hudson: Sloan Publishing LLC.

Nation, J. R., & Woods, D. J. (1980). Persistence: The role of partial reinforcement in psychotherapy. *Journal of Experimental Psychology: General, 109*(2), 175–207.

O'Heare, J. (2007). *Aggressive behavior in dogs*. Ottawa: DogPsych Publishing.

O'Heare, J. (2011). *Empowerment training: Training for creativity, persistence, industriousness, resilience & behavioral well-being*. Ottawa: BehaveTech Publishing. [This book was written to provide guidance on training companion animals for creativity and persistence and the help rehabilitate animals in which disempowerment has been conditioned.]

O'Heare, J. (2013). The least intrusive effective behavior intervention (LIEBI) algorithm and levels of intrusiveness table: A proposed best-practices model. associationofanimalbehaviorprofessionals.com. [The most current version of this article will be linked to from the page on professional practice guidelines.]

O'Heare, J. (2015). *Changing problem behavior (2nd edition)*. Ottawa: BehaveTech Publishing.

Overmier, J. B., & Seligman, M. E. (1967). Effects of inescapable shock upon subsequent escape and avoidance responding. *Journal of Comparative and Physiological Psychology, 63*, 28-33.

Pierce, W. D., & Cheney, C. D. (2008). *Behavior analysis and learning* (4th ed.). Mahwah: Psychology Press.

Pierce, W. D., & Cheney, C. D. (2013). *Behavior analysis and learning* (5th ed.). New York: Psychology Press.

Pryor, K. (1999). *Don't shoot the dog! The new art of teaching and training*. New York: Bantom Books.

Schlinger, H. D., & Blakely, E. (1994). The effects of delayed reinforcement and a response-produced auditory stimulus on the acquisition of operant behavior in rats. *The Psychological Record, 44*, 391–409.

Schlinger, H. D., & Blakely, E. (1987). Function-altering effects of contingency-specifying stimuli. *The Behavior Analyst, 10(1)*, 41–45.

Sidman, M. (2001). *Coercion and its fallout* (Revised ed.). Boston: Author's Cooperative, Inc. Publishers. [This is an excellent book length treatment on the side effects of aversive stimulation.]

Seligman, M. E. P. (1975). *Helplessness: On depression, development, and death.* San Francisco: W. H. Freeman and Company.

Seligman, M. E., Maier, S. F., & Geer, J. H. (1968). Alleviation of learned helplessness in the dog. *Journal of Abnormal Psychology, 73*(3), 256-262.

Skinner, B.F. (1938/1991). *The behavior of organisms.* Cambridge: B.F. Skinner Foundation.

Skinner, B.F. (1957/1992). *Verbal behavior.* Cambridge: B.F. Skinner Foundation.

Vargas, J. S. (2009). *Behavior analysis for effective teaching.* New York: Routledge.

Vargas, J. S. (2013). *Behavior analysis for effective teaching* (2nd ed.). New York: Routledge.

Weatherly, J. N., McSweeney, F. K., & Swindell, S. (2004). Within-session rates of responding when reinforcer magnitude is changed within the session. *Journal of General Psychology, 131*(1), 5–16.

Wilde, N. (2003). *It's not the dogs, It's the people!* Santa Clarita: Phantom Publishing

INDEX